Nerve Cells and Animal Behaviour

Extensively revised, the third edition of this introduction to neuroethology – the neuronal basis of animal behaviour – is written for zoology,
biology and psychology students. It focuses on the roles of individual
nerve cells in behaviour, from simple startle responses to complex behaviours, such as route learning by rats and singing by crickets and by birds.
The text begins by examining the relationship between brains and behaviour and shows how study of specialised behaviours reveals neuronal
mechanisms that control behaviour. Information processing by nerve
cells is introduced using specific examples, and how the roles of neurons
in behaviour are established is described for a predator–prey interaction,
toads versus cockroaches. Much new material is included in chapters: on
vision by insects, which describes sensory filtering; hunting by owls and
bats, which describes sensory maps; and on rhythmical movements including swimming and flying.

PETER SIMMONS regularly publishes research on insect neurobiology, especially on the ocellar and compound eye visual systems and their role in
controlling flight, and on the physiology of synaptic transmission. He is
currently Director of the Zoology Degree at Newcastle University.

DAVID YOUNG has undertaken research and teaching on the link between
neurobiology and behaviour in insects, looking at both sensory and motor
systems. A special interest has been the mechanisms of sound production
in crickets and cicadas. He is also the author of *The Discovery of Evolution*
(2007) published by the Natural History Museum/Cambridge University
Press.

Nerve Cells and Animal Behaviour

Third edition

Peter J. Simmons
Newcastle University

David Young
University of Melbourne

CAMBRIDGE UNIVERSITY PRESS
Cambridge, New York, Melbourne, Madrid, Cape Town, Singapore,
São Paulo, Delhi, Mexico City

Cambridge University Press
The Edinburgh Building, Cambridge CB2 8RU, UK

Published in the United States by Cambridge University Press, New York

www.cambridge.org
Information on this title: www.cambridge.org/9780521899772

First edition published 1989
Second edition published 1999
Third edition published 2010
3rd printing 2012

Printed and bound by MPG Books Group, UK

A catalogue record for this publication is available from the British Library

Library of Congress Cataloguing in Publication data
Simmons, Peter J. (Peter John), 1952–
Nerve cells and animal behaviour / Peter J. Simmons, David Young. – 3rd ed.
   p. cm.
ISBN 978-0-521-89977-2 (hardback)
1. Neurobiology. 2. Neurons. 3. Animal behavior. 4. Neurophysiology.
I. Young, David, 1942 II. Title.
QP356.Y68 2010
573.8′61–dc22 2009052764

ISBN 978-0-521-89977-2 Hardback
ISBN 978-0-521-72848-5 Paperback

Contents

Preface

Our aim in this book is to introduce university students to research on nervous systems that is directly relevant to animal behaviour, and to do so without assuming a detailed knowledge of neurophysiology. We concentrate on examples of studies in neuroethology that illustrate clearly how the activity of nerve cells is linked with animal behaviour. The level of the book is for advanced undergraduate students, particularly those studying zoology, biology or psychology, but we hope it will also be useful to students in other disciplines and to postgraduates.

Each chapter is given a title in two parts: a description of its general area and then usually the specific topics to be described. We begin with a consideration of how animal behaviour and brains are organised. Chapter 2 is an introduction to the nuts and bolts of how nerve cells work, and we approach this by referring to specific examples that illustrate concepts without delving into detailed cellular physiology. The next two chapters describe some clear examples where the roles of particular neurons in predator–prey interactions have been established. We then describe two different types of sensory systems in which roles of specific neurons in behaviour have been recognised – vision in insects and hearing in owls and bats – followed by a chapter on the control of rhythmical movements. Chapter 8 describes research on changes in behaviour, including learning. The final chapter describes signalling behaviours, and shows how the ways that nerve cells are involved in relatively complex behaviours can be studied.

At the end of each chapter are features designed to help students consolidate and extend their knowledge. The purpose of the questions is to stimulate thought and discussion: they are not meant to be the kinds of questions that have definite right or wrong answers. Further reading often provides avenues for further explanation of the topics covered in the chapter, as well as useful reviews. Partly in response to demand from students, each chapter ends with a summary – but we hope these summaries will not discourage students from making their own notes as they read. Many new terms are set in **bold** type the first time they appear in the book in order to help a reader to come to grips with unfamiliar terms and concepts, and the page on which each term is first described are also indicated in bold in the index.

In this third edition we have made some significant changes to content and arrangement. All of the chapters include new material, and some are almost completely new, and some of the figures from the previous editions have been modified. Some topics, such as singing by crickets, that were omitted from the second edition are now included again with new research stories. Questions and summaries for each chapter are new. Also new are photographs which introduce some topics and should help to bring some of the behaviours we describe to life.

Neuroethology is an active area of research, and there are many interesting topics we cannot cover within the limits of our book. Some, such as animal navigation, are outside the scope of a book on nerve cells and behaviour. Others, of which learning by *Aplysia* is a prime example, are covered well by other neurobiology or general behaviour texts. Inevitably, some of the topics covered by the book – bird song is a likely one – will come to be in need of updating fairly soon. But it is striking that a number of sections in the first edition, written by David Young over 20 years ago, still provide solid foundations in this subject. Some people argue that textbooks are now out of date, superseded by web-based provision of learning material. We would not have embarked on writing a new edition of this book if we agreed with that. There are advantages in a coherent approach offered by one or two authors in a book, and although there are some excellent websites in neuroethology, coverage as a whole is extremely patchy.

Much of the revision was by Peter Simmons, and he is grateful to many people for helpful comments and for encouragement. These include a number of undergraduate and postgraduate students, and members of the Newcastle University Centre for Evolution and Behaviour. Colleagues he would like to thank include Claire Rind, Alan Roberts and Natalie Hempel de Ibarra, as well as others who have been very generous in providing photographs for us to use.

Organisation of animal behaviour and of brains: feeding in star-nosed moles and courtship in fruit flies

What is special about animal behaviour? Many people like watching animals behave, and an understanding of animal behaviour has been vital throughout human history, enabling people to hunt, to farm, and to understand something about themselves. More recently, understanding how the brains of animals work has given important information about how the human brain works, and why it sometimes malfunctions. But although animal behaviour can be complex and even sometimes seems mysterious, it can be understood and appreciated by the same scientific approaches that are used to study other aspects of the structure and function of living organisms. It is shaped by evolution in the same way as anatomical characters, and natural selection acts on animal behaviour by shaping the ways in which nervous systems work.

The ways in which the internal workings of brains control behaviour is the subject of this book and we shall illustrate them by using examples drawn from many different animal groups. There are several reasons for this catholic approach, but two are particularly important. First, some animals have nervous systems that are especially favourable for study. For example, nervous systems of invertebrates usually contain smaller numbers of nerve cells than those of mammals, so it is much more possible to trace the flow of signals from cell to cell within these simpler nervous systems. In some cases, there are particularly large nerve cells that are especially easy to study, such as the giant neurons involved in escape responses described in Chapters 3 and 4. Second, animals that specialise in particular behaviours often have parts of their brains that reflect that specialisation, making it feasible to relate the function of those brain parts to particular identifiable behaviours. We describe two different examples of specialisations for particular behaviours in this chapter, but there are many more examples in the book including echolocation by bats (Chapter 7) and bird song (Chapter 9). The two examples for this chapter are hunting by the star-nosed mole, in which specialisation of brain areas is associated with their extraordinary nose; and courtship by fruit flies, in which expression of a particular gene labels some neurons with the specialisation of playing a role in that behaviour.

The scientific study of natural behaviour in animals is called **ethology**. Many of the fundamental approaches to studying ethology arose from pioneering observations by Konrad Lorenz and experimental work by Niko Tinbergen that started in about 1930 (Tinbergen, 1951; Bolhuis and Verhulst, 2008). Tinbergen (1963) described four principal questions about a particular behavioural trait that still form the basis of a structured approach to the study of animal behaviour. Our book is mainly concerned with one of these, about the nervous mechanisms that control it; the others are about its function, its development, and its evolution. **Neuroethology** describes the field of science that is concerned with the way that nervous systems control animal behaviour.

Elements of behaviour

A study in which Lorenz and Tinbergen (1938) collaborated provides an excellent introduction to the way we can analyse behaviour into basic elements. The study was on egg retrieval behaviour in greylag geese, *Anser anser*, but various other birds perform it too. These birds nest on the ground, and while an adult is incubating its eggs it is common for an egg to roll away from the nest. The adult goose extends its long neck to the egg and then retrieves it, drawing the head towards its body and using the bill to gently roll the egg (Fig. 1.1). To identify the stimuli that trigger this response, Tinbergen made models, each of which had some of the features of an egg. One of the models was a cardboard Easter egg; others included cardboard boxes of various shapes, and model eggs of different sizes and colours. A stimulus is described as **releasing** a behaviour if it usually triggers that response by the animal, and by comparing the effectiveness of different model eggs in triggering retrieval behaviour, Lorenz and Tinbergen could compare the releasing values of different stimulus features. The most important feature that enables a goose to recognise an egg as an egg is its shape, but colour and pattern also matter. Larger eggs than normal are particularly

Fig. 1.1 Egg retrieval behaviour by a goose, *Anser*. To retrieve an egg, a goose extends its head to the egg, and then rolls it towards its nest guiding the rolling movements of the egg with side-to-side movements of its beak. (Redrawn from Lorenz and Tinbergen, 1938.)

attractive for retrieval – they are supernormal stimuli. A natural stimulus that releases a particular behaviour is termed the **sign stimulus** for that behaviour. An essential step in discovering how a nervous system controls a behaviour is to identify the mechanisms in sensory pathways that recognise the sign stimulus. The process of recognition is often referred to as **sensory filtering**: the nervous system selectively filters out and retains some features of the original stimulus, but ignores and discards others. An excellent example that has been analysed in some detail is the way toads recognise worm-like objects as potential food, described in Chapter 3.

Within the nervous system, there needs to be a link between the processes of stimulus recognition and of triggering the muscle activity responsible for the behaviour. Tinbergen and Lorenz coined the phrase '**innate releasing mechanism**' for this link. It can be likened to a lock. When the correct sign stimuli are filtered, they act as a key in the lock, to release the appropriate behavioural response. In some cases, single nerve cells play this role – for example, giant neurons of crayfish that trigger startle responses (Chapter 4) – but the exact cells involved in innate releasing mechanisms are usually very hard to pinpoint. The word 'innate' was important to Lorenz and Tinbergen because they were particularly interested in the heritability of patterns of behaviour. However, there is no reason why some releasing mechanisms should not be learned by an animal – for example, the sight of Konrad Lorenz walking would release following behaviour by geese that he reared as hatchlings.

It is hard to roll an egg with a beak or a pencil because the egg often escapes sideways. When a goose is retrieving an egg and it escapes, the goose does not immediately re-extend its head towards the egg, but instead it first retracts the head to the nest before re-extending the neck. The movement of flexing the neck towards the body is always completed before the next behaviour pattern starts – the goose completes the action pattern of neck extension and retraction. Such a sequence of movements that goes to completion once initiated is called a **fixed action pattern**, but the word 'fixed' does not mean that the behaviour pattern is completely stereotyped in form – a goose does steer an egg by side-to-side movements of the beak while trying to retrieve it, for example. The observation that some sequences of movements tend to be completed rather than interrupted mid-flow is significant because it implies that central nervous systems generate sequences of instructions, a bit like computer programmes, that specify the order in which different muscles should be used. An alternative would be that sequences of movements are broken down into small elements arranged in a chain, so completion of one element triggers the next. Mechanisms within a nervous system that generate sequences of actions are called **pattern generators** (Chapter 7). Some action patterns can be learned – for example, bird song and human music often include very fast sequences of movements that are almost certainly laid down as kinds of programmes within the central nervous system.

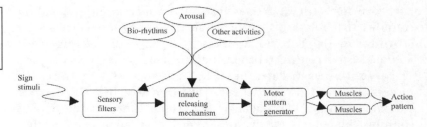

Fig. 1.2 Schematic diagram to show the steps involved in producing a behavioural response to a sign stimulus.

The idea of programmes for behaviour was important in the thinking of Tinbergen and Lorenz, and is still significant today, although terms such as fixed action pattern or innate releasing mechanisms are not much used. The idea provides impetus for neuro-ethologists to search for networks of particular neurons in the animal nervous system that create the programmes. In terms of the life history of an animal the idea of programmes is important, too. It is normal for an animal to encounter specific kinds of stimuli at particular times in its lifetime, which enables the brain to be programmed according to environmental needs as the animal develops.

The relationship between basic concepts about the mechanisms for the control of behaviour is illustrated in Fig. 1.2. Thinking in this way about how behaviour is organised enables neuroethologists to frame experiments designed to determine how nervous systems control behaviour. Although putting names inside boxes might indicate that there are identifiable regions of a brain that correspond to sensory filters, innate releasing mechanisms or motor programme generators, this is unlikely to be the reality. Neurons involved in a particular operation are often scattered in different regions, and some neurons fulfil more than one job. As indicated in the diagram, behaviour is subject to many factors that act within the nervous system to influence its performance, including daily patterns of wakefulness and sleep or yearly reproductive cycles.

Nerve cells and networks

Neurons, or nerve cells, are thickly interwoven within a nervous system. Distinguishing one neuron from its neighbours is very much like trying to make out the detail of a single tree in a dense wood. During the last part of the nineteenth century, however, a histological method was discovered that enabled the intricate detail of single neurons to be traced using the light microscope. In this method, silver is deposited onto individual neurons; but the method is very selective in that only a small proportion of the neurons in a block of nervous tissue are stained. The silver staining method is sometimes called Golgi staining, after its discoverer, and it was used to great effect by the Spanish doctor Santiago Ramon y Cajal. One of Ramon y Cajal's drawings (Fig. 1.3a) illustrates one of the main types of neuron in the mammal brain, a pyramidal neuron. The name pyramidal neuron is given for the shape of the shape of the **cell**

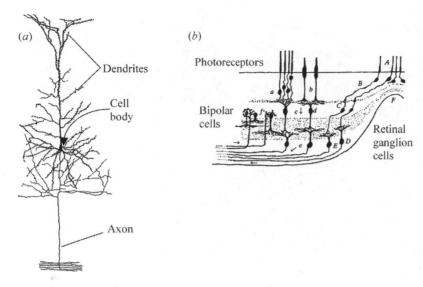

(a)

Dendrites

Cell body

Axon

(b)

Photoreceptors

Bipolar cells

Retinal ganglion cells

Fig. 1.3 Drawings of silver-stained neurons in mammalian nervous systems. (*a*) A single pyramidal cell from the brain. The cell body, about 20 μm (0.02 mm) across, gives rise to two different kinds of processes: dendrites, and an axon. At the bottom of the figure, the axon is shown joining a bundle of axons from similar cells. (*b*) Some of the types of neurons in the primate retina, including sensory neurons (rods and cones, *a, b, B, A*), bipolar cells (*c, D, C*) and the retinal ganglion cells (*e, E, D*). *F* indicates the fovea, where bipolar and retinal ganglion cells are displaced from their receptors. Arrows indicate pathways through which Ramon y Cajal thought that signals passed. (From Ramon y Cajal, 1911.)

body, the region of cytoplasm that contains the cell nucleus. Ramon y Cajal thought that neurons are dynamically polarised, by which he meant that some parts have the function of collecting signals and other parts pass signals onwards. Although the experimental techniques were not available to him to test his ideas, and Golgi vehemently disagreed with them, this view of a functional separation between different parts of a neuron is fundamental to a modern understanding of how they work. In the case of the pyramidal neuron, the **dendrites** collect signals from other neurons, and the **axon** carries signals on to target cells. The way in which information is received, processed and transmitted by individual neurons will be explained in Chapter 2.

Ramon y Cajal also had the insight to propose that individual neurons connect with each other to form networks that process information in specific ways. One of his drawings proposing interactions between neurons in a vertebrate retina is shown in Fig. 1.3*b*, which also illustrates some of the great diversity of form between different neurons. For example, in the retina neurons of the innermost layers called ganglion cells have long axons that convey signals into the brain, whereas the photoreceptors and bipolar cells are involved in much more local operations and lack long axons. In studies of neuroethology, considerable effort goes into attempts to trace the routes by which information flows between different neurons, and into understanding how it is collected, transformed and transmitted to control animal behaviour. These routes are sometimes referred to as 'circuits', although the pathways rarely form a complete loop around which the signals travel.

Nervous systems

Jellyfish, sea anemones and starfish have relatively simple nervous systems in which neurons are connected into a nerve net. In this

A ganglion that controls the forewings and middle pair of legs in a cockroach thorax. The ganglion was stained with a dye called toluidine blue that sticks to nucleic acids, so the cell bodies and nuclei of neurons stain darkly. The outline of the ganglion and some of the major nerves, which do not stain with this dye, have been drawn in. (Photograph by Peter Simmons.)

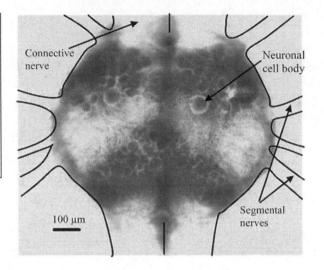

Fig. 1.4 Invertebrate central nervous systems. (*a*) A snail such as the pond snail, *Lymnaea*. Above is a dorsal view of the major ganglia; below, a diagram of a snail from the side, showing positions of ganglia and connective nerves. Three of the ganglia are labelled. (*b*) A locust, *Schistocerca*, with its central nervous system below. (*c*) Cut-away view of the head and thorax of a locust showing the segmental nerves innervating some of the large flight muscles. (*d*) A fly such as the blow fly (*Calliphora*) or *Drosophila*. Note that the thoracic ganglia are fused into one. (Part (*a*) redrawn from Kandel, 1979, and Bullock and Horridge, 1965; (*b*) based on Burrows, 1996; (*c*) redrawn from Wilson, 1968; (*d*) redrawn from Bullock and Horridge, 1965.)

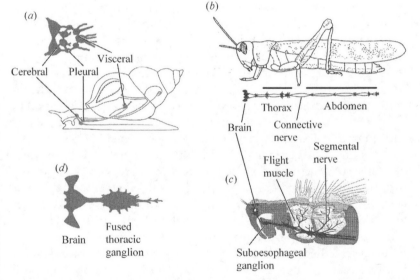

net, there are no or relatively few centralised points for collecting information, so that if one of these animals is prodded the resulting signal spreads out in rather a diffuse manner. Most animals are more highly organised for action, with bilateral symmetry, appendages for locomotion and a clear head end that carries special sensory structures. In most invertebrates, the central nervous system is composed of discrete **ganglia** connected with sensory structures and with muscles by nerves. Usually, each ganglion serves a particular area of the body. For example, in a snail, different individual ganglia are associated with the muscular foot, with the feeding muscles and other body parts (Fig. 1.4*a*). In clearly segmented animals such as crayfish and insects, there is usually one ganglion associated with each body segment (Fig. 1.4*b*, *c*) but complex segments, such as the hind two thoracic segments that bear wings and legs in most adult insects, tend to have ganglia that are larger and contain more neurons than those of more simple segments. Nerves called **connectives** run between different ganglia, although fusion between ganglia occurs in several arthropods such as flies (Fig. 1.4*d*), bees, moths and crabs.

The brain, which originated by fusion of different ganglia, is located above the gut, and includes several specialised areas dedicated to particular functions such as senses of sight or smell. One area of the insect brain, called the mushroom body, is important for learning and memory, and organisation of complex behaviour patterns. The invertebrate central nervous system is solid and, apart from the brain, is located near the ventral surface of the body. Nerves are essentially bundles of axons, which function to transmit signals over some distance. Each ganglion contains the cell bodies and dendrites of neurons and tracts of axons. In areas of **neuropile**, dendrites and axon branches form a tangled network, and in these areas neurons communicate with each other and process information. Often the cell bodies are clustered in areas distinct from the neuropile.

In vertebrates, the central nervous system includes the spinal cord and brain, which are hollow and dorsally located. Muscles and sense organs are connected with the spinal cord by segmental nerves (31 pairs in humans) and with the brain by cranial nerves. The ancestral segmentation of the brain is still apparent in some of the brain stem structures that connect the brain with the spinal cord, especially reticulospinal neurons (Chapter 4).

The vertebrate brain can be divided into three major structural areas: the hindbrain, midbrain and forebrain. Figure 1.5 shows brains

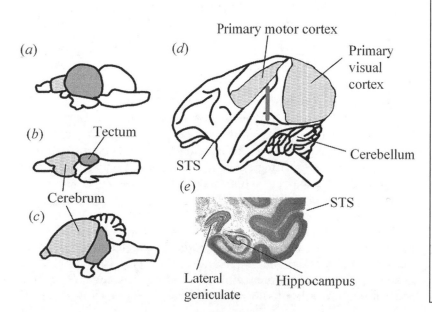

(a)

(b) Tectum

Cerebrum

(c)

(d) Primary motor cortex

Primary visual cortex

Cerebellum

STS

(e)

STS

Lateral geniculate   Hippocampus

Fig. 1.5 Vertebrate brains. Side views of brains of (*a*) teleost fish, (*b*) frog, (*c*) pigeon, (*d*) macaque monkey. (e) A section through a macaque brain in the region indicated by the vertical grey bar in (*d*). This shows the dark grey, folded cortex, which contains six layers of cell bodies, dendrites and axons. The paler white matter beneath it contains mainly axons. Structures in this section that are deeper in the brain include the lateral geniculate body, which receives information from the optic nerves and distributes it to the visual cortex; and the hippocampus, which is involved in spatial memory. STS in (*d*) and (e) is the superior temporal sulcus. (Part (e) drawn from www.brainmaps.org (accessed 2 September 2009).)

of various different vertebrate classes. In fish the hindbrain is most significant and in mammals the forebrain has become greatly enlarged, including the cerebral cortex. The midbrain also has a different proportion relative to other brain regions in different classes. It includes the tectum (indicated in the diagram of an amphibian brain, Fig. 1.5b), which is where visual neurons described in Chapter 3 are found. Birds have a much smaller cerebral cortex than mammals, but these two groups of animal are recognised as having similar cognitive abilities, with many birds able to learn complex spatial relationships or songs (Chapter 9). During evolution, birds have followed an alternative pathway of brain sophistication to that followed by mammals.

In terms of function, the mammalian cortex is divided into various sensory regions, motor regions and associative regions (Fig. 1.5d). To reach the sensory regions, signals from sense organs are processed in a series of brain regions, each dedicated to particular aspects of sensory stimuli. Most sensory information first passes through a region called the thalamus from where different aspects are parcelled into separate pathways that act in parallel, before signals from the different pathways reconverge at higher levels. Similarly, motor areas of the cortex control muscle actions by relaying information through a series of processing stations. Areas of white matter are distinguishable from areas of grey matter in both the brain and spinal cord: white matter is composed of bundles of neuronal axons, and grey matter contains cell bodies and dendrites, so is where neurons communicate with each other and process signals. Many areas of the brain are covered with distinct layers of grey matter. In fish that are considered related to ancestral vertebrates, there are three layers; some parts of the mammalian cortex have six or seven layers. The area of grey matter in these areas is sometimes greatly increased by folds in the surface, such as those that are a familiar feature of the human cerebral cortex. These folds on the surface of the cerebral cortex of a macaque monkey are shown in Fig. 1.5d. A slice through part of the macaque brain is shown in Fig. 1.5e (the region is indicated by the vertical grey line in Fig. 1.5d; notice that the superior temporal sulcus occurs in both parts). The slice shows the layered nature of the grey matter of the outer cortex, and indicates how the folds increase the number of nerve cells that can be accommodated in the grey matter. A human brain is estimated to contain 10^{12} neurons, each potentially connecting with thousands of others. Despite this complexity, many regions of the cortex are arranged as repeating modular columns, each of which is thought to process information in essentially the same manner. Deeper regions of the brain have significant impacts on behaviour – for example, the hippocampus (Fig. 1.5e; Chapter 8) is important in some types of learning; and the amygdala is responsible for processing emotional aspects, such as fear.

In the rest of this chapter, we explore two different approaches to discovering which neurons are involved in particular behaviours. The first approach is to study an animal with clear anatomical and behavioural specialisations, which are often reflected in obvious adaptations of parts of the nervous system. The second is to use genetic mutants to correlate changes in behaviour with neurons or parts of the nervous system.

An animal with a specialised nervous system: the star-nosed mole

A spectacular specialisation is the star-shaped nose of a species of mole, *Condylura cristata*, that lives in wetlands in the eastern parts of USA and Canada. Its nostrils are surrounded by 22 fleshy and mobile appendages or rays, a star-shaped structure that gives the mole its name. Although the rays are part of the nose and look a bit like fingers of a hand, the structure is actually a highly specialised touch-sensitive organ. It plays a vital role enabling the star-nosed mole to detect and catch food. This mole needs to sustain a high rate of energy input, and lives in an environment where there are large numbers of relatively small prey, including small worms, and this mole is an expert at eating large numbers of small food items rather than relying on finding larger single bites to eat (Catania and Remple, 2005). It can identify and take into its mouth up to ten pieces of earthworm within 2.3 seconds, making it the fastest mammal at food handling. Like other moles, this species spends most of its time

The nose of the star-nosed mole, *Condylura cristata*, is a specialised sensory structure that enables the mole to detect and identify worms and other food very quickly. Its surface is represented by three separate maps in the animal's brain. Inset is a scanning electron micrograph of part of the surface of the nose, showing individual touch-sensitive dome-shaped Eimer's organs, each about 40 μm across. (Photographs provided by Kenneth Catania, Vanderbilt University.)

in underground burrows and has poor eyesight. Investigation of the nose by Ken Catania and various colleagues led to an appreciation that it is a unique sensory structure, served by specialised regions of the cerebral cortex. The mole identifies the worm as food and distinguishes it from inedible objects using touch-sensitive structures on its star-shaped nose, although it is not yet known exactly how an edible object feels different from an inedible one. The rays of the nose are covered by touch-sensitive structures called Eimer's organs. On the surface, each is a dome-shaped projection of skin about 40 μm across, and below the surface is an array of flat skin cells regularly arranged, a bit like an onion. A number of different nerve endings supply the organ, and these are exquisitely sensitive to touch over a very small area of the ray. All moles have Eimer's organs on their noses, but the star-nosed mole has many more than other species.

When a star-nosed mole is feeding, it explores the ground with the rays of its nose, rapidly applying rays to a patch of substrate and then moving them to another patch. Up to 13 different patches are sampled each second. If the mole misses a chunk of worm with its nose rays, even by a very small distance, it ignores it. But if one of the longer rays hits a worm, the mole immediately moves its nose to examine the worm in greater detail with the shortest nose ray, the one numbered 11 on each side. A mole hardly ever bites and swallows a piece of food it has not first inspected with ray 11. This behaviour, of first noticing a potentially interesting object and then using a specialised region of a sense organ to examine it, is familiar to humans in our sense of sight. While our eyes make us aware of objects anywhere within a wide area, particularly if they are moving, a relatively small part of a retina called the **fovea** is able to examine an object in detail. While reading, the eyes move so that the images of words under attention are directed to the fovea, but more peripheral parts of the retina have insufficient acuity to distinguish letters from each other. In a similar way, a star-nosed mole uses ray 11, and to a lesser extent ray 10, to examine the feel of potential food.

So what is special about ray 11? It is smaller than other rays; and has 900 Eimer's organs, compared with 1500 that larger rays have. But each of its Eimer's organs is innervated by more sensory axons: ray 11 on average has 7.1 sensory axons per Eimer's organ compared with 4 for rays 1–9 and 5.6 for ray 10. This greater number of axons probably increases the sensitivity of individual Eimer's organs in ray 11 compared with other rays. But it is in the brain that the most significant difference between ray 11 and the others was found by Catania and Kaas (1997). Using small electrodes, they recorded responses from individual neurons in the sensory cortex of anaesthetised moles to touches of the nose and other parts of the body. As in other mammals, the star-nosed mole has a somatosensory cortex, a region of cortex that is dedicated to receiving and processing touch to the body surface. A light touch anywhere on the body surface is registered somewhere within the somatosensory cortex. But the area of cortex in which Catania and Kaas recorded responses when

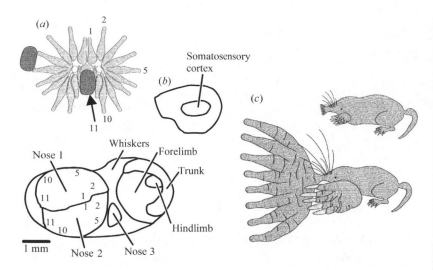

Nose 1

Whiskers Forelimb

Trunk

Hindlimb

Nose 2   Nose 3

1 mm

Somatosensory cortex

Fig. 1.6 The somatosensory cortex of the star-nosed mole. (a) Front view of the nose, which includes 22 fleshy rays on either side, numbered as indicated on the right. If the mole encounters an object with rays 1–9, as on the left, it very quickly moves the nose so rays 10 and 11 can inspect it in more detail, as shown. (b) The small drawing (above) shows the location and size of the somatosensory cortex in the left cerebral cortex, and below is a diagram of the organisation of the somatosensory cortex. Much of it is dedicated to mechanosensory information from the nose surface, with three separate maps of the nose in which particular areas are dedicated to each ray (numbers). (c) Anatomical drawing of the star-nosed mole (top) together with a 'molunculus', in which body parts are rescaled according to the area of somatosensory cortex dedicated to each. (Parts (a) and (c) redrawn from Catania, 1999; (b) redrawn from Catania and Kaas, 1996.)

the nose was touched was much larger than the area dedicated to other parts of the body. This is illustrated by the star-nosed 'molunculus' (Fig. 1.6c), in which the body parts of the animal are drawn scaled in proportion to the amount of cortex dedicated to them.

So the mole dedicates a particularly high proportion of its available brain power to processing information from touch receptors on its nose. The parts of the cortex associated with the nose were further identified by using a stain for the enzyme cytochrome oxidase. This method is called activity-dependent staining, because it picks out cells that have been most active in a particular region of tissue. When applied to the somatosensory cortex of a star-nosed mole after experiments in which the nose rays have been touched, distinct stripes are seen in the somatosensory cortex, with one stripe for each ray. As is usual for both sensory and motor cortex, the left side processes information from the right nose rays and vice versa. The stripes are arranged in an orderly way, corresponding with the way the rays are arranged around the nose, but nose ray 11 has a larger area of cortex associated with it than the other rays: each sensory axon from ray 11 is associated with more than twice the area of cortex surface compared with rays 1–9. This kind of arrangement is quite common for other touch-sensitive sensory systems in mammals. One that has been particularly studied is the whiskers of rats, each of which corresponds with a barrel-shaped ensemble of cortex in the rat's brain. So the surface of the body is represented in a map-like manner on the surface of the cortex.

In the star-nosed mole, there is not just one map of the nose rays, however, but three distinct maps. A nose-like array of dark staining is found in three different, closely spaced areas of the brain's cortex. It is likely that each of the three maps is dedicated to different types of sensory features, so that the different features are processed in parallel, enabling a mole to distinguish extremely rapidly between edible and inedible objects that its nose has touched. The cortex of the star-nosed mole, then, is specialised to deal with fast food.

Identifying neurons involved in behaviour: fruit fly courtship

If there is a way that individual neurons can be made to light up whenever a particular behaviour is performed, that would provide important clues about where to look for the neurons that control the behaviour. Brains can be examined with various scanners and imaging techniques that indicate, often through measures of metabolic activity, which areas are most active during different behavioural or thought tasks. Those techniques are often important in medicine, but their application to study animal behaviour is limited because they usually require an animal to keep still during a scan, and the areas and time periods they sample are usually quite large. Another technique is to attach tags to mark neurons according to the genes that they express. This technique is applied particularly successfully to the fruit fly, *Drosophila melanogaster*. Not only has the genome of this insect been mapped, but it is also fairly routine to create different strains that are genetically engineered to incorporate into the genome tags of different kinds to label the products of particular genes. Sometimes the engineering is applied to the whole animal, but it can also be restricted to particular parts of the nervous system. Some tags enable the cells expressing the genes of interest to be stained, and so identified on maps of the nervous system. Others enable the neurons expressing those genes to be selectively activated or inactivated by an experimenter, adding weight to an argument that those neurons are involved in a particular behaviour.

A behaviour that is extensively investigated using this technique is courtship in *Drosophila*. These flies tend to congregate near rotting fruit, attracted by its smell, a strategy that increases the chance that males and females will find each other. Once a male and female are close enough to exchange signals, a courtship ritual follows in order to confirm that they both belong to the same species and are ready to mate. Male and female *Drosophila* emit their own scents, or pheromones, and the smell of a female combined with sight increases a male's urge to court. Courtship normally lasts several minutes and involves a number of distinct action patterns, normally occurring in orderly sequences. Initially a male follows and orients towards a female, and taps her with a foreleg, perhaps using chemosensors to taste her or delivering his own pheromone onto the female. Then the male extends either his left or right wing and vibrates it (Fig. 1.7), using some of the same muscles that power the wing during flying. Wing vibration produces sound that acts as a love song, and *Drosophila* makes two types of song: one in which the wing is oscillated regularly for a few seconds; and the other in which the wing is vibrated in regular, short pulses. The interval between pulses differs between species, and females find pulses of males of their own species more attractive than pulses of others. In *Drosophila melanogaster* the inter-

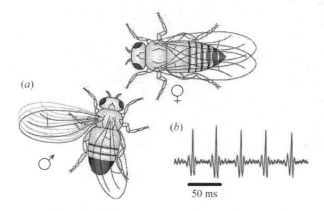

Fig. 1.7 Courtship behaviour in *Drosophila*.(*a*) Drawing of a male singing to a female. The male extends one wing, and vibrates it up and down. (*b*) Recording of sound pulses of a male's song. (Part (*a*) redrawn from Sokolowski, 2001; (*b*) redrawn from Clyne and Miesenböck, 2008.)

50 ms

pulse interval is 32–38 ms, drifting in a regular way that might also be a species-specific cue. The female hears the song with her antennae, and is also stimulated by male pheromones. In addition to singing, the male licks the female's genitalia. If the song and licking engage the interest of the female, she consents to copulate when the male curls his abdomen while approaching and mounting. Otherwise she tries to move away from and avoid the male. Males can learn to recognise signs of rejection, so experienced males leave unreceptive females alone much earlier in courtship than inexperienced males.

The *fruitless* gene and courtship

In the early 1960s, a gene that affects courtship and its consummation was identified in *Drosophila*. It is called '*fruitless*', named because one allele causes males to court both sexes indiscriminately but never to consummate relationships. It acts together with other genes, especially one called '*doublesex*' that has a similar counterpart in mammals, to ensure males and females differ from each other. *Fruitless* is an extremely large gene, and neurons are the only cells, apart from one small muscle, in which the proteins it encodes are expressed. Its gene products are found in about 2% of the 130 000 neurons throughout the body. Its expression normally occurs during the pupal stage of the fly's life history, during which the body of the maggot is extensively rebuilt to generate the adult stage that has wings, legs and reproductive organs. *Fruitless* codes for four different proteins, each of which is a regulatory gene whose protein has the function of activating other genes, directing cells to take particular developmental pathways. One of its products is processed in one of two different ways depending on the sex of the animal: fru^m is expressed in male cells, and fru^f in female cells (in *Drosophila*, an embryo with one X chromosome becomes male, and an embryo with two X chromosomes becomes female).

A number of different lines of experimental evidence have been accumulated by several different research groups that link *fruitless* with courtship behaviour, and in particular fru^m with male-specific courtship behaviour (Manoli *et al.*, 2005, 2006; Demir and Dickson, 2005; Stockinger *et al.*, 2005). Apart from some specific brain areas, the fru[m] protein occurs in neurons throughout the nervous system including particular subsets of neurons in the visual and olfactory pathways, both of which are involved in triggering courtship behaviour. There are a few neurons in male *Drosophila* that are not found in adult females: probably these neurons die during pupal development in the female, but are protected in males by the fru[m] protein. Some brain neurons have a different anatomy in a male compared with a female adult fly, but in many cases the differences between male and female neurons seem to be very subtle or absent.

If flies are created with mixtures of patches of male and female tissue, only those that are male in particular brain regions will sing courtship songs. Interfering with the normal expression and function of fru[m] alters a male's normal behaviour. For example, if a small number of neurons in a male's olfactory system are engineered in a way that stops them making the fru[m] protein, such males become indiscriminate in their courtship and will persist in courting other males as well as females for long periods. A likely reason for this is that these engineered males cannot distinguish between male and female pheromones. In another genetic engineering manipulation, male neurons that expressed fru[m] were tagged with a temperature-sensitive enzyme pathway. Warming the flies for a short time to 31 °C selectively inactivated the neurons that contained fru[m], but not others in the nervous system. These flies would court normally at lower temperatures, and would walk, fly and feed normally at 31 °C, but would not perform courtship at this temperature.

Does the fru[m] protein instruct a developing male nervous system to construct networks that are absent in an adult female, or are there particular key neurons that the male has but the female lacks? Dylan Clyne and Gero Miesenböck (2008) quite literally threw light on this by genetically engineering neurons that could be excited by brief ultraviolet pulses. In the strains of flies they studied, neurons in which the *fruitless* gene was expressed also made a protein that is not normally found in *Drosophila* neurons, but which made these neurons sensitive to ultraviolet. Because only the neurons that expressed the *fruitless* gene expressed the light-sensitive protein, most neurons in the body were not excited by ultraviolet pulses. In a male with this additional protein, shining a 10 ms long pulse of ultraviolet light reliably causes him to shake one wing in the same way that is characteristic of courtship. This shows that the thoracic ganglion contains the neuronal pattern generator responsible for the programme for this aspect of courtship. Courtship song is only triggered by the light pulse if the nerves that connect the brain with the thorax had been cut; the same light treatment applied to intact engineered male flies rarely caused courtship movements. This is not surprising because the light might well excite simultaneously some neurons in the brain that switch on wing shaking, but others that switch it off – a bit like pressing a car accelerator and brake at the same time.

Clyne and Miesenböck went on to test females that had also been genetically engineered to make the light-sensitive protein in their *fruitless*-expressing neurons. They found that, as in males, as long as the brain was disconnected from the thorax, the females would also often respond to a strong pulse of ultraviolet by shaking either the left or right wing to make a song. This shows that the network that generates singing normally lies dormant in the female and does not need instruction from fru^m, and suggests that the female nervous system lacks some kind of switch to activate the network.

The courtship song by the engineered females was not as good as that made by the engineered males. This was shown by a rather sneaky kind of experiment. Recordings were made of the songs made by males and females induced to sing by light pulses and then were played to receptive female flies. Each female was paired in a container with a male that was normal except that his wings had been cut off so that when he tried to court he was not successful because he sang silently. But if the recorded experimental song from a male was played, these pairs of flies usually consummated their courtship: the recorded song plus tactile stimulation from the wingless male combined to persuade the female to copulate. But the songs recorded from light-stimulated females were ineffective at persuading females to mate. So, although the female contains most of the neuronal elements needed to generate a song programme, the pattern generator does not quite work well enough to be functional.

In the thorax, the number of neurons that express the *fruitless* gene in one form or another is just under 220 in females, and slightly more than 220 in males. The extra neurons might be needed to fine-tune the song. Clyne and Miesenböck (2008) showed that the fru^m protein really is responsible for directing a network of neurons to develop into a fully functional courtship programme. They made a strain of female flies in which neurons expressed fru^m rather than fru^f. Recordings of these songs were effective at ensuring success by the courting, wingless males.

A key site where male and female nervous systems differ from each other is likely to be in the brain. This is clearly indicated by these experiments, and by earlier work. When a strain of flies in which a small subset of brain neurons were engineered to prevent them making the fru^m protein, courtship behaviour was drastically affected (Manoli and Baker, 2004). These males certainly had not lost the urge to mate, but had lost their patience to do so. Instead of generating an orderly sequence of distinct behaviour patterns that could last minutes, they often omitted initial stages of orienting towards and tapping the female, but tried to sing, and lick and copulate with the female all at the same time. The region of the brain affected is called the median bundle (Fig. 1.8), and about 60 neurons here express fruitless proteins. Perhaps they function as parts of a command centre that organises the sequence of programmes to be played out by the thoracic nervous system. Neurons of the median bundle send axons to the suboesophageal ganglion, which

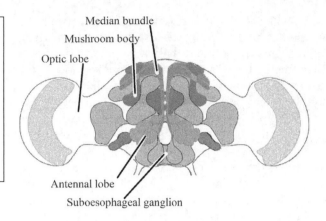

Fig. 1.8 *Drosophila* brain showing some of the most significant areas of neuropile. The median bundle contains neurons known to be important in the release of male courtship behaviour. The gut runs through the hole between the brain and suboesophageal ganglion. (Redrawn from http://flybrain. neurobio.arizona.edu/ (accessed 2 September 2009).)

is just below the brain, where they enter regions that probably combine signals from a number of different sense organs in a way that is necessary to co-ordinate behavioural sequences appropriately. Neurons of the median bundle also send axons to another region of the brain, the protocerebrum, from where other neurons send axons to the thoracic motor centres, and could provide the messages to switch on or terminate the motor programmes.

It may not seem surprising that the nervous systems of males and females are wired differently. But an electrical engineer designing control networks for courtship might create completely separate networks, rather than include a functional singing control system in females without the switch that activates it in males. But we have to remember that neuronal networks in *Drosophila* were not designed for a predefined purpose but have evolved through mutation and natural selection. Also, although the set of proteins coded by *fruitless* identifies neurons that are responsible for courtship behaviour, the neurons are likely to be involved in other activities as well. For example, some of the neurons that control wing movements are probably also members of networks that control flying. The function of the fru^m protein is not just restricted to courtship behaviours either, but is significant in other sex-specific behaviours. *Drosophila* will fight each other, for example squabbling over food, and sex-specific differences in fighting style correlate with expression of different forms of *fruitless* (Vrontou *et al.*, 2006). Some of the manoeuvres used in fighting are common between the sexes, but others are sex specific. Females use shoving and head butting when fighting other females whereas males use lunging and boxing when fighting other males, and males also form dominance hierarchies that specify rank in mating rights. So expression of fru^m or fru^f does not just lay the foundations for courtship behaviour, but also directs sex-specific aspects of fighting.

It could be dangerous to make general extrapolations about the control of behaviour from studies on a single species. *Drosophila* has several unusual features, and it could be a peculiarity of this fly that the female possesses but does not use networks needed to produce song. Studies of mice, however, suggest that mammals adopt a similar approach to generating different courtship behaviours in males and females (Kimchi *et al.*, 2007). In mice, pheromones are important in sexual and other social behaviours, and they are detected by the vomeronasal organ, an olfactory organ that most mammals other than humans have in addition to the nose. Pheromone detection depends on a protein called Trpc2, which is part of the mechanism enabling sensory cells to respond to these chemicals. Males that lack Trpc2 fail to discriminate between males and females, and perform poorly in aggressive encounters with other males. Females that lack this protein behave in an even more unusual manner. When paired with normal females, they perform many behaviour programmes that are characteristic of male mice, including mounting, thrusting and making a complex series of ultrasonic squeaks. Surgically removing the vomeronasal organ in normal female mice also causes them to behave in this unusual male-like manner. So female mice, like female fruit flies, possess neuronal programmes for generating behaviour patterns that are normally characteristic of males. In mice, these behaviours are usually held in check through the action of pheromone-activated pathways. In fruit flies, by contrast, normal females lack the means to activate the male behaviour programmes.

Conclusions

The scientific study of animal behaviour was put on a sound footing with the classical ethological work of Tinbergen, Lorenz and colleagues. They pioneered techniques for investigating the natural behaviour of animals, including the use of simple experiments in conjunction with field observations. In the course of this work, they developed a number of key concepts that have helped to guide efforts to understand the mechanisms of behaviour. These concepts provide a starting point for a physiologist to look for neurophysiological mechanisms that relate to behaviour.

One concept is that of the motor pattern, in which aspects of an animal's movements are clearly recognised. An ambition of physiologists can then be to understand how the programme for co-ordinating that sequence of movements is generated in the nervous system. Another concept is that of the releasing mechanism, which can be thought of as a kind of neural filter that extracts specific sign stimuli and ensures appropriate action is taken. The nervous system must be organised so as to sort out different stimuli and to make decisions about which motor pattern to put into action at any one time. The egg-retrieval response in nesting birds provides a good example of the development of these two concepts, although the neurophysiological bases of the egg retrieval motor programme or of recognition of eggs have not been specifically investigated in geese. The usefulness of these classical ethological concepts will be apparent in many of the examples described in the book, starting with prey capture by toads (Chapter 3), which is an excellent example of the physiological basis of an innate recognition mechanism.

The first step in discovering the neuronal basis of a particular behaviour is usually to identify the individual neurons involved. To establish that a specific neuron is involved in a particular behaviour is, in practice, quite hard to do unambiguously, partly because the needs to investigate signals in individual brain neurons and to ensure an animal can behave as naturally as possible are usually incompatible with each other. Once neurons involved in a behaviour have been identified, the way they are organised into networks with specific functions in behaviour, such as filtering out sign stimuli or generating a particular motor pattern, can be investigated. This work involves tracing the flow of signals from one nerve cell to the next, and so is most easily done where particular nerve cells can be uniquely identified.

Diversity is important in neuroethology research. As well as using a great variety of techniques to study brains, neuroethologists study many different invertebrate and vertebrate species. Often, particular specialisations in an animal's nervous system provide a way in to investigate a problem of general significance in animal behaviour. The star-nosed mole provides an excellent example where anatomical, behavioural and neuronal specialisations involved in prey handling correlate well with each other. Study of the way information from the nose is represented as maps of the nose surface in the cortex is a clear illustration of the way information is represented in a mammalian cortex. One of the specialisations of this mole is that it has not just one but three different nose surface maps in its cortex. What are the relative functions of these maps; and what happens during development to create three maps rather than a single one? In *Drosophila*, the *fruitless* gene is a different kind of specialisation, and is a rare example where a single gene can be unambiguously related to clearly defined behaviour patterns. Because its expression labels the neurons that are involved in courtship, and also enables the neurons to be manipulated, it provides an invaluable route by which to explore the neuronal basis of behaviour. However, it is technically extremely challenging to study the flow of signals between individual neurons in *Drosophila* because of its small size. The sea hare, *Aplysia*, has some neurons so large that a *Drosophila* would almost fit into one, and networks made by its neurons are more amenable for studying how networks of neurons process electrical signals.

Questions

What is special about animal behaviour compared with other aspects of biology?

Is there likely to be a gene for egg retrieval behaviour by geese?

Summary

Animal behaviour

- An aim of ethology, the scientific study of animal behaviour, is to analyse behaviour into basic elements such as: sign stimuli; innate releasing mechanisms; and fixed action patterns. These elements can be performed by neuronal processes such as sensory filtering and pattern generation.

Neurons and nervous systems

- Electrical signals are carried and processed in two types of process in a neuron: dendrites, which receive signals; and axons, which carry output signals.
- An aim of neuroethology is to trace the neuronal routes by which information is collected, transformed and transmitted to control animal behaviour.
- In many invertebrates, neurons are aggregated into ganglia. Axons connect ganglia with each other, with sense organs and with muscles.
- In vertebrates, the surface of the brain in many regions consists of 5–7 layers of cells. This is particularly large in the cerebral cortex of mammals, which is divided into various sensory, motor and association regions.
- One approach to identify the neurons involved in a particular behaviour is to study an animal with clear anatomical and behavioural specialisations. Another is to use genetic mutants to correlate behaviour with neurons or parts of the nervous system.

Star-nosed mole

- Eats many pieces of food per second and identifies food by touch with its nose that includes 22 motile rays.
- Ray 11 is a fovea and is always used to examine food in detail before a bite. It is the smallest ray, but has the greatest number of sensory axons.
- A particularly large proportion of the somatosensory cortex is dedicated to the nose, especially ray 11. There are three distinct maps of the nose.

Drosophila courtship

- A sequence of events, including a male singing by vibrating one wing.
- Males with a *fruitless* gene mutation court indiscriminately but never finish.
- The *fruitless* gene can be tagged to allow the neurons that express it to be seen. It is expressed in 2% of neurons in the pupa, spliced to code for four different regulatory proteins.

- The fru^m protein is associated with male courtship. Interfering with it alters normal courtship; experimentally exciting fru^m neurons switches on courtship.
- Female thoracic ganglia contain neuronal networks that can create courtship behaviour, but their normal activation probably requires male-specific fru^m brain neurons.

Further reading

Catania, K. C. (1999). A nose that looks like a hand and acts like an eye: the unusual mechanosensory system of the star-nosed mole. *J. Comp. Physiol. A*, **185**, 367–372. A review of the experimental work that established the mapping of the nose onto the cortex of this unusual-looking mammal.

Greenspan, R. J. (2007). *An Introduction to Nervous Systems*. New York: Cold Spring Harbor Laboratory Press. This book is about the roles that single neurons play in behaviour, but takes a different point of view from this present volume. An excellent introduction to invertebrate nervous systems, with a lot to think about.

2

Signals in nerve cells: reflexes in mammals and insects

Individual neurons receive and transmit information in the form of small electrical signals. Most significantly, they also **integrate** the signals they receive, combining inputs from different sources and over time to determine their own outputs. Integration enables neurons to work together and form networks that perform the kinds of operations needed to control behaviour, such as sensory filtering or motor pattern generation. In this chapter we first provide the information about how neurons work that is needed to understand the later chapters in the book, starting with a description of how a simple sense organ works. We then illustrate the ways in which neurons receive and integrate signals in simple behavioural responses in two different animals: a mammal, and an insect. Sometimes these movements are called reflexes; they are almost automatic reactions to simple stimuli that affect a limited part of the body, and involve pathways of only a few neurons. They illustrate very well the basic principles of neuronal physiology.

The shortest neuronal pathways in arthropods and vertebrates include just two connected neurons: a sensory receptor, and a motor neuron. **Sensory receptors** are cells specialised to receive a particular form of environmental stimulus – such as light energy, mechanical movement, or chemical molecules in an odour – and to respond with an electrical signal. Some sensory receptors provide information about positions or movements of an animal's body parts, and receptors that do this are called **proprioceptors**. Many proprioceptors are placed within muscles, tendons and joints and play important roles in ensuring the animal's movements are executed smoothly as circumstances change, such as a change in terrain. A **motor neuron** is a nerve cell that connects with and controls muscle fibres. The sensory neuron connects with the motor neuron, and the motor neuron connects with muscle fibres at junctions between cells that are specialised to transmit electrical signals, called **synapses**. The neuron that is passing a signal on is called **presynaptic**; and a neuron receiving signals from others is called

postsynaptic. In most synapses, the presynaptic and postsynaptic neurons are separated by a small gap, and communication is effected by a chemical neurotransmitter, but there are also electric synapses where electric current flows directly between two neurons. In most neuronal pathways sensory neurons do not make synapses directly with motor neurons, but communicate through additional neurons, collectively called **interneurons** (or sometimes association neurons or relay neurons), which greatly outnumber motor neurons. Some interneurons connect neurons in distant parts of the nervous system, whereas others have a much more restricted extent, for example being confined to a small region of one ganglion.

The signals that neurons produce and process are localised changes in the electrical potential, or voltage, across the cell membrane. The nature of these signals was first revealed at about the same time as ethologists were developing their scientific approach to studying animal behaviour, in the 1940s, when electronic devices were developed that allowed the small electrical signals produced by individual neurons to be picked up, amplified and displayed. It is useful to distinguish types of signals: spikes and graded potentials. **Spikes** or action potentials are the most familiar signals that neurons produce and they travel by an active process of propagation, which provides a means for long-distance communication in nervous systems. They are pulse-like, 'all-or-none' signals, sometimes compared to digital signals in electronic systems. **Graded potentials** are much smaller voltage signals than spikes. They are not fixed in amplitude, and can be summed together, allowing neurons to combine signals in time and space. Graded potentials originate at synaptic junctions between neurons, or within sensory receptor cells, and they are responsible for controlling spike production.

A simple neuron: crayfish stretch receptor

We shall introduce the topic of signals in nerve cells by describing the operation of a sensory cell called a muscle receptor organ in crayfish, which is relatively simple and has been thoroughly investigated. It consists of a thin strand of muscle in which a sensory cell is embedded (Fig. 2.1a). Dendrites of the sensory cell branch from the cell body and mingle with the fibres of the slim receptor muscle. The cell body is also connected to the axon that runs into the central nervous system. The receptor muscle runs alongside a larger muscle that spans the joint between two segments in the abdomen, and when the larger muscle contracts it straightens the abdomen. The receptor organ is a proprioceptor and is part of a reflex pathway that allows the crayfish to hold its abdomen straight when an external force tries to flex it. The receptor organ's task is to **transduce** mechanical stretch, converting the mechanical energy into an electrical signal in the dendrites. The way this happens is that the mechanical stretch of the dendrite membrane has the effect of causing channels embedded in it to open and

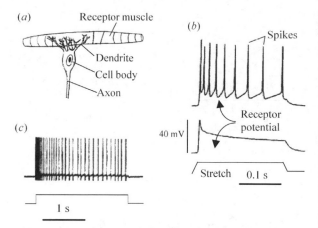

Fig. 2.1 Structure and function of a crayfish muscle receptor organ. (*a*) Diagram showing how the sensory neuron's dendrites are embedded in a thin strand of receptor muscle. (*b*) Two different recordings made with an intracellular electrode in the cell body to a stretch (bottom trace). The middle trace shows a recording of the receptor potential. In the upper trace, both the receptor potential from the dendrites and spikes from the axon were recorded. (*c*) Sensory adaptation in a crayfish muscle receptor organ in response to a sustained stretch (lower trace). Spikes were recorded with extracellular electrodes (top trace); spike rate is initially high, but declines to a lower rate during the stimulus. (Redrawn from Rydqvist *et al.*, 2007.)

allow sodium ions to pass through. The ions carry a net positive charge to the inside, creating a voltage change in the sensory cell that is called the **receptor potential**. The size of the receptor potential is proportional to the strength of the stretch.

The cell body sums together the receptor potentials originating in all the dendrites. But the receptor potential is only a few millivolts (mV, thousandths of a volt) in amplitude, and does not travel very far along the axon before it fades away. At the start of the axon, the receptor potential is converted into another form of electrical signal: **spikes**. A spike is a brief, pulse-like change in voltage across the cell membrane and has an amplitude much larger than the receptor potential, about 100 mV. Its amplitude is fixed, and it travels along the axon at a speed of a few metres per second by a process called propagation in which, like a spark travelling along a trail of gunpowder, a spike in one stretch of axon triggers the next stretch to generate a spike. The larger the receptor potential is, the greater the rate at which spikes are produced. As a result, the strength of a stretch is encoded as the rate, or frequency, of spikes, measured in spikes per second. The sequence of events from mechanical stretch to receptor potential to spikes is shown in Fig. 2.1*b*. Note that the order of events flows from the bottom trace towards the top trace in this figure. As far as possible, we shall stick to this bottom-towards-top order throughout the book. In some cases, we have altered trace order from the original research papers as there is no universally recognised order for arranging traces from recordings.

When a stretch is maintained, the rate of spikes is initially high, but declines to a more sustained level (Fig. 2.1*c*). This reduction of response during a maintained stimulus is called **sensory adaptation**, and it is important in allowing the nervous system to ignore constant stimuli and to concentrate on changes in stimuli, which are most likely to be significant and require a behavioural response. Different sensory cells adapt at different rates. Crayfish have two pairs of stretch receptors in each segment in the abdomen, and one type adapts much more quickly than the other. The more rapidly adapting receptors are termed **phasic**; they respond briskly but briefly to any increased

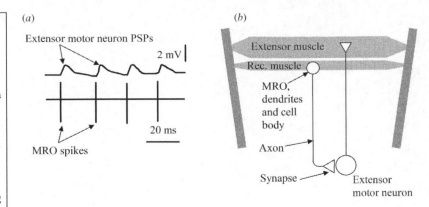

Fig. 2.2 Synaptic transmission between a muscle receptor organ (MRO) and an extensor motor neuron in a crayfish abdomen. (*a*) When the MRO is gently stretched, each MRO spike causes a postsynaptic potential (PSP) in the motor neuron. MRO spikes were recorded with extracellular electrodes on the nerve that contains its axon; an intracellular electrode recorded the motor neuron EPSPs. (*b*) Diagram showing the two neurons involved, which are drawn using symbols as described in the text. The neurons are shown associated with the extensor muscle and receptor (Rec.) muscle, which are attached to cuticle plates in adjacent segments. (Part (*a*) redrawn from Wine, 1984.)

stretch. The receptors that give a more sustained level of response are termed **tonic**, and can perhaps provide over time a more accurate measure of stretch. In addition, crayfish can exert control over the sensitivity of their stretch receptors: the thin receptor muscle can contract in response to spikes in its own motor neuron; and the stretch receptor itself receives synapses from central neurons.

When a muscle receptor organ is stretched, it generates a train of spikes that travel within a few milliseconds into the abdominal ganglion that serves its segment. Signals are then conveyed to various target neurons by way of synapses: functional connections between neurons. A spike in a presynaptic neuron is followed after a short delay by an electrical response in the postsynaptic neuron, called the **postsynaptic potential** (PSP). A train of muscle receptor organ spikes and the PSPs they set up in an extensor motor neuron are shown in Fig. 2.2*a*. Like a receptor potential, a PSP is usually a relatively small electrical signal of the order of less than 1 to 10 mV in amplitude, and it fades within a short distance from its origin. Often, symbols are used to represent neurons and synapses in diagrams of neuronal networks. The simple network of muscle receptor organ and extensor motor neuron is shown in Fig. 2.2*b*, which summarises some of the conventions adopted in this book. Usually, dendrites and cell bodies are indicated together by a circle and axons by lines. In Fig. 2.2*b*, the presynaptic terminals of the receptor organ are indicated by an open triangle.

In the next few sections, we describe how electrical signals are recorded from neurons and give an introduction to some of the important properties of these signals that is needed to understand later sections of the book.

Spikes: how they are recorded and some of their properties

A spike travels as a wave of excitation along an axon, enabling long axons to convey signals between distant parts of a nervous system. Some of a spike's main characteristics are illustrated in Fig. 2.3, which shows in part (*a*) an arrangement for an experiment on a small length

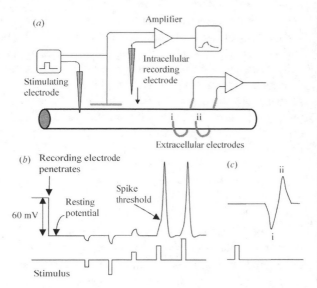

Fig. 2.3 Spikes and axon excitability. (*a*) Arrangement for an experiment on a length of axon. Two intracellular electrodes are shown, one for stimulating the axon and the other for recording its membrane potential. On the right, a pair of silver wire hooks act as extracellular electrodes. The triangle is the electronic symbol for an amplifier. (*b*) At the start of the experiment, the recoding intracellular electrode penetrates the membrane and records the resting potential. Then five short stimulating pulses of current are delivered. The axon responds passively to the two first, negative pulses. Note that the voltage waveform of the response changes more slowly than the square stimulus pulses; this is due to the electrical capacitance of the axon membrane. A small, positive stimulus pulse also causes a passive response, but larger pulses are sufficient to excite the membrane to the threshold for a spike. (*c*) A spike recorded by the extracellular electrodes.

of axon. Two glass intracellular **microelectrodes** are shown, each consisting of a fine-bore glass capillary tube that has been heated to near melting point along a small length and then drawn out. The electrode tip is extremely sharp, and has a small opening, less than 1 μm across. The lumen of the electrode is filled with a conducting salt solution such as 2 M potassium acetate, and a silver wire enables it to be connected to electrophysiological equipment.

Microelectrodes can be used to record from, to stimulate and to inject dye into a neuron. The microelectrode on the left of Fig. 2.3*a* is connected to a stimulator, and the other microelectrode is connected to an amplifier that enables the experimenter to measure the voltage across the cell membrane. The amplifier measures the voltage at the electrode tip with reference to a silver wire placed outside the axon. In the drawing, the tip of the recording electrode is outside the axon, and at the beginning of an experiment it would be carefully advanced with a micromanipulator to penetrate the axonal membrane. The glass tip of the electrode tip seals into the membrane, so the salt solution inside is in electrical contact with the cytoplasm. As the electrode penetrates the membrane, the voltage that is measured jumps downward to a level, usually 60 to 70 mV (thousandths of a volt) negative to the extracellular fluid. This sustained voltage across the membrane is called the **resting potential** and is characteristic of most animal cells, not just neurons.

The stimulating electrode is used to deliver brief pulses of electrical current to the inside of the axon, which alters the voltage across the membrane. The trace from the recording electrode (upper trace in Fig. 2.3*b*) shows that a pulse of negative current briefly increases the polarity of the membrane, or **hyperpolarises** it, and that the voltage change increases with stimulus strength. Small pulses of positive stimulating current briefly decrease the polarity of the membrane, or **depolarise** it. The first three stimulus pulses in Fig. 2.3*b* cause passive responses by the axon membrane, but the fourth and fifth are sufficiently strong to trigger an active response – a spike. A spike is triggered when the membrane is depolarised to a threshold value. The membrane potential then changes rapidly and its polarity briefly reverses and then reaches about 50 mV positive inside before returning to the original resting potential. The whole event is usually over in a fraction of a millisecond, depending on temperature. Increasing the size of the stimulus pulse beyond threshold, as in the fifth compared with fourth pulse in Fig. 2.3b, does not affect the size of the spike although it takes a little less time to reach threshold. When a patch of membrane has produced a spike, there is a brief **refractory period** when it is unable to be excited again immediately, so that spikes remain discrete and do not fuse into each other.

As well as two glass intracellular electrodes, Fig. 2.3b shows a pair of extracellular electrodes consisting of two silver hooks positioned quite close to each other and around part of the axon. These hooks are connected to an amplifier that responds to changes in the difference in voltage recorded at the two hooks. When a patch of axon is excited to make a spike, small electrical signals flow into the extracellular fluid, and it is those that the extracellular electrodes pick up. As shown in Fig. 2.3c, the waveform of a spike recorded in this way typically has a trough when the spike is nearest one electrode (i), and a peak when the spike is nearest the other electrode (ii). The order of trough and peak depends on how the electrodes are connected to the amplifier; reversing the connections to the amplifier would change the response from electrode (i) from a trough into a peak. From this experiment, the conduction velocity or speed of travel of the spike along the axon could be measured by dividing the distance between the electrodes by the delay between the trough and the peak in the recording. Mammalian motor neuron axons generally have conduction velocities of several tens of metres per second (m/s), and the largest insect axons up to about 10 m/s.

It is easier to record spikes from axons using extracellular compared with intracellular electrodes. Also, extracellular electrodes can be placed around bundles of axons or nerves and record spikes from several different axons. The size of the recorded spike varies between axons, depending on the strength of the current that travels from outside the axon to the electrodes so axons that are either wide or near to the electrodes produce the largest spikes in extracellular recordings. The amplitude of a spike from one axon is quite constant, but is very much smaller than the voltage recorded by an intracellular electrode. As with intracellular electrodes, extracellular electrodes can also be used to stimulate axons.

Conduction of signals in axons and dendrites

Axons are relatively poor conductors of electric signals compared with insulated metal cables. An appreciation of their passive

(a)

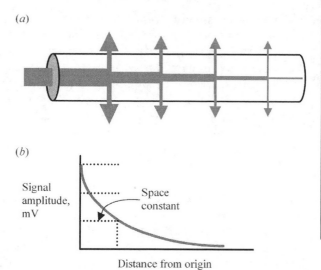

(b)

Signal
amplitude,
mV

Space
constant

Distance from origin

Fig. 2.4 Decay in passive signal strength along an axon. (a) Diagram to illustrate how the strength of electric current (indicated by arrow width) flowing within an axon declines because current is progressively lost by flowing through the membrane. (b) A graph to show the exponential decline in the amplitude of an electrical signal across the membrane with distance from its origin. The space constant is the distance over which the signal reduces to roughly one third of its original value.

electrical properties explains why spikes are needed for communication over distance, and is also useful to understand the function of dendrites. Axons or dendrites are narrow, and their salty cytoplasm does not conduct anything like as well as a copper wire. Also, the cell membrane is a poor insulator compared with plastic, so some of the electric signal that travels along the inside of an axon escapes through the membrane into the salty, conductive extracellular fluid (Fig. 2.4a). Consequently, a voltage signal across an axon membrane becomes progressively smaller with distance from its point of origin. The decline in signal strength with distance has a negative exponential form (Fig. 2.4b). A measure of signal travel is provided by the **space constant** of the axon, which is the distance over which the signal declines to roughly one third of its original value (it is not possible to pinpoint exactly where the signal declines to zero). The space constant of an axon depends on its width, and a 0.5 mm wide squid giant axon has a relatively large space constant of about 5 mm, so a 100 mV spike at one point in the axon will be accompanied by a simultaneous 33 mV signal 0.5 mm away.

A depolarisation of 33 mV is more than sufficient to bring a patch of axon membrane to threshold, so a spike in one patch of membrane will trigger a spike in another patch that is somewhat more than 0.5 mm distant. The wider an axon, the greater the distance along which excitation spreads, and so the faster the conduction velocity of a spike. Note that the passive electrical signal travels at the speed of light; the delay in spike propagation is caused by the time it takes for the cell membrane to trigger a spike at each successive location. Although excitation spreads passively in both directions away from a patch of excited membrane, the spike only travels in one direction because a recently excited patch is in its refractory period.

Some axons have improved insulation, which increases their space constant and their conduction velocity. These include many axons of vertebrates, which are wrapped by a myelin sheath of lipids made by a special class of glial cells (there are several types of glial cells in central nervous systems, and they provide mechanical and nutritive support to neurons). A few neurons in a few invertebrates, including some prawns, have axons that are insulated in a similar way. At intervals, there are short nodes where an axon is not enclosed by myelin, and a spike travels by jumping from node to node. Myelination, together with a large width and warm body temperature, gives mammalian axons particularly fast spike conduction velocities, approaching 100 m/s.

Dendrites conduct passive electrical signals in much the same way as axons, but with a few important differences, associated with their function in combining signals together rather than ensuring they are relayed from one end to the other. They are usually considered to be inexcitable, lacking the signal boosting capability that an axon has in propagating spikes. Consequently, signals in dendrites decline in size from their point of origin. Because dendrites can be extremely narrow, they have relatively short space constants. Signals in dendrites usually originate at synapses, and although a signal might be several millivolts in amplitude at the synapse, the signal can have declined to a fraction of a millivolt in the cell body or axon origin. Dendrites often form complex branching structures, and the ways in which passive signals flow from one part of a neuron's dendritic tree to another can be quite complex.

Ions, channels and signals in neurons

The way a neuron generates its resting potential, and the potential changes that underlie electrical signals, including spikes and synaptic or receptor potentials, is by a controlled flow of charged ions through proteins called **ion channels** that are embedded in the cell membrane. Each channel is selective for particular ions, such as sodium, potassium, chloride or calcium, and allows just those ions to pass through when the channel is open. Each type of channel opens or closes in response to a particular trigger. For example, in an axon that has reached spike threshold, voltage-sensitive sodium-conducting channels open, followed very soon by voltage-sensitive potassium-conducting channels. Another type of potassium channel is responsible for the negative resting potential of neurons: potassium tends to leave the inside of a neuron because it is maintained at a higher concentration inside than outside the neuron. Each potassium that leaves the inside carries a positive charge, so the inside of the neuron develops a negative charge, and an equilibrium is established in which the concentration gradient across the membrane for potassium is balanced by the electrical gradient, or voltage. The same balance of forces affects the way other ions flow through their conducting channels when they are open.

Neurons typically have many tens of distinct types of channels in their membranes, with each region having a particular complement that suits its function. Axons have voltage-sensitive potassium and sodium ion channels; postsynaptic sites on dendrites are where chemical-sensitive channels are located. At axon terminals, another type of voltage gated channel plays an important role, a calcium channel, which has the function of regulating the release of chemical neurotransmitters from presynaptic terminals. **Neurotransmitters** are specific chemicals that are released by a presynaptic terminal in response to rising intracellular calcium levels. The released molecules diffuse rapidly across the short extracellular gap to postsynaptic sites on the receiving neuron, where they bind with receptor sites on chemical-sensitive ion channels, usually causing these channels to open. So when a spike travels to the terminals of an axon, it depolarises the terminals and triggers calcium channels to open, which leads to the release of a burst of neurotransmitter. When the transmitter binds to receptor sites on the outside surface of chemical-sensitive channels in the postsynaptic terminal, this causes the channels to open for a short time. The resulting flow of ions sets up the postsynaptic potential (PSP). Although the ion channels open and shut rapidly, the waveform of a PSP rises and falls with a fairly smooth, rounded shape. The same occurs when square electrical current pulses are injected into a neuron through a microelectrode (Fig. 2.3b). The reason for this smoothing effect is that the cell membrane has appreciable electrical capacitance. Capacitors act as containers that can store electrical charge, and take time to fill and empty. The smoothing effect the membrane has on voltage changes is important in enabling dendrites to sum PSPs together.

There are many kinds of chemical-sensitive ion channels, characterised by the chemical nature of the neurotransmitter they respond to, and by the identities of the ions they are permeable to when open. Some excite their neurons by enabling ion fluxes that depolarise the postsynaptic regions, making it more likely the postsynaptic neuron will spike. Others inhibit their neurons, usually by increasing permeability to chloride or potassium ions.

Functional description of a neuron: mammalian motor neuron

A mammalian motor neuron serves well to illustrate the general principles by which a neuron receives, integrates and transmits information. In fact, study of the way synaptic potentials are generated and integrated in a cat spinal cord motor neuron, by J. C. Eccles and colleagues in the early 1950s, provided many of our fundamental concepts in synaptic physiology. In overall structure, a spinal cord motor neuron (Fig. 2.5) is similar to the pyramidal cell in the brain, introduced in Chapter 1 (Fig. 1.3a), in that two types of processes extend from the cell body of the neuron: dendrites and an axon. As shown in the drawing by Ramon y Cajal (Fig. 2.5b), most dendrites

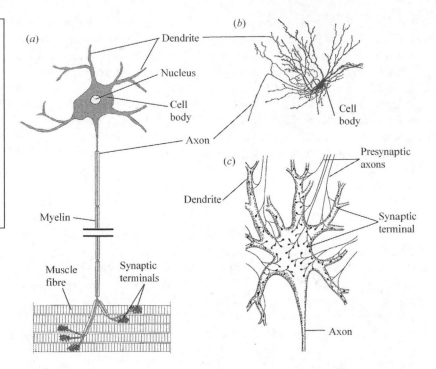

Fig. 2.5 Structure of a vertebrate motor neuron. (*a*) Diagram of the principal structural components. Only a few dendrites are drawn here. (*b*) Drawing by Ramon y Cajal of a silver-stained motor neuron in a cat spinal cord. (*c*) Diagram of cell body and some dendrites showing the way that individual presynaptic terminals branch to each make a few discrete synaptic terminals onto the motor neurons. (Parts (*a*) and (*c*) redrawn from Eccles, 1977; (*b*) redrawn from Ramon y Cajal, 1909.)

branch several times, with the result that most of the total area of the neuron's cell membrane is contained within dendritic branches. Dendrites receive signals from other neurons at synapses, and the cell body also receives synapses. Each synapse is very small, and a presynaptic axon branches to make about half a dozen individual, anatomical synaptic contacts with the motor neuron (Fig. 2.5*c*). A motor neuron typically receives synapses from thousands of different presynaptic neurons, each one contacting the motor neuron at around half a dozen different discrete sites. The received signals are typically small electrical signals, a few millivolts across the cell membrane at most. They are transmitted passively within the motor neuron's dendritic tree, and their sum at the origin of the axon from the cell body determines the neuron's output signals. The axon origin, therefore, acts like a point of decision for the neuron. In the case of a motor neuron, outputs are transmitted along the axon to its terminals at synapses made with the surface of muscle cells, and regulate muscle cell contraction. The cell membrane of a motor neuron axon is excitable, unlike that of the dendrites, a property that enables it to conduct signals along its length without a decrease in their amplitude.

Synapses in action: a mammalian spinal reflex

When a muscle in a mammal is stretched, for example if a limb is moved in a particular way, a reflex pathway comes in to play to make the muscle shorten and relieve the stretch that is imposed on it

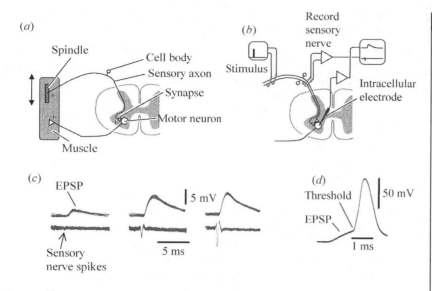

(a) Spindle, Cell body, Sensory axon, Synapse, Motor neuron, Muscle

(b) Record sensory nerve, Stimulus, Intracellular electrode

(c) EPSP, 5 mV, Sensory nerve spikes, 5 ms

(d) Threshold, EPSP, 50 mV, 1 ms

Fig. 2.6 Synaptic basis of a reflex in the spinal cord. (*a*) Diagram, including spinal cord in transverse section, to indicate the relationship between components in the reflex pathway. Symbols for neurons and synapses are the same as introduced in Fig. 2.2*b*. (*b*) Set-up for an experiment to measure postsynaptic potentials from a motor neuron. (*c*) EPSPs recorded at three different stimulus intensities. Several traces for the same stimulus intensity are superimposed. A weak stimulus (left) evokes a very small extracellular spike (arrow) from a single spindle axon. As stimulus intensity is increased (middle and right recordings), an increasing number of spindle axons are recruited so the extracellular spike increases in amplitude as does the response from the motor neuron. Note the scale bars, which indicate the timescale, and the amplitude of the EPSPs in the intracellular recordings (extracellular spikes are not measured). (*d*) A strong shock activates a sufficient number of synapses to depolarise the motor neuron to its spike threshold. Note the difference in scale bars compared with (*c*). (Parts (*c*) and (*d*) redrawn from Eccles, 1977.)

(Fig. 2.6*a*). This is what usually happens when a doctor taps a patient's leg just below the knee, and the leg jerks to straighten. The stretch is detected by the muscle spindles, proprioceptors whose axons enter the spinal cord by a dorsal nerve root. They function in a similar way to the crayfish muscle receptor organ, to convert stretch of a small bundle of specialised muscle fibres into a train of spikes whose rate depends on the strength of the stretch. All tetrapod vertebrates have muscle spindles in their skeletal muscles; a handful in the smallest muscles and several tens in the largest. Their cell bodies are in a dorsal root ganglion, near to the spinal cord. The cell bodies of the motor neurons are towards the ventral horn of the grey matter of the spinal cord (Fig. 2.6*a*), and their axons leave by the ventral nerve root. Each muscle in a vertebrate is innervated by several motor neurons, sometimes several hundred. The sensory neuron axons branch and make synapses with motor neuron dendrites, also in the ventral horn of the cord.

The set-up for an experiment to record synaptic responses from a motor neuron, like one of the experiments Eccles and colleagues performed, is shown in Fig. 2.6*b*. Two pairs of metal electrodes are placed in close contact with a dorsal nerve root, one to deliver brief electrical stimuli and the other to record the spikes the shocks trigger in spindle receptor axons. An intracellular microelectrode inserted into a motor neuron cell body records its membrane potential.

The weakest stimuli elicited no responses because they were not strong enough for any of the spindle axons to reach the threshold for spiking, but as stimulus strength was gradually increased, they were strong enough to elicit spikes in a few axons in the nerve that were nearest to the electrodes, indicated by a very small deflection picked up by the recording electrodes (Fig. 2.6*c*). The intracellular electrode recorded a small positive deflection: an **excitatory postsynaptic potential** (**EPSP**), caused by the excitatory action of the synapses made by the spindle axon terminals on dendrites of the motor neuron. In these recordings, several repetitions at the same stimulus strength are superimposed to indicate the consistency of the responses. The slight delay between the axon spike recorded from the nerve and the start of the EPSP is partly accounted for by the time taken for the spikes to reach the ventral horn of the cord, but mainly by the synaptic delay, which is mostly the time between arrival of a spike at an axon terminal and the release of neurotransmitter into the synaptic cleft.

As stimulus strength was increased further (middle and then right records, Fig. 2.6c), an increasing number of sensory axons were recruited as the stimulating currents spread further from the electrodes. Accordingly, the number of excitatory synapses onto the motor neuron that were activated by each stimulus increased. The different EPSPs summed with each other, producing a larger, compound EPSP. This is one kind of integration: **spatial summation** of postsynaptic potentials originating at different synapses onto the motor neuron. The EPSPs in Fig. 2.6c are all subthreshold and did not elicit a spike in the motor neuron. However, increasing the stimulus strength further would cause the EPSPs to sum beyond the threshold for a spike, as in Fig. 2.6d.

A second kind of integration, **temporal summation**, is illustrated in Fig. 2.7. Here, two stimuli follow each other in short succession, each triggering an EPSP in the motor neuron. The second one starts before the first has decayed, so the second EPSP piggybacks on the first, and the peak on the second reaches a higher amplitude than the first. The peak amplitude depends on the delay between the two EPSPs. In the recordings in Fig. 2.7, a very slight decrease in the delay between the two EPSPs was sufficient to enable the second EPSP to trigger a spike in the motor neuron. Temporal summation is made possible by the smoothing effect that membrane capacitance has in enabling the dendrites to store charge and prolong the time course of an EPSP.

The chain of excitatory events in this reflex pathway is summarised in Fig. 2.8. Stretching the muscle spindle causes spikes in the sensory axon, and each spike causes an EPSP in the motor neuron. The EPSPs sum in time. The force with which the spindle is stretched is encoded in spike rate, and the summation of the EPSPs caused by the spikes depends on how closely spaced they are. EPSPs originating in different spindle axons converging onto the motor neuron sum in space, and because the spikes in different sensory axons are not synchronous, the jagged peaks of EPSPs from single spindle axons are smoothed in the summed response from the motor neuron. The summed EPSP regulates the production of spikes in the motor neuron, and each motor neuron spike excites and causes a twitch in its muscle fibres. In a process of signal decoding, twitches caused by individual motor neuron spikes sum, increasing the muscle contraction force. The muscle contraction relieves the stretch on the muscle spindle, which, along with sensory adaptation, deactivates the reflex pathway.

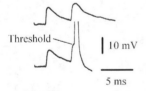

Threshold

10 mV

5 ms

Fig. 2.7 Temporal summation of EPSPs in a mammal motor neuron. The two traces each show an intracellular recording with two EPSPs, produced by stimulating a sensory nerve with two shocks. The second EPSP sums with the first. The interval between the shocks was slightly longer in the upper than the lower trace. In the lower trace, the second EPSP sums with the first sufficiently to exceed the spike threshold in the motor neuron (the top of the spike is omitted). (Redrawn from Eccles, 1957.)

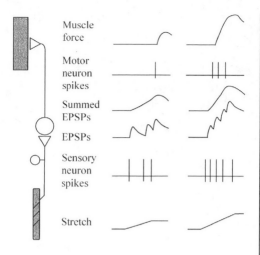

Muscle
force

Motor
neuron
spikes

Summed
EPSPs

EPSPs

Sensory
neuron
spikes

Stretch

Fig. 2.8 Summary of signal transformation in a mammalian spinal reflex. The pathway is summarised on the left, and responses at different stages in the pathway to a weak and to a stronger stretch are shown. The strength of each stretch is encoded in a train of spikes in a spindle sensory neuron, and each spike causes an EPSP in the motor neuron. The EPSPs from a single spindle neuron show temporal summation. The EPSPs from all the spindles sum to generate in the motor neuron a smoother waveform than the EPSPs from a single sensory neuron. The summed EPSPs regulate spike production in the motor neuron axon and these spikes regulate the strength with which the muscle contracts (spikes are not shown in the intracellular recordings). (Redrawn from Eccles, 1977.)

Inhibitory synapses

A reflex pathway could cause injury if it activated a muscle to contract while an antagonist muscle was also contracting. Pathways incorporating synaptic inhibition act to ensure this does not occur. For example, if a muscle that extends a limb is stretched, the reflex pathway described above comes into play, causing the extensor muscle to contract. A second pathway then inhibits motor neurons of the limb flexor pathway. In addition to exciting extensor motor neurons, the sensory neurons excite interneurons in the grey matter which make inhibitory synapses with the flexor motor neurons (Fig. 2.9a). The arrangement is reciprocal, so if the flexor muscle is stretched, its spindle cell axons excite its motor neurons and inhibit the limb extensor motor neurons. Each spike in an interneuron causes a hyperpolarising **inhibitory postsynaptic potential** (**IPSP**) in its target motor neurons. The IPSP and an EPSP sum in a similar way to the spatial and temporal summation of EPSPs described above. In this case, however, the hyperpolarising effect of an IPSP and the depolarising effect of an EPSP tend to cancel each other out. Eccles and colleagues showed how the EPSPs and IPSPs arriving separately at a motor neuron interact (Fig. 2.9b).

The interaction between excitatory and inhibitory synapses is, in practice, usually more complex than simple addition of potentials. One reason in this case is that the excitatory synapses are made onto the motor neuron dendrites, but the inhibitory synapses are mainly made onto the motor neuron cell body and so are strategically located near the main control point, the origin of the axon where spikes are initiated. A second reason is that besides causing small PSPs, neurotransmitters usually open ion channels in the postsynaptic membrane, which reduces the membrane's electrical resistance. This decrease in resistance has an effect rather like an electrical short-circuit, and reduces the amplitudes of voltage signals in nearby regions of the neuron.

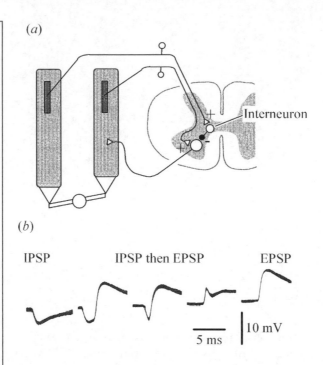

Fig. 2.9 An IPSP and its interaction with an EPSP in a spinal motor neuron. (*a*) Diagram similar to Fig. 2.5*a*, but including two antagonistic muscles with muscle spindles. A motor neuron of the muscle on the right is shown, and it is directly excited by a sensory neuron of a spindle in its own muscle. It is also inhibited by a pathway from a spindle in the other muscle (which is being stretched), by way of an interneuron in the grey matter of the spinal cord. The inhibitory synapse is indicated by a filled circle. (*b*) Intracellular recordings from a spinal cord motor neuron showing PSPs in response to stimulating two different branches of a sensory nerve, one carrying spindle axons from its own muscle and the other carrying spindle axons from an antagonistic muscle. Stimulating the first caused an EPSP (right recording), and stimulating the second caused an IPSP (left recording). In the three middle recordings, an IPSP and then an EPSP were evoked, with different intervals between them. (Part (*b*) redrawn from Eccles, 1977.)

Locust wing reflex

A good example of a reflex in an invertebrate is from a study by Malcolm Burrows (1975) on flight machinery of the locust. His intracellular recordings from flight motor neurons reinforce and enlarge on the principles of synaptic integration from the cat spinal cord. Each wing of a locust has a number of different types of proprioceptors in its base, one of which is a stretch receptor. The stretch receptor consists of a single sensory cell with a cell body and dendrites at the wing base, and an axon that runs in a nerve to the central nervous system. It is not associated with a particular muscle, but its dendrites are stretched and the axon spikes when the outstretched wing is moved upwards. The rate of spikes depends on the extent of wing elevation (Fig. 2.10*a*).

Within its ganglion, a stretch receptor axon branches and intermingles with dendrites of motor neurons that control the wing muscles (Fig. 2.10*b*). A synaptic connection between the two neurons in this case consists of a few hundred discrete anatomical synapses – many more than is typical in a mammalian nervous system. Unlike in a vertebrate motor neuron, the dendrites in an arthropod motor neuron do not branch off from the cell body, and the cell body is attached to the dendritic tree by a slender process. The cell body of a locust motor neuron does not receive any synapses and is not in the pathway for signal flow, but rather is to one side of it. Because the cell body is quite large – up to about 50 μm across for a large flight motor neuron – it is a good target for a microelectrode; but all the electrical signals recorded from it have conducted passively from their sites of

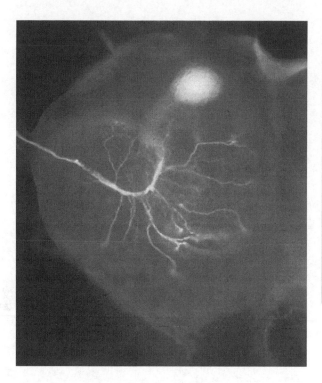

A flight motor neuron responsible for controlling a muscle that depresses a forewing. The neuron was injected with a dye called lucifer yellow that fluoresces brightly when illuminated with blue light. This photograph shows well how the neuron has a three-dimensional structure. The cell body is below the pane of focus, as are many of the smaller dendrites. The main process that collects input signals from different dendrites is sharply focused, and is continuous with the axon, which leaves the ganglion in one of the segmental nerves on the left. (Photograph provided by Berthold Hedwig, Cambridge University.)

origin, so are smaller and more rounded in the cell body than in the dendritic tree or axon. For example, a spike in the axon would be 100 mV in amplitude and last less than 1 ms, but in the cell body it might be represented as a 10 mV signal lasting several milliseconds.

Burrows devised a way to make intracellular recordings from wing motor neurons while moving the wing and recording stretch receptor spikes. The locust was fixed upside down to a clamp, and a wing was extended and held in a device that would move it up and down (Fig. 2.10c). A window was cut into the ventral cuticle of the thorax and a metal platform was placed under the thoracic ganglia, to hold them still during an experiment. The cell bodies of most motor neurons lie near to the ventral surface of the ganglia, and particular motor neurons have their cell bodies in characteristic locations. To record from a motor neuron, the electrode tip was positioned over its cell body, and the micromanipulator holding the electrode was tapped gently, causing the electrode to penetrate the cell body. The identity of the motor neuron was confirmed by recording which muscle it controlled.

A single spike in the stretch receptor is followed, after a short delay, by an EPSP a few millivolts high in wing depressor motor neurons (Fig. 2.11a), and closely spaced EPSPs show temporal summation, adding in a time-dependent manner. The stretch receptor on its own rarely excites a motor neuron sufficiently to bring it to the threshold for spiking, but if it did, the motor neuron spike would cause its muscle to twitch, depressing the wing tip and reducing stretch receptor excitation. As well as exciting wing depressor

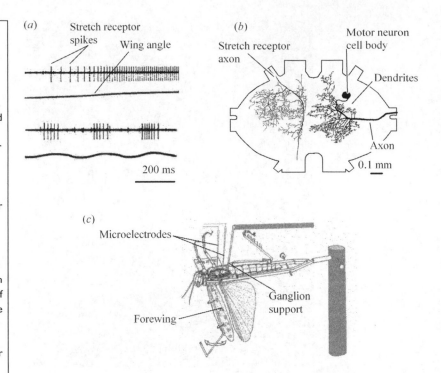

Fig. 2.10 Wing hinge stretch receptor and a motor neuron of a locust. (*a*) Two recordings of responses to imposed wing movements. Extracellular recordings of spikes were made using extracellular electrodes placed beneath the nerve that carries the stretch receptor axon. In the upper record, the wing tip was elevated slowly; in the lower, it was moved repeatedly up and down. (*b*) Structure of the stretch receptor axonal branches and of a wing depressor motor neuron in the thoracic ganglion that controls the forewings. The stretch receptor on the left side, and a motor neuron on the right side are shown. Note that if both neurons on the same side were drawn, there would be very many overlaps between axonal branches of the stretch receptor and of motor neuron dendrites, places where synaptic connections could occur. (*c*) Diagram of the arrangement of an experiment to study the reflex between stretch receptor and motor neurons. (Redrawn from Burrows, 1975.)

motor neurons, spikes in the stretch receptor cause IPSPs in wing elevator motor neurons (Fig. 2.11*b*). The delay from a stretch receptor spike to an IPSP is a little longer than the delay to an EPSP, and so it is likely that rather than making inhibitory synapses directly with elevator motor neurons, the stretch receptor excites interneurons which then inhibit the elevators, in a similar manner to the way spindle sensory neurons inhibit antagonistic motor neurons in the spinal cord. However, these interneurons in the locust have not been identified.

Normally, motor neurons receive a barrage of PSPs from many different neurons, and in the recording from the depressor in Fig. 2.11*b* there are IPSPs from unknown presynaptic neurons as well as EPSPs from the stretch receptor. When the locust is flying, the motor neurons receive strong synaptic input from interneurons responsible for generating the regular rhythm of up and down wing movement (Chapter 7). The stretch receptor spikes in each wing beat, and the PSPs it causes are integrated together with all the others the motor neurons are receiving.

Some of the motor neurons receive synapses from brain neurons, enabling the locust to make appropriate steering movements during flight. For example, there is a brain neuron that receives signals from the eyes, and responds specifically to a fast-approaching object such as another locust in a swarm, or a swooping predatory bird, by producing a train of spikes that increase in rate. This visual inter-neuron excites a particular wing elevator motor neuron, and an intracellular recording from the wing elevator neuron provides a

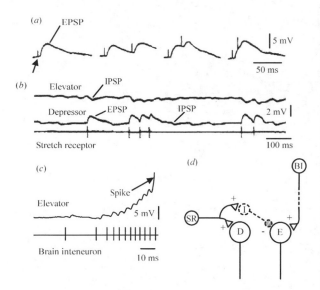

Fig. 2.11 Postsynaptic potentials in locust flight motor neurons. (a) Temporal summation of EPSPs shown by delivering a single and then pairs of stimuli to trigger spikes in the stretch receptor. Small line (arrowed left) indicates stimulus time. (b) Recordings from a wing elevator and a wing depressor motor neuron. Each stretch receptor spike caused an EPSP in the depressor and an IPSP in the elevator, and the motor neurons received additional PSPs. (c) Intracellular recording from a wing elevator motor neuron showing summation of EPSPs from a presynaptic brain interneuron (spikes in lower trace). The EPSPs in the motor neuron sum to trigger a spike (peak not shown). (d) Diagram of connections between neurons in this figure, using the same symbols for synapses introduced in Figs. 2.5 and 2.8. SR, stretch receptor; D and E, wing depressor and elevator motor neurons; BI, brain interneuron; I, possible interneuron between a stretch receptor and elevator motor neuron. (Parts (a) and (b) redrawn from Burrows, 1975; (c) redrawn from Santer et al., 2006.)

good illustration of temporal summation of EPSPs leading to a motor neuron spike (Fig. 2.11c). In this case, the electrode was inserted into the dendritic fan of the motor neuron, near to the origin of the axon rather than in the cell body, and EPSPs at this location rise and fall more sharply than in the cell body. A diagram of connections between these neurons is shown in Fig. 2.11d.

Locust legs and local interneurons

Jointed limbs, such those of a mammal or an insect, are used in a wide variety of ways. Besides being involved in locomotion, they are capable of finely controlled movements such as grooming, where the end of the limb is brought precisely to a particular spot on the animal's surface. Touching a hindleg lightly at any point on its surface causes the locust to move the leg away (Fig. 2.12a). In a locust, these local reflexes involve flow of signals through four different layers of neuron in the third thoracic ganglion. Between the sensory and motor neurons are two types of local interneuron – called local because they are confined to one region of their ganglion. They do not carry signals between one part of the body and another, but only within a restricted part of their ganglion. One type produces spikes, and the other type normally operates and communicates without producing any spikes, so they are called '**non-spiking interneurons**'.

About 10 000 sensory neurons project into each half of the third thoracic ganglion, and many of these originate on the surface of the leg, so there is a great deal of convergence as information flows from sensory neurons to motor neurons, of which there are about 100 involved in leg movements. The spiking local interneurons mostly receive inputs from various kinds of sensory neurons,

Fig. 2.12 Components in local movements of a locust leg. (*a*) Gently touching a hindleg at the point indicated causes the locust to flex the leg to a new position (shaded dark). (*b*) Axon terminals of a hair sensory neuron in the third thoracic ganglion. (*c*) Structure of a local spiking interneuron that is involved in local leg withdrawal movements. (*d*) Recordings showing EPSPs in a local spiking neuron (upper trace) caused by stimulating individual hairs on the leg (spikes in lower trace) in the locations on the leg shown in the drawing. In each case, several recordings are superimposed. (Parts (*a*), (*c*) and (*d*) redrawn from Burrows, 1992, (*b*) redrawn from Newland, 1991.)

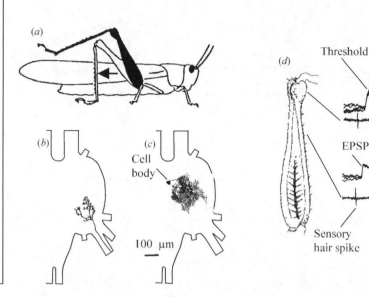

including small hairs up to 0.8 mm long which each have a single sensory neuron at the base. It is possible to record from and stimulate these hairs by cutting the tip off then placing a glass microelectrode with the tip broken off over the tip-less hair. Gently moving the electrode stimulates the receptor, and the electrode records spikes from the neuron. The sensory neurons project in an orderly way into the ganglion so that it contains a spatial map of the leg surface – for example, a hair near the foot projects to a relatively posterior region in the ganglion (Fig. 2.12*b*), whereas a hair on the femur near to the knee projects more anteriorly.

Each local, spiking interneuron has two layers of branches. The ventral layer of one interneuron is drawn in Fig. 2.12*c*, and the cell body is also situated ventrally. The ventral layer of branches receives synapses from sensory and other neurons and is connected by a narrow process to the second, dorsally located layer of branches that mainly act as presynaptic elements to other neurons in the ganglion. It is likely that the narrow connecting process is an axon, conducting spikes from the ventral to the dorsal region. Each local spiking interneuron responds to stimulation of hairs and other sense organs on a particular region of the surface of the leg (Burrows and Siegler, 1985; Burrows and Newland, 1993). Some interneurons respond to stimuli on particular parts of one segment of the leg, but others are interested in broader regions that extend in both directions from a joint. Typically, stimuli to one small part of the leg evoke larger EPSPs and so stronger responses than stimuli to other parts of the leg. The recordings in Fig. 2.12*d* illustrate this. Using the electrode-over-cut-hair technique, two different hairs were stimulated while EPSPs were recorded from a spiking interneuron. Spikes from the hair on the knee joint usually produced an EPSP that was large enough to trigger a spike, whereas spikes from hairs further towards the body produced EPSPs that were below spike threshold. When two sensory neuron spikes occur in quick succession, the two EPSPs sum in time; however, the second is often smaller than the first. Changes in effectiveness of synapses with use are fairly common, underlying many plastic changes in behaviour (Chapters 8 and 9), and here it helps to enhance responsiveness by the locust to new, rather than maintained, tactile stimuli (Burrows, 1992).

The non-spiking interneurons provide a major source of synaptic input to leg motor neurons. They branch quite profusely within a local region of the ganglion, and possess no single process that can be called an axon. Studies using electron microscopy show that input and output synaptic sites occur close to each other on many of the branches. As their name suggests, these interneurons do not spike when they are excited. Instead, the rate at which they release neurotransmitter is controlled by their membrane potential. There is a smoothly graded relationship between the membrane potential of a non-spiking interneuron and of a motor neuron that it drives (Burrows and Siegler, 1978), which is seen when one electrode is used to inject current into a non-spiking interneuron while another one is used to record the membrane potential of a postsynaptic motor neuron. The same relationship between presynaptic and postsynaptic potentials is found at all chemical synapses, but the all-or-nothing nature of a spike often obscures it. As for a spike, there is a threshold potential for synaptic transmission. Consequently, the membrane potential of a non-spiking neuron needs to be depolarised at least to the threshold level for any transmission to occur at its output synapses.

Figure 2.13a shows responses from two different motor neurons to stimuli applied to a non-spiking interneuron by injecting positive current into it through an electrode. The two different stimulus strengths in the figure show that the more strongly the interneuron was depolarised from its resting potential, the greater the size of the PSPs in postsynaptic neurons. This relationship arises because the amount of neurotransmitter released depends on how strongly the interneuron is excited. In this example, the rate of neurotransmitter release remained constant at a particular level of stimulus, so PSPs lasted as long as the stimulus. The interneuron excited a motor neuron that extended the hindleg, and inhibited a motor neuron that flexed it, an arrangement that makes functional sense. It is likely that the interneuron made inhibitory synaptic connections directly with the flexor motor neuron, but excited the extensor motor neuron indirectly by way of the pathway shown by shaded cell bodies in Fig. 2.13b, by inhibiting a second non-spiking interneuron that itself inhibited the extensor motor neuron. The effect of exciting the first interneuron would be to reduce the on-going inhibition of the motor neuron by the second interneuron – a mechanism called disinhibition.

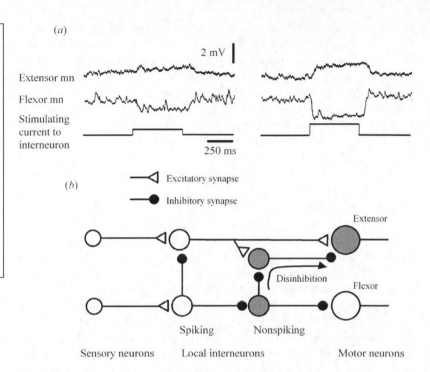

Fig. 2.13 Non-spiking local interneurons controlling leg motor neurons in the locust. (a) Intracellular recordings that show the graded nature of transmission from a non-spiking interneuron. (b) Diagram to show some of the kinds of synaptic interaction in the networks involved in local reflex movements. The pathway through the three shaded neurons illustrates excitation through a process of disinhibition: one non-spiking interneuron inhibits a second which in turn inhibits the extensor motor neuron. (Redrawn from Burrows, 1989.)

Figure 2.13b summarises the kind of connections found among the network of neurons controlling local reflex movements of the leg. The non-spiking interneurons are each excited by particular sense organs including the hairs on the surface of the cuticle and various internal proprioceptors that monitor movements and angles of joints. Non-spiking interneurons can sustain neurotransmitter release, exerting a steady synaptic drive on their postsynaptic targets, with the result that they can set a general level of excitability in motor neurons in a way that depends on the animal's posture. To illustrate how this works, consider a pathway in which a sensory neuron excites a non-spiking interneuron which, in turn, excites a motor neuron. If the balance of synaptic inputs to the interneuron does not excite it by much, its potential will be below the threshold for transmitter release and its output synapses will effectively be switched off. As a result, the interneuron will not contribute to the membrane potential and excitation of the motor neuron. If the angle of the joint is altered, the new combination of sensory inputs onto the interneuron might excite it sufficiently for it to release transmitter, which will in turn excite the motor neuron with a steady, depolarising potential. The non-spiking interneurons, therefore, can act as central control points, where signals from various sense organs sum to regulate the excitability of motor neurons.

This kind of graded synaptic transmission is found in many neuronal networks. It is particularly clear in the early stages of visual processing; both in invertebrates and vertebrates, the first two layers of retinal neurons are non-spiking neurons. Some neurons have a mixed control over their synapses, transmitting both smoothly and gradually changing levels of potential and spikes. The exact advantages of one mode of transmission over the other can be hard to pinpoint. Spikes are certainly needed for long-distance communication along axons from one part of the body to another. But the situation is not as simple as a distinction between long-distance and local communication. In the locust thoracic ganglia, non-spiking interneurons communicate over similar distances to spiking local interneurons. One use for a spike is that it can provide a precise timing signal, whereas timing a graded amplitude signal might be more ambiguous.

Conclusions

Signals in dendrites and the axon originate as tiny electrical currents created by ions flowing through specific channels that open or close in response to particular triggers, either changes in voltage across the membrane, or neurotransmitters released by other neurons. The input and output connections of each neuron, reflected in its specific branching pattern, control the flow of information and create functional circuits in the nervous system. The process of synaptic integration allows a neuron to combine information from different sources, and it underlies the way in which a nervous system ensures that behaviour patterns are selected and controlled in appropriate ways according to sensory stimuli and other factors. In synaptic integration, EPSPs reinforce each other, and their effect is reduced by summing with IPSPs.

A mammalian motor neuron in which dendrites collect and sum PSPs and the resulting graded potential regulates spike initiation serves as a good basis for understanding neuronal function. But additional characteristics are often significant in the integrative activity of neurons and their networks. First, as illustrated by the networks that control locust leg movements, not all neurons operate with spikes. Small and variable changes in presynaptic voltage can be effective at regulating the release of neurotransmitter, and of conveying an analogue type of signal from one neuron to another. Second, particularly in non-spiking neurons, there is not necessarily a physical separation between input and output regions of a neuron. Sometimes dendrites make output synapses as well as receiving inputs, so their input and output synapses are intermingled and information is processed in a localised region. Third, some dendrites have in their membranes voltage-sensitive channels that can boost the amplitude of their input signals. One place where this has been studied is in output neurons of the thalamus of the vertebrate brain, which have the job of relaying sensory inputs to particular processing areas in the cortex. These examples serve to illustrate that networks of neurons are not designed and built according to a limited set of operational rules, but have been shaped by an evolutionary process of natural selection.

Questions

The action of a neurotransmitter is determined by the nature of the ion channels it causes to open or close, so a neurotransmitter such as acetylcholine can exert a wide variety of diverse postsynaptic effect. Why, then, are so many different types of neurotransmitter found in nervous systems?

When a person, or a locust, experiences an itch they can move a finger or claw accurately and quickly to the itchy part of skin. What are the challenges in designing a network of neurons that could accomplish this task?

Summary

- Neurons receive, transmit and integrate electrical signals.
- Types of neuron include motor neurons, sensory neurons and interneurons.

Crayfish muscle receptor organ
- Dendrites transduce stretch into a small receptor potential.
- Receptor potential amplitude depends on stretch strength.
- Receptor potential does not travel far in axon.
- Axon converts receptor potential to spikes that travel by propagation as stereotyped, brief pulse-like signals.
- Stretch strength is coded as spike frequency (rate).
- With a sustained stimulus, responses adapt – strongly in phasic receptors, less strongly in tonic receptors.
- Communicates with other neurons at synapses.

Neuronal signals: recording and mechanisms
- Microelectrodes are used to make intracellular recordings, for stimulation and for staining.
- Neurons have a negative resting potential, usually -60 to -70 mV.
- A spike is triggered when excitable membrane is depolarised above a threshold.
- Graded potential signals in dendrites are smaller, not fixed in amplitude and can be summed.
- The cell membrane is electrically leaky, so graded potentials in dendrites or subthreshold potentials in axons do not travel far, and has capacitance, which prolongs (smooths) signal waveforms
- Electrical signals result from flow of ions, driven by electrochemical gradients, through voltage- and neurotransmitter-sensitive ion channels.

Mammalian motor neuron

- Excited by synapses from muscle spindles in its own muscle.
- EPSPs from different presynaptic neurons are integrated by spatial summation.
- Over time, EPSPs are integrated by temporal summation.
- During reflex activation, motor neurons of antagonistic muscles are inhibited. The antagonist spindle excites inhibitory interneurons in the spinal cord.
- By spatial summation, IPSPs are integrated with EPSPs.

Locust thoracic ganglion neurons

- Integrate EPSPs and IPSPs by spatial and temporal summation, in a similar manner to mammal motor neurons.
- Involved in control of leg movements are local spiking interneurons, which are excited by sense organs.
- Non-spiking interneurons, in which small changes in potential effect graded changes in postsynaptic potential, are major components of networks that control motor neurons.

Further reading

Burrows, M. (1992). Local circuits for the control of leg movements in an insect. *Trends Neurosci.* **15**, 226–232. A review of the organisation of neurons involved in local reflex movements of the hindlegs of a locust.

Yuste, R. and Tank, D. W. (1996). Dendritic integration in mammalian neurons, a century after Cajal. *Neuron* **16**, 701–716. A review of different types of integrative processes that contribute to the function of a neuron.

3

Neuronal mechanisms for releasing behaviour: predator and prey – toad and cockroach

Imagine a damp forest floor with a toad sitting motionless by a log. An insect scuttles from under the log, moving too fast to identify, and immediately the toad lunges towards the insect, flicking its tongue towards it. The toad misses its meal this time and the insect swivels away from the lunge of the toad and runs for cover; it was a cockroach (Fig. 3.1). These two behaviours, prey capture by toads and escape running by cockroaches, are excellent case studies in neuroethology because both show how it is possible to identify the roles played by individual nerve cells in recognising significant stimuli and triggering appropriate behavioural responses. In these cases, the stimuli require immediate action on the part of the animal. A toad will in fact try to catch and eat any small animal that moves along the ground in front of it; it probably does not hunt for cockroaches in particular. Likewise a cockroach will turn and run away from rapidly accelerating air currents, such as those produced by the sudden strike of any predator including a toad. Toads and cockroaches are not specifically adapted to detect each other, but natural selection has shaped the evolution of effective neuronal mechanisms that enable toads to recognise scuttling insects as a potential meal and cockroaches to escape from predatory assaults. In each behaviour, the animal needs to assimilate sensory information rapidly and to organise its motor response appropriately. Here we shall focus on identifying the roles that individual neurons play in recognising sensory stimuli of particular significance; in later chapters we focus on other examples where the flow of signals between different neurons has been studied in more detail.

The common European toad, *Bufo*, preys on small animals like beetles, earthworms and millipedes as well as cockroaches. Jörg-Peter Ewert (1985, 1987) identified the neuronal basis of this prey recognition in toads by first using classical ethological tools to isolate the salient features that a toad uses to recognise a potential meal. Ewert then applied neurophysiological and anatomical techniques to

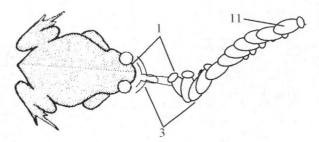

Fig. 3.1 Unsuccessful predatory attack by a toad on a cockroach. The animals were filmed from above using a camera running at a rate of 64 frames per second. Outlines were drawn from successive frames: numbers 1–3 for the toad, and 1–11 for the cockroach. At frame 1, just before the attack started, the cockroach was facing towards and to the left of the toad. At frame 2, the toad's tongue is visible and the cockroach had already begun its turn away. The turn continued in frame 3, allowing the cockroach just to escape the sticky tongue tip, and then run away from the toad. (Redrawn from Camhi and Tom, 1978.)

pinpoint the parts of the visual system involved in this recognition. By recording from, stimulating and eliminating small areas of the brain he established that the releasing mechanism resides in a particular class of neurons. Jeffrey Camhi and colleagues have used similar techniques to show that cockroaches can escape capture by toads by reacting to the air currents produced as a toad starts its attack. A small number of large, identifiable interneurons in the cockroach nerve cord form the core of the neuronal pathway that is responsible for enabling cockroaches to escape from attacks.

Stimuli that release prey-catching behaviour in toads

The prey-catching behaviour of toads illustrates very well some of the classical concepts about the way a nervous system is involved in controlling behaviour. Particular sign stimuli release the prey-catching behaviour. Toads that have been reared in isolation are good at catching small prey, and the way a toad recognises potential prey is a good example of an innate releasing mechanism. In terms of neuronal hardware, the releasing mechanism includes neurons called feature detectors that are specifically excited by the sign stimuli that release behaviour patterns. An insect scuttling near to a toad will almost always cause the toad to start a series of discrete movement patterns, or action patterns, and Ewert with colleagues showed that prey capture by toads consists of a sequence of action patterns, each triggered by its own releasing mechanism. The first action, released when a small insect first comes to the attention of a toad as a movement in the periphery of its visual field, is for the toad to turn to face towards the insect. Next, with the insect fixated in front of the toad, it approaches towards the insect, tracking it if it moves away from the centre of the toad's visual field. If the toad manages to approach within striking distance, the sight of an insect within range releases a rapid lunge towards the prey. The toad extends its tongue during the lunge, followed by a swallow and often a wipe of the mouth by a front flipper, a sequence that continues even if the toad misses the insect, or if a model stimulus is removed from site. Because this sequence tends to be completed once it has started, it is a fixed action pattern. The experiments that Ewert and his colleagues performed allowed them to identify a particular class of neurons in the toad's brain as feature

A frame from a high-speed video showing a toad, *Bufo woodhousii*, capturing a cricket with its tongue. Prey catching is released by stimuli with the appropriate combination of visual characteristics. (Photograph provided by Kiisa Nishikawa, Northern Arizona University.)

detectors that release the toad's behaviour of tracking potential prey by turning to follow it as it moves.

In the visual world of a frog or toad, just a few simple criteria serve to categorise moving objects as prey, enemy or lover. Ewert quantified the sign stimuli by which the prey is recognised by measuring the releasing values of different aspects of model prey, which is possible because toads are easily deceived by simple, cardboard models. In an experiment, a hungry toad (*Bufo*) was confined in a glass vessel, from which it could see a model circling around (Fig. 3.2*a*). If the toad interpreted the model as a prey animal, it tried to bring it into the frontal visual field, and in doing so turned around jerkily after the moving model. The number of orientating turns per minute elicited by a given model, compared with the number elicited by others, was therefore a measure of that model's releasing value (Fig. 3.2*b*). From the toad's point of view, the higher the releasing value, the better the resemblance between that model and prey.

A small 2.5 mm square did not interest the toad much, eliciting only a few movements by the toad to turn and follow the moving shape. But increasing the horizontal length of the cardboard model in a stepwise manner increased its effectiveness at releasing the turning behaviour. In these experiments, the height of the shape was always 2.5 mm, and shapes with progressively greater horizontal lengths were tested. So that the distance between the toad and the stimulus can be ignored, the size of each shape is expressed as the angle that it occupied within the toad's visual field rather than the shape's absolute size, and in Ewert's experiments a 2.5 mm square had an angular size of 2°. The number of turns per minute that the toad made to follow the moving shape increased proportionately to the length of the stimulus shape until, with a 16° long stimulus, the toad had reached the limit of its ability to make rapid turns, so increasing stimulus length to 32° did not elicit a greater response from the toad. Thus, elongation of the model in the same direction as its horizontal movement increased its resemblance to prey, up to a certain limit. Ewert called this stimulus pattern the 'worm configuration'. If the small, square shape was elongated in the vertical dimension at right angles to its direction of movement, its

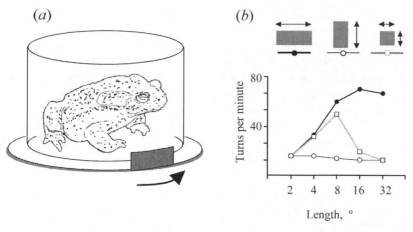

(a)

(b)

Fig. 3.2 Experiments to analyse the sign stimuli that release prey capture in toads. (*a*) A toad sat inside a glass container, and a turntable rotated simple cardboard shapes around it. If the toad was interested in a shape, it turned to follow it. The number of turns per minute is a measure of the releasing value of a particular model stimulus. (*b*) Shows responses to three different stimulus configurations: worm, antiworm and square. In each case, the average response from 20 experiments is plotted. (Redrawn from Ewert, 1985.)

releasing value decreased to zero. In fact, the toads often interpreted this stimulus as a threat and froze in a defensive posture and so Ewert called it the 'anti-worm configuration'. If both dimensions of the model were lengthened equally, so that the toad was presented with squares of increasing size, the prey-catching activity initially increased but then declined rapidly to zero. This is probably the result of activating some parts of the brain that detect horizontal (worm) and other parts that detect vertical (antiworm) edges.

The toad's ability to distinguish between worm and antiworm did not vary with other stimulus parameters, such as the colour of the model or its direction of movement. Within certain limits, increasing the speed of a worm stimulus increased its releasing value. If the models were moved past the toad in a vertical direction, the vertical stripe elicited prey catching and the horizontal stripe elicited either no response or a defensive posture. The distinction between worm and antiworm is based on the combination of just two simple stimulus parameters, the elongation of the object in relation to its direction of movement. These parameters are the sign stimuli that release prey-catching behaviour in a hungry toad, and it is obvious that they correspond to a real worm only in a very rough way. Nevertheless, they normally enable a toad to distinguish correctly between potential prey and inedible objects in its natural environment.

The retina

The function of the retina is to convert the optical image into a neuronal representation that the brain can act upon. Rod and cone photoreceptors transduce light into receptor potentials, and signals pass first to bipolar cells and then to retinal ganglion cells, which have axons that travel into the brain. Each photoreceptor samples a small patch of the image, collecting light from a small cone-shaped region of space with an angle about 1/35 of a degree. In toads, about 18 photoreceptor cells would sample the region of sky occupied by a full moon. The region of the visual field that a neuron is interested in is termed its

receptive field, as illustrated in Fig. 3.3*a*. Photoreceptors greatly out-number retinal ganglion cells, and each photoreceptor feeds signals to more than one ganglion cell, so retinal ganglion cells have much larger receptive fields than photoreceptors, between 1.5 and 15 degrees in different types. Also, most ganglion cell receptive fields have a more complex structure than the simple circular shapes of photoreceptor receptive fields. This is a consequence of processing within the retina, in which signals are modified by two layers of horizontally oriented cells: horizontal cells and amacrine cells. The arrangement of the basic cell types in a vertebrate retina is shown in Fig. 3.3*b*.

Details of the way signals are processed in the retina were first studied in another amphibian, the mudpuppy salamander *Necturus*, an animal that has particularly large neurons in its retina from which it is relatively easy to make intracellular recordings (Dowling and Werblin, 1969; Werblin and Dowling, 1969). Most of the principles of retinal processing learned from *Necturus* apply to other vertebrates, and are similar to the kinds of operation that image processing software pro-grammes perform, for example to improve the quality of photographs. Some processes ensure that the retina can work well over a wide range of ambient lighting conditions, from bright sunlight to dark night, and others sharpen contrasts in the image. The horizontal cells are respon-sible for creating a two-part receptive field structure, called a centre-surround organisation, in retinal bipolar cells. In one type, whereas photoreceptors in the central part of a bipolar cell's receptive field act to increase its response, those in a peripheral ring act to decrease its response, connecting with the bipolar cell via horizontal cells. So a small spot of light in the central part of the receptive field causes an increase in the response of the bipolar cell; but a spot of light in a ring-shaped region surrounding the central part causes a small decrease in its response – the central excitatory part and surrounding inhibitory ring in the receptive field work in an antagonistic manner (Fig. 3.3*c*). This means that the most effective stimulus for a bipolar cell is a spot of light that just fills the central part of its receptive field; a larger spot causes a smaller response because it strays into the inhibitory sur-round region. The balance between the centre and surround is such that they cancel each other out if the whole region of the receptive field is filled. A functional advantage of this arrangement is that large, uniform areas of equal shading in the image do not cause large signals in bipolar cells or retinal ganglion cells, so these cells are much more interested in features such as spots or edges in the image, which are likely to provide the animal's brain with information that is useful in recognising particular objects. This is a good example of stimulus filtering; only the informative aspects of the stimulus are preserved for further analysis in the brain, so it does not have to be concerned with a large amount of irrelevant information, which is discarded.

The amacrine cells modify the signal that is transmitted from the bipolar cells to the ganglion cells in various ways. Many are involved in processing moving images, so that some retinal ganglion cells are also most responsive to moving images. In frogs and toads, about half a

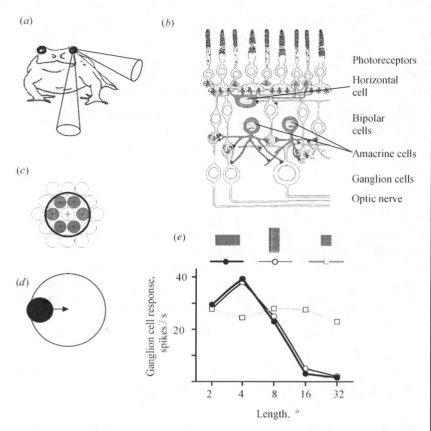

Photoreceptors

Horizontal cell

Bipolar cells

Amacrine cells

Ganglion cells

Optic nerve

Fig. 3.3 A toad's retina and retinal ganglion cells. (*a*) The receptive fields of two different retinal neurons are shown, each being a cone-shaped region in which the neuron will respond to visual stimuli. Most receptive fields are much narrower than those drawn here. (*b*) The five principal types of neuron in a vertebrate retina, drawn as if a thin section cut perpendicularly to the retina surface were viewed with a microscope. Neurons of the vertical pathway from photoreceptors to bipolar to ganglion cells are unshaded, while horizontal and amacrine neurons are shaded. (*c*) Diagram illustrating the principle of a centre-surround organisation of a bipolar cell's receptive field. The dark ring describes the extent of the receptive field of the bipolar cell, and the smaller rings describe the receptive fields of individual photoreceptors. The photoreceptor in the centre (+) excites the bipolar cell (just one is drawn for simplicity); those surrounding the centre (- and shaded) inhibit it. (*d*) A type 2 retinal ganglion cell of a toad responds whenever a small dark blob moves into, or within, its receptive field. (*e*) Responses by a toad type 2 retinal ganglion cell to the three different stimulus configurations used in behavioural experiments. (Part (*b*) redrawn from Dowling and Boycott, 1966; (*e*) redrawn from Ewert, 1985.)

dozen distinct types of ganglion cells can be recognised and are categorised according to the way their receptive fields are organised. Some are most sensitive to objects of a given angular size, others to objects moving in a particular direction or to the difference in brightness between adjacent areas of the image. Some types are found much more commonly than others. Type 2 ganglion cells have a strong centre-surround organisation, becoming excited when a small dark blob enters the centre of their receptive field and inhibited as it moves into the surrounding part (Fig. 3.3*d*). This type of retinal ganglion cell was first reported from the frog retina in the late 1950s, and was named a 'bug detector' because it seemed to be adapted to pick out small, moving bug-shaped objects. In toads, Ewert and colleagues found that this and two other types of ganglion cells provide the brain with information about basic parameters of stimuli that might interest toads, such as angular size, velocity and darkness. However, none of them distinguished between worm, antiworm and square stimulus configurations (Fig. 3.3*e*). Therefore, his search for the neurons that first distinguish prey from predators led him to look deeper into the toad's visual system.

Most retinal ganglion cells are connected to the optic tectum, a layered region of the midbrain that is visible as a large bulge on either side (Fig. 3.4). A small proportion of ganglion cells are connected to the thalamus, which is the most prominent part of the posterior

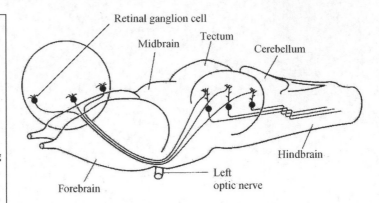

Fig. 3.4 Layout of the main visual pathways in the brain of a toad that are concerned with prey detection. The axons of most ganglion cells travel from the eye to the optic tectum on the opposite side of the brain via the optic nerve and connect with feature-detecting neurons such as T5(2) neurons in the tectum. These feature-detecting neurons send their axons to the motor regions of the contralateral hindbrain.

forebrain, and to the pretectal areas of the midbrain. The connections to the tectum are spread out in an orderly manner in its superficial layers, with each ganglion cell keeping the same relative position with respect to its neighbours that it has in the retina. Consequently the relative different locations of different parts of the visual image are preserved in a map-like fashion in the tectum, with the receptive field of a particular neuron being adjacent to and overlapping those of its neighbours.

Neuroethology of a releasing mechanism

Ewert and colleagues recorded the responses of tectal neurons by probing different layers of the optic tectum with a microelectrode (Fig. 3.5). Like the ganglion cells, the neurons of the thalamus and tectum can be divided into different classes according to their patterns of response, and those with similar properties are usually found in the same layer as each other. Of the thalamic and tectal neurons that have been investigated, at least three classes showed different responses to moving stimuli of worm and antiworm configurations. The thalamic Class TH3 neurons responded best to squares; stripes with the antiworm configuration elicited a lesser response and the worm configuration elicited the least response of all (Fig. 3.5b). In the optic tectum, the Class T5(1) neurons also gave their most vigorous responses to squares, but when tested with stripes they preferred the worm to the antiworm configuration.

Another class of tested neurons, Class T5(2), distinguished much more clearly between the worm and antiworm configurations, with the worm configuration eliciting the greatest response, the squares a lesser response and the antiworm by far the least response. Each T5(2) neuron has a large receptive field with an angle of 20–30°, and it spikes vigorously when a worm-like stimulus enters its receptive field, but not when an antiworm stimulus enters it. Among all the neurons tested, the response pattern of the T5(2) neurons shows the best correlation with the sign stimuli for prey-catching behaviour (compare Figs. 3.2b and 3.5b). Like the behaviour, brisk responses by T5(2)

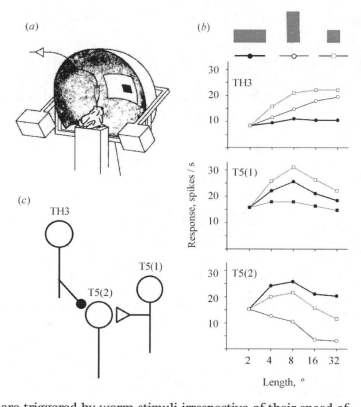

Fig. 3.5 Responses by tectal neurons to visual stimuli. (*a*) Set-up for recording the responses of neurons in the brain to moving visual stimuli. The toad is held in a fixed position and its brain is probed with a microelectrode for recording the spikes in single neurons. Each stimulus is moved in front of the toad by means of the perimeter device. (*b*) Response of three different classes of neuron, TH(3) in the thalamus and T5(1) and T5(2) in the tectum, to the same three stimulus configurations used in the behavioural tests (Fig. 3.2). (*c*) Pattern of connections onto a T5(2) neuron from a TH3 and a T5(1) neuron that could explain the selectivity of T5(2) neurons for worm stimuli. Triangle indicates excitatory synapse; closed circle an inhibitory synapse. (Part (*a*) redrawn from Ewert, 1985; (*b*) redrawn from Ewert, 1980.)

neurons are triggered by worm stimuli irrespective of their speed of movement, their orientation, or their shade against the background. The match between neuronal response and behaviour was not a perfect one, however: both gave maximum response for worms 16° long, but the behaviour showed little decline from this maximum when toads were tested with longer worms, although the neuronal responses did decline. Nevertheless, these results indicate strongly that the T5(2) neurons are an essential component of the releasing mechanism, and this was borne out by other experiments. Some of this evidence that the T5 layer is important in distinguishing worms from antiworms came from experiments in which the relative activity in different brain regions following long exposure to different patterns was revealed using activity-dependent stains (Finkenstädt and Ewert, 1988).

It is evident from Fig. 3.5 that the responses of the Class T5(2) neurons are more selective for the worm stimulus configuration compared with the antiworm or square than responses of TH3 or TH(1) neurons. The kind of interaction shown in Fig. 3.5*c* can explain this: each T5(2) neuron receives excitatory input from Class T5(1) neurons and inhibitory input from Class TH3 neurons. The fairly strong response to the worm configuration in Class T5(1) neurons would be minimally inhibited by the poor response to it in the Class TH3 neurons, resulting in a strong response in the Class T5(2) neurons. Similarly, the poor response to the antiworm configuration in Class T5(1) would interact with the moderate response in Class TH3 to give a very poor response in Class T5(2).

The possibility that these interactions shape the selectivity of TH(2) neurons has been tested by removing the input from the Class TH3 neurons, which can be accomplished by severing the pathway that is known to run from the thalamus to the optic tectum. Whether this lesion was done permanently by microsurgery or temporarily by local application of anaesthetic, the effect on Class T5(2) neurons was dramatic. Their responsiveness to all kinds of visual stimuli increased and selectivity was lost, with the neurons responding best to squares and failing to distinguish clearly between stripes in worm and antiworm configurations. This shows that the normal selective response of the Class T5(2) neurons is dependent on inhibition from thalamic neurons, including the Class TH3 neurons.

When a toad with a pretectal lesion was allowed to recover from surgery and was tested behaviourally, its responses closely paralleled those of the altered T5(2) neurons: each operated animal responded vigorously to all shapes, preferring squares, and failed to distinguish clearly between worm and antiworm configurations of stripes. Such a close correspondence between the responses of the Class T5(2) neurons and of the whole animal suggests that these neurons are directly involved in prey detection and hence in releasing prey-catching activity. This was confirmed by means of a small device fitted to a toad's head that enabled the experimenter to record from and stimulate single neurons in the optic tectum. The device was radio operated, so the toad was not impeded by wires as it moved around. Recordings made with this system show that activity of Class T5(2) neurons preceded and continued during the orientation of the toad towards the prey. Once the location of the receptive field of a particular T5(2) neuron had been established by recording its pattern of excitation, the neuron was stimulated experimentally by passing a tiny current through the microelectrode. This experimental excitation consistently elicited orientating movements by the toad that were directed to the receptive field of the T5(2) neuron that was stimulated. Stimulating a second T5(2) neuron with a different receptive field would cause the toad to look towards the second neuron's receptive field.

If they are involved in prey detection in this way, the Class T5(2) neurons would be expected to be connected, directly or indirectly, with the motor circuits in the hindbrain of the toad. A variety of histological methods demonstrate that a number of the connections arriving in the motor regions on one side of the hindbrain do come from the opposite side optic tectum (see Fig. 3.4). The inclusion of the Class T5(2) neurons amongst those that make such connections was confirmed by physiological methods. Localised stimulation of the appropriate neural tract in the hindbrain sends signals travelling back up to the optic tectum, where they can be recorded in individual T5(2) neurons with a microelectrode.

In summary, there are a number of very good lines of evidence showing that the T5(2) neurons play an essential role in prey recognition and capture. First, there is good correlation between their responses and behaviour because their response properties to worm, antiworm and square stimuli differ in the same way as the behavioural responses. Second, their removal or inactivation alters the toad's responses to worm stimuli. Finally, artificially stimulating an individual T5(2) neuron causes a toad to behave in the same way that it would when the neuron was excited by a visual stimulus.

The Class T5(2) neurons provide an excellent example of specific brain cells that are involved in releasing a simple, important behaviour pattern. On the basis of the stimulus parameters selected by the retinal ganglion cells, the neurons of the optic tectum are able to respond to specific parameter combinations that carry information relevant to the toad's way of life. The specific combination of visual parameters to which the T5(2) neurons respond carries information enabling the toad to distinguish between its natural prey and inedible objects, although they do not distinguish between worms, slugs or cockroaches. However, the toad's behaviour is not completely automatic and it will not necessarily respond to every scuttling insect. Whether or not a toad tries to catch an insect or worm depends on how hungry it is, and during the breeding season toads are less likely to react to worm stimuli. The behaviour is flexible in other ways, too. A toad can be trained to turn towards, catch and eat antiworm stimuli (although not to prefer antiworms to worms); and if a prey animal disappears from sight the toad alters its movement patterns in a way that is appropriate to continue pursuit.

The startle reaction of a cockroach

Cockroaches are extremely difficult to catch, partly because they are exquisitely sensitive to very small air currents to which they respond with a fast turning and running response that is accurately directed away from the source of wind. In their natural habitat, which for the American cockroach *Periplaneta americana* is leaf litter on the forest floor, cockroaches have been shown to be able to evade the strikes of one of their predators, the toad, by detecting the movement of air caused by movement of the toad's tongue before the tongue has left the mouth (Camhi and Tom, 1978). Cockroaches are particularly sensitive to sudden increases in the velocity of air currents, which is how they can distinguish an attack by a toad from meteorological wind (Plummer and Camhi, 1981). This is another good example of sensory filtering, enabling an animal to respond to the most significant stimuli but to ignore less significant, background events. The first movement a cockroach makes is triggered rapidly, within 50 ms of the start of a wind stimulus, and is a rapid turn away from the direction of attack (Fig. 3.6a) that involves the co-ordinated movement of all six legs at the same time in a pattern distinct from the way they are used in walking or running (Fig. 3.6b). Each leg pushes or pulls, so that the cockroach swivels to face away from the direction from which the air current is coming. It then runs forward, using the usual tripod gait mode of locomotion in which the legs are moved in two sets of three.

Photograph of the hind end of a cockroach, *Periplatenta americana*. The arrow points to hair-like filiform sensilla that hang beneath each cercus and provide an early-warning system by responding to sudden, small air currents. (Photograph by Peter Simmons.)

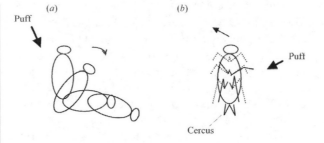

Fig. 3.6 Startle response of the cockroach. (*a*) The turning response to a wind puff delivered from the front left of the animal as recorded with a video camera looking down on the animal. The outline of the body and head are traced from every second frame. (*b*) Leg movements during an escape turn caused by a puff of wind coming from the right and slightly to the front. The cockroach was held so that its body could not move, but its legs slipped over a lightly oiled glass plate. The initial positions of the legs are indicated with dotted lines; the final positions of the middle and hindlegs are drawn as solid lines. The arrow above the cockroach indicates the direction in which the animal would have faced if its body had been free to rotate. (Part (*a*) redrawn from Comer and Dowd, 1993; (*b*) redrawn from Ritzmann, 1993.)

Sensory coding and filiform sensilla

The sense organs that detect air currents are called filiform sensilla, located on the ventral surfaces of a pair of sensory appendages, the cerci, that a cockroach bears on its last abdominal segment. Each cercus carries several types of sense organs, some of which are bristle-like, but the filiform sensilla are sufficiently slender and delicate to be deflected by the most gentle of breezes. If the filiform sensilla are immobilised by coating the ventral surface of a cercus with a thin layer of petroleum jelly, the cockroach no longer turns away from puffs of air, and careful removal of the coating restores this response. So the filiform sensilla are responsible for the cockroach's ability to defect air puffs. Each sensillum is attached to the dendrite of a single sensory cell (Fig. 3.7*a*). The cell body of the sensory cell is located at the base of the sensillum, and gives rise to a long axon that lies alongside axons of its neighbours in the sensory nerve that connects the cercus with the last abdominal ganglion. In the last abdominal ganglion the axon terminates in a series of fine branches that are restricted to one side of that ganglion. The sensilla are arranged in 14 columns, most of which run the length of the cercus, and each column is usually represented by one filiform sensillum on each of the 20 segments that an adult cockroach cercus has. The arrangement for two adjacent cercal segments is shown in Fig. 3.7*b*.

Each filiform sensillum is most easily deflected in one particular direction, and shares that with others in its row. When a puff of air

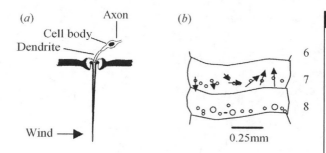

Fig. 3.7 Filiform sensilla on the cercus of a cockroach. (*a*) Diagram of the structure of a single filiform sensillum. (*b*) Best directions of deflection for exciting sensory neurons of the filiform sensilla of the 7th segment of the right cercus, indicated by arrows. Each circle shows the location of one filiform sensillum on the ventral surface of segment 7 or 8; the diameter of the circle depends on the length of the hair. (Part (*a*) redrawn from Keil, 1997; (*b*) redrawn from Dagan and Camhi, 1979.)

deflects the sensillum, it responds by producing spikes that travel by propagation along its axon to the last abdominal ganglion. The greatest number of spikes are produced when the wind blows in the direction in which the sensillum moves most easily; wind at right angles to this will not cause any spikes. So the sensillum and its sensory neuron have directional sensitivity, and information about the direction from which a puff of air is coming is available to the cockroach by determining which row of sensilla give the most vigorous response to that puff.

Giant interneurons

In the last abdominal ganglion, the axons of filiform sensilla make synaptic connections with the dendrites of seven pairs of interneurons that have axons that run in the ventral nerve cord. Because the axons of these interneurons are considerably wider than others in the nerve cord, ranging from 20 to 60 μm across, they are called 'giant interneurons'. Anatomically, two separate groups of giants can be distinguished; one with axons located dorsally and the other with axons located ventrally in each connective nerve. The ventral group are larger than the dorsal group and are the group that trigger escape running. The structure of one ventral giant, number 1, is shown in Fig. 3.8*a*. The three lines of evidence for the involvement of an individual neuron in a particular behaviour show that the giant interneurons play roles initiating and directing the cockroach's turn. First, the way the giants respond to air currents suggests they are suitable for this role. Second, selectively killing one giant interneuron reduces the cockroach's responsiveness to wind from that giant's direction (Comer, 1985). Finally, stimulating a single giant electrically affects the turning behaviour (Liebenthal *et al.*, 1994). Stimulating a single giant rarely initiates running, but these neurons act together to control the escape behaviour.

Carefully conducted experiments have revealed that each giant responds most vigorously to air currents from particular directions (Kolton and Camhi, 1995), and this directional sensitivity arises because each giant interneuron is excited by particular columns of filiform sensilla. To determine the directional selectivity of a giant,

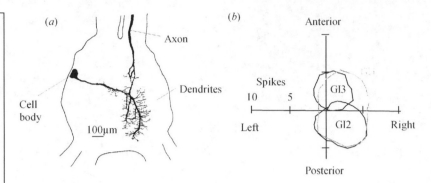

Fig. 3.8 Giant interneurons and sensory analysis of wind puff direction in the cockroach. (a) Drawing of right giant 1 in the last abdominal ganglion. (b) Polar plot of the directional sensitivities of the three ventral giants that have their axons in the right connectives, constructed as explained in the text. (Part (a) redrawn after Harrow et al., 1980; (b) redrawn from Kolton and Camhi, 1995.)

spikes were recorded from its axon using a microelectrode that contained a stain so that the neuron could later be identified from its anatomy. Controlled puffs of air, always with a peak velocity of 0.85 m/s, were directed at the right cercus from a nozzle that could be rotated to stimulate it from different directions. Air currents were applied from an angle of 45° above the cockroach, which is the kind of angle from which an attack might be made. In each stimulus, spikes were counted for the first 50 ms following the start of the stimulus; this is roughly the time over which a giant neuron would respond before a cockroach starts to move in response to an air current. The results are expressed as a polar plot (Fig. 3.8b). The origin of the plot represents the cercus, and each response is plotted as a point whose direction from the origin represents the direction of the stimulus, and the distance from the origin represents the number of spikes.

In Fig. 3.8b, each point was the mean of several stimuli in experiments on five different cockroaches. For each interneuron, the different points are joined by a line, and the resulting shape gives an immediate impression of its receptive field, and the direction of wind that it responds most vigorously to. All three ventral giant interneurons respond preferentially, but not exclusively, to stimuli coming from the same side as their axons (the right side, in Fig. 3.8b), and interneuron 1 responds well to stimuli from the front or from the rear. There is a little overlap between the receptive fields of interneurons 2 and 3, but if we look at the receptive field to determine which stimulus direction elicits the best responses in these two neurons we find that the directions for the two giants are almost exactly at right angles to each other; interneuron 2 prefers stimuli from behind and to the side of the animal whereas interneuron 3 responds best to stimuli from the front and the side. In fact, the best direction for interneuron 3 corresponds to the direction in which the cercus normally points, and the best direction for interneuron 3 is at right angles to this. Thus, a cockroach can determine whether an air current is coming from the left or from the right by comparing the responses of its left and right giants; and it can distinguish stimuli coming from the front from those coming from the rear by the relative excitation of interneurons 2 and 3. If a current of air is coming directly from the side, these two giants will respond with almost equal vigour.

A cockroach giant interneuron can generate 25 spikes or more between the times when it starts to respond to wind and when the cockroach starts to move. By stimulating single giants to generate extra spikes during responses to wind, Liebenthal *et al.* (1994) showed that the most critical parameter in the response is the number of spikes in left compared with right giants. No single giant interneuron responds exclusively to air currents from the left or from the right, yet the cockroach always makes quite accurately oriented turns that take it away from a source of danger. This means that information from different giants must be combined together in some way. Rafael Levi and Jeff Camhi (2000a) showed that excitation in different giant interneurons is combined by a kind of collaborative summing mechanism, with no indication that one giant can veto the input from another. They delivered wind puffs to the cerci of a cockroach that was fixed in place, but could move its legs and make steering movements on an oily glass plate. At the same time, they used a microelectrode both to record from and stimulate a single giant axon, number 3, which has a narrower receptive field than the other two. They compared turning movements to wind puffs alone and then to the same puffs when extra spikes were added electrically to a single axon. Change in the direction of turn was directly related to the number of extra spikes. On average, each extra spike in giant interneuron 3 added an extra 7.4° to the cockroach's attempted turn. This was found whether they studied interactions between left and right, or between the front and rear on the same side.

One way in which signals from different giants could contribute to steering is by what Levi and Camhi (2000b) call a 'steering wheel' mechanism. In this, each of the three giants on one side would collaborate as if they all pull in the same direction on a steering wheel, while the three giants on the other side pull in the opposite direction. However, adding extra spikes to giant interneuron 2, which responds to wind from one side and the rear of the cockroach, does not bear this mechanism out. Extra spikes to the right giant 2 increase the strength of a cockroach's leftward turn in response to a wind puff directly from the right of the cockroach. So each giant interneuron seems to work independently, attracting the cockroach to turn towards the direction of its own receptive field. A plausible explanation is that each giant drives a different set of interneurons in the thoracic ganglia that determine the way the legs move.

Conclusions

The T5(2) neurons of toads and giant interneurons of cockroaches play significant roles in the lives of these animals, enabling one to catch prey and the other to escape capture. Each type of neuron is tuned to generate vigorous responses to particular stimuli within a defined receptive field. We can understand how both the stimulus configuration and the receptive field are established in terms of the synaptic connections made by neurons in the pathways that converge on these interneurons. In the case of T5(2) neurons, the image that falls on the toad's retina is processed at successive stages in the visual pathway in a way that enables tectal neurons to filter out particular combinations of stimulus features, such as shape and movement, while disregarding others, such as colour. A cockroach giant interneuron filters out sudden air currents from a particular direction as a result of the particular columns of cercal filiform sensilla that excite it.

The roles of these neurons in controlling behaviour have been demonstrated with three different types of test. First, each neuron can be shown to be suitable for this role by correlating the way it responds to appropriate stimuli with the occurrence of behavioural responses to the same stimuli. Second, artificially exciting a neuron through an electrode either triggers or modifies the behavioural response under study. Finally, removal of a neuron or, in the case of T5(2), a population of neurons alters behavioural responses in a predictable way.

Questions

Would you expect that eliminating a single T5(2) neuron in the toad's tectum would affect its ability to detect prey?
What would a cockroach's reaction be if two jets of air, from different directions, were simultaneously directed at it?

Summary

- Prey capture and evasion from predation provide good examples of behaviour in which roles of individual neurons have been established.
- In establishing a role for a neuron in a behaviour: electrical signals should correlate with the behaviour; stimulating the neuron should elicit or affect the behaviour; and removing the neuron should prevent or alter the behaviour.

Toads
- Prey catching is released by stimuli that have a worm configuration, being elongated in the direction of movement, independent of colour, speed and other features.
- The retina processes the visual image, particularly emphasising borders. Each retinal neuron has a receptive field, within which it responds to particular stimulus features.
- Retinal ganglion cells send trains of spikes to the thalamus and tectum in the brain. None responds specifically to worm-like stimuli.
- The excitation of one class of tectal neuron, type 2 in tectal layer 5, matches the releasing characteristics of the behaviour well, although not perfectly.

- T5(2) neurons are feature detectors: they distinguish between worms and antiworms or other stimuli.
- Further evidence for their role in the innate releasing mechanism for prey catching comes from experiments in which stimulating one causes a toad to gaze towards its receptive field. Additional evidence comes from selective lesions.

Cockroaches

- Detect a threat as a sudden increase in wind velocity flowing over filiform sensilla on the cerci, and turn accurately away from the wind direction.
- Each row of filiform sensilla filters out wind from a particular direction.
- They synapse with three giant interneurons on each side of the last abdominal ganglion.
- Each giant has a different receptive field. Wind direction can be determined by comparing responses by different giants.
- Number of spikes each giant generates is significant in determining the turning response.

Further reading

Dowling, J. E. (1987). *The Retina: An Approachable Part of the Brain*. Cambridge, MA: Harvard University Press. A book that explains clearly how the vertebrate retina is organised and processes visual information, including the way that receptive fields of different neuron types are established.

Libersat, F. (2004). Maturation of dendritic architecture: lessons from insect identified neurons. *J. Neurobiol.* **64**, 11–23. An account that focuses on cockroach giant interneurons. It shows how the shapes of their dendritic trees influence their responses, and how these change during development.

4

Neuronal pathways for behaviour: startle behaviours and giant neurons in crayfish and fish

When an animal is startled by a sudden attack from a predator, it must respond with great urgency if it is to escape, and neuronal pathways that initiate such an escape response must be both straight-forward and reliable in order to fulfil their biological function. Straightforward pathways are essential to ensure speed in initiating the escape and they must be reliable not only to make sure the response occurs when needed but also to avoid false alarms. These qualities of simplicity and reliability, which are of great survival value to the animal, are also of service to the neuroethologist exploring the roles that nerve cells play in behaviour. Consequently, several startle responses have been studied in detail and they provide valuable insight into the flow of information through the nervous system from sensory inputs to muscular output.

As we have already shown in our description of cockroach escape behaviour in the previous chapter, the neuronal pathways responsible for startle responses often involve neurons that are exceptionally large and so are called giant neurons. The unusual width of the axon of a giant neuron enables it to conduct spikes rapidly along the animal's body. For an experimenter, the size of giant neurons also makes it relatively easy to insert microelectrodes into them, although because any small movements will dislodge an intracellular microelectrode, it is not possible to make intracellular recordings from neurons in freely moving animals. But in extracellular record-ings, giant neuron spikes are much larger than those of other neurons and various techniques are used to make extracellular recordings while an animal swims, walks or flies, impeded only slightly by long flexible leads attached to the electrodes.

Two main functions must be carried out by the neuronal pathway that initiates a behaviour pattern such as startle. First of all, a decision to initiate the activity must be made. This includes processing incoming sensory information in a way that extracts particular stimulus features, emphasising the significant features so that the animal only responds to truly threatening stimuli, and sensory processing must be done rapidly to give the startled animal a chance to escape. The second function for the neuronal pathway, once the decision to initiate an escape has been made, is an executive one. The pathway must ensure that muscle actions needed to achieve escape are not hindered by other muscles responsible for incompatible movements. To achieve this function, escape pathways include inhibitory connections responsible for shutting down on-going movements so that escape proceeds smoothly.

Early work on startle responses in this area gave rise to the idea that there may be a special class of high-order interneurons that are responsible for delivering a clear command signal for escape. Keis Wiersma reported in 1947 that direct stimulation of a single giant axon in a crayfish consistently causes a vigorous tail flip. Later, Wiersma and Ikeda (1964) found that electrical stimulation of single interneurons elicited predictable, co-ordinated movements of the abdominal ventilatory and swimming appendages of crayfish called swimmerets. Crucially, the pattern of swimmeret movements depended on which interneuron was stimulated, and not much on the pattern of stimulus such as the rate of spikes elicited. They introduced the term **command neuron** for such interneurons. Essentially, a command neuron acts as a point on which sensory information converges, and makes a decision to elicit a particular response on the basis of the information it receives. If the balance of sensory information drives the command neuron to spike, the excitation is delivered to divergent pathways that lead to activation of a particular set of muscles. The idea of a command neuron was effective at triggering much debate and experimental work. In an influential review, Kupferman and Weiss (1978) suggested that the term should be reserved for neurons that have been shown to be both necessary and sufficient to initiate a particular, naturally occurring, behaviour pattern. At the extreme, it would mean a catalogue of recognisable behavioural acts would be represented in the nervous system by a library of corresponding command neurons.

However, as already shown for cockroach giant interneurons, the way nervous systems control behaviour is usually not so straightforward. First, it is usual for neurons to act in synergy with each other, so that the flow of information from sensory input to motor action is rarely funnelled through a single interneuron. Second, the pattern of excitation does matter, so that a weak discharge of spikes in a particular axon may be ineffective at triggering a behavioural response, whereas a more intense burst of spikes will be effective.

The first topic for this chapter is the lateral giant interneuron of the crayfish. It provides an excellent introduction to the mechanisms for the control of behaviour and we probably know more about the way that this neuron works in controlling behaviour than any other.

Study of the lateral giant interneuron illustrates well how ethological and neurophysiological methods of analysis can mutually reinforce one another during a focused study of a single system. But it is important to bear in mind that this neuron has some unusual features, one of which is that it does fulfil the strict criteria for being a command neuron. The second topic is another large and unique interneuron, the Mauthner neuron of bony fish. Like the crayfish lateral giant interneuron, the Mauthner neuron is involved in triggering rapid escape movements but it normally works alongside others to ensure the fish directs its escape in an appropriate direction, so it does not fulfil the criteria for a command neuron as strictly as the crayfish lateral giant. These two neurons only ever produce a single spike each time their startle responses are initiated, providing a clear sign of a behavioural decision. This is an unusual feature because most neurons with long axons carry information as trains of spikes, with spike rate often coding stimulus intensity.

Giant neurons and the crayfish tail flip

Freshwater crayfish, such as the American red swamp crayfish, *Procambarus clarkii*, escape from the strike of a predator by means of a rapidly executed tail flip, produced by flexing and re-extending the whole abdomen. The abdomen is able to move the crayfish effectively and rapidly because its last two (6th and 7th) segments are modified to form a paddle-like tail fan (Fig. 4.1a). A single flip of the tail fan is capable of moving the animal several centimetres through the water. The power for this movement is provided by the fast flexor muscles, which occupy much of the space within each abdominal segment. These are called fast muscles because they produce rapid twitch contractions, in contrast to a set of much smaller and slower contracting muscles that produce graded postural movements of the abdomen. Each abdominal segment also contains fast and slow extensor muscles, which are much less substantial than the flexor muscles.

About 10 large motor neurons send axons to the flexor muscles on each side of each abdominal segment. One large motor neuron is exceptionally large and is called the motor giant. It sends an axon branch to every fast flexor muscle fibre. Another motor neuron is an inhibitory motor neuron that also innervates every muscle fibre. The remaining motor neurons are simply known as fast flexor motor neurons. Each of these innervates only a restricted group of fibres within the fast flexor muscles, and this arrangement enables a crayfish to use different parts of the flexor muscles independently of each other in some types of movements. Two pairs of exceptionally large interneurons in the nerve cord are able to initiate tail flips by exciting the motor giant and fast flexor motor neurons. These are called the lateral and the medial giant interneurons (Fig. 4.1c), and their influence

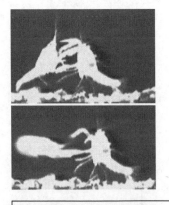

Two frames from a high-speed film of an encounter between a dominant crayfish on the right and a subordinate on the left. The dominant punches the subordinate with its claw, and the subordinate escapes by shooting backwards. This escape response is triggered by a large interneuron called the medial giant. The two frames from a high speed video are 70 ms apart. (Photographs provided by Donald Edwards, Georgia State University.)

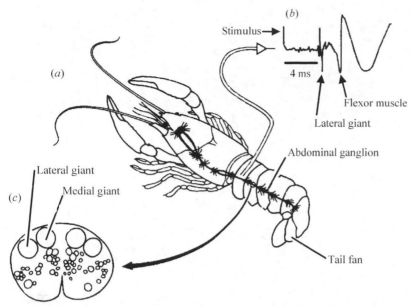

(b)

Stimulus→

4 ms

Flexor muscle

Lateral giant

Abdominal ganglion

(a)

Lateral giant

Medial giant

(c)

Tail fan

Fig. 4.1 Giant interneurons involved in crayfish (*Procambarus*) startle behaviour. (*a*) Crayfish showing the location of the central nervous system (in solid black), a chain of ganglia. Also shown are electrodes implanted to record neuronal activity in the freely moving animal. (*b*) Activity recorded by these electrodes during a startle response: a tap on the abdomen (stimulus) is followed by a spike in a lateral giant and excitation of the abdominal flexor muscles. (*c*) Transverse section of the connectives between two abdominal ganglia, showing the locations of the lateral and medial giants. (Part (*a*) redrawn from Schramek, 1970; (*b*) redrawn from Krasne and Wine, 1975; (*c*) redrawn from Krasne and Wine, 1977.)

is so strong that a single spike in either giant interneuron will inevitably trigger a tail flip.

The most striking feature of the initial tail flip is its speed: within 50 ms, abdominal flexion is completed and the animal has usually moved some distance through the water. The mean delay between the stimulus to the onset of flexor muscle potential is 6 ms (Fig. 4.1*b*). This speed of reaction is needed because the tail flip is probably a response to a predator that is extremely near to or touching a crayfish. In the nervous system, the speed is partly achieved by extensive use of neurons with wide axons, which conduct spikes more rapidly than narrow axons, and of electrical synapses. At the point of synaptic contact between a lateral and motor giant, the motor giant branches over the surface of the interneuron's axon in a characteristic manner, which is readily recognised when the motor axons are filled with intracellular dye (Fig. 4.2*a*). Historically, this synapse provided the first physiological demonstration that neurons can communicate between each other by way of electrical synapses, which overturned a commonly held assumption that neurons always communicate by way of chemical synapses. When intracellular recordings were made simultaneously from these two neurons, a negligible delay was found between the start of a spike in a lateral giant and a motor giant (Fig. 4.2*b*).

Both the lateral and medial giant interneurons extend along much of the length of the central nervous system, but they differ considerably in structure. The left and right medial giants have their cell bodies and dendrites in the brain, where they receive sensory input, and their axons extend down to the last abdominal ganglion. In contrast, the lateral giants are segmentally repeated structures on both sides of the body, formed from separate cells linked end-to-end. Each segment contains a cell body, dendrites and

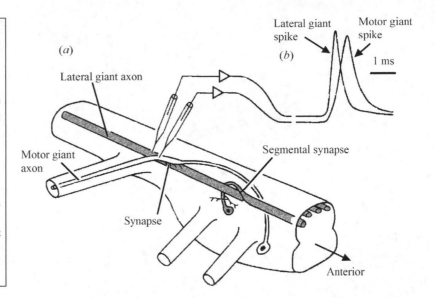

Fig. 4.2 Electrical synapses between neurons involved in crayfish startle behaviour. (*a*) Drawing of an abdominal ganglion to show the relative positions of the lateral giant and motor giant neurons, the synapse between them, and the arrangement for recording from each side of this synapse. The segmental synapse between successive lateral giants is also shown. (*b*) Simultaneous intracellular recordings from the lateral giant and motor giant neurons close to the synapse, demonstrating the negligible delay between the start of a spike in the interneuron and in the motor giant. (Redrawn from Furshpan and Potter, 1959.)

a length of axon that abuts against the corresponding axon in the next segment. As shown in Fig. 4.2*a*, where adjacent lateral giant axons abut, they communicate with each other by an electrical synapse that transmits spikes between the two so the whole chain of lateral giant interneurons acts as if it were a single neuron extending from the tail to the head of a crayfish. The lateral giants receive sensory input only in the abdominal segments.

A detailed series of studies, using a combination of neurophysiological and high speed filming techniques, showed that the lateral and medial giants initiate different patterns of behaviour (Fig. 4.3). In these studies, wire electrodes were implanted into the crayfish to record electrical responses from giant neurons and muscles. The electrodes had long leads, so the animal was able to move relatively unimpeded within its aquarium. Activation of the medial giants elicits contraction of the fast flexor muscles in all abdominal segments, and this produces a uniform curling of the abdomen that propels the animal straight backwards. Activation of the lateral giants elicits flexor contraction in the anterior segments of the abdomen but not in the posterior segments. The posterior part of the abdomen remains straight and so the thrust is directed mainly downwards, pitching the animal into a somersault-like movement. This turns the animal over so that it faces the source of annoyance, and is in a position where subsequent swimming movements produced by the abdomen and tail fan will take the crayfish away from potential danger. The two kinds of tail flip are well adapted to the different sorts of stimuli that excite the two types of giant interneuron. The lateral giants are triggered only by sudden mechanical stimuli that originate posteriorly, such as a sharp tap to the abdomen, and a lateral giant flip prepares the animal for escape away from potential danger approaching its abdomen. Similarly, the medial giants are triggered only by

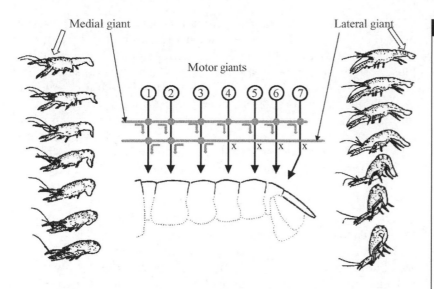

Medial giant   Motor giants   Lateral giant

Fig. 4.3 The different kinds of tail flip produced by the medial and lateral giant interneurons. The precise pattern of movement was correlated with activity in the giant interneurons by filming animals with electrodes implanted (as in Fig. 4.1*a*). Tracings from these high-speed films show (left) a medial giant flip elicited by a tap on the head and (right) a lateral giant flip elicited by a tap on the abdomen. Diagram of the pattern of synaptic connections between the two giants (grey) and the motor giant neurons in the abdominal segments. Direct synaptic connections are represented by filled plus arrow circles and the absence of synaptic connections is indicated by an x. (Redrawn from Wine, 1984.)

stimuli applied to the head region and a medial giant flip carries the animal backwards. In this way, each kind of tail flip removes the animal from the source of the stimulus.

The differences between the two kinds of tail flip can be explained by differences in the synaptic connections between the giant interneurons and the motor giant neurons in the abdominal segments (Fig. 4.3). In the anterior segments of the abdomen, both lateral and medial giants receive these synaptic branches from the motor giants, but in the posterior segments the synaptic branches to the lateral giant are clearly missing, while those to the medial giant are present. This distribution of synapses has been confirmed by testing for post-synaptic responses by recording with a microelectrode: responses to medial giant activity can be obtained in all abdominal segments, but responses to lateral giant activity are only obtained in the anterior segments. In the thorax, the situation is reversed: the medial giant makes no output connections to motor giants, but the lateral giant does connect in the more posterior segments (Heitler and Fraser, 1993). Contraction of flexor muscles in the posterior part of the thorax helps bend the body into the jack-knife shape that propels it rapidly forwards. Hence, the consistent difference in the pattern of abdominal flexion is brought about by differences in the synaptic connections between the respective controlling interneurons and a shared motor output system.

Sometimes the startle response consists of a single tail flip, but at other times the tail flip is followed by a series of up and down movements of the tail fan in rapid succession. These movements are produced by alternating flexions and extensions of the abdomen, repeated at a frequency of 10 to 20 per second. The behaviour is termed escape swimming.

Following a medial giant response, swimming tends to simply continue the backward movement away from the threat at the head end. But often steering movements are incorporated into the swimming movements, and a lateral giant response is usually followed by one or two pitching flips, which turn the animal in a complete somersault so that it lands on its back with its head facing the stimulus. Then two or three twisting flips turn the animal dorsal side up again, and a series of flips carry it backwards away from the stimulus. In addition, the swimming system is able to act upon directional information in the stimulus that is ignored by the giant systems. The medial giants, for example, always generate a bilaterally symmetrical response that carries the animal straight backwards, regardless of whether the stimulus comes directly from behind or from one side. But the path followed by subsequent swimming has a lateral component that steers the animal away from a stimulus delivered to one side of the abdomen. So escape swimming is well adapted to exploit the initial advantage gained by a fast giant-mediated tail flip.

Donald Edwards and colleagues (Herberholz *et al.*, 2004) established that crayfish really do use their giant interneurons and tail flip responses to evade predators, and that these startle responses are effective means of escape. They placed individual juvenile crayfish, about 2 cm long, together with slightly larger dragonfly larvae in small aquaria. Dragonfly larvae are voracious predators and catch prey by suddenly extending a prehensile structure, the labium, to grab the prey and bring it to the mouth. In all 38 attacks they observed, the crayfish responded with a tail flip. By using electrodes in the aquarium water, they could record and distinguish spikes from the two types of crayfish giant interneuron as well as muscle potentials from both animals. Most attacks were to the front or thorax of the crayfish, and most of those evoked medial giant spikes. A few evoked tail flips and swimming with no giant involvement. Attacks directed at the abdomen consistently evoked both lateral giant spikes and jack-knife tail flips. In 45% of attacks, the crayfish escaped any contact with the dragonfly, and the dragonfly larvae were successful in obtaining a meal in only one in every five strikes. This type of study is important as it ties conclusions about how neuronal pathways work together with measures of the value of the behaviours they control in the natural life of the animal.

The following account will focus on the lateral giant response and its relation to subsequent swimming. A lateral giant interneuron acts as a central channel in the startle response, providing a rapid route for the flow of information from sensory to motor systems. This is illustrated in Fig. 4.4, in which part (*a*) represents neurons and synapses schematically and part (*b*) includes more realistic drawings of some of the neurons. Information converges onto the lateral giant through two different pathways, and it also distributes information through two pathways. Some receptor neurons connect directly with the lateral giant, but most sensory information is routed to the lateral giant through sensory interneurons. From the lateral giant, information diverges into two routes. The most direct is through the motor giant neurons, each of which synapses with all the fast flexor muscle fibres in its half

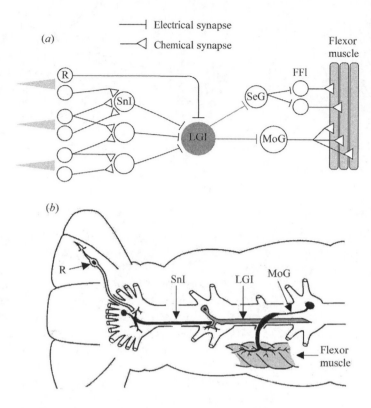

Fig. 4.4 Neuronal circuit for startle behaviour mediated by the lateral giant interneuron. (*a*) Schematic representation of the pathway from the mechanoreceptors to the flexor muscles. Labelled circles represent: R, receptor neurons; SnI, sensory interneurons; LGI lateral giant; SeG, segmental giant; MoG, motor giant; and FFI, fast flexor motor neurons. (*b*) Drawing to show the arrangement of some of these neurons within the nervous system. They are not drawn to scale; and only one segment of the lateral giant is shown. (Redrawn from Wine and Krasne, 1982.)

segment. In the other route, the lateral giant excites a segmental giant interneuron which sends excitation to all of the fast flexor motor neurons in its half ganglion. Each fast flexor motor neuron excites a subset of flexor muscle fibres. The connections between a lateral giant and a motor giant, between a lateral giant and segmental giant, and from a segmental giant to individual fast flexor motor neurons are all electrical, so there is minimal delay in transmission of excitation towards the muscles.

The trigger for a tail flip

It is clear that a lateral giant has a crucial location in the pathway that generates a tail flip. It has been shown that the lateral giant is essential to the production of a tail flip by hyperpolarising the neuron so that it cannot produce spikes, and when a lateral giant is prevented from firing in this way, stimuli that would normally cause a tail flip no longer do so (Olson and Krasne, 1981). For a given stimulus intensity, the amount of hyperpolarisation needed to abolish the tail flip is exactly that needed to prevent the lateral giant from firing.

A lateral giant interneuron responds to sudden water movements over the surface of the abdomen and to touch of the tail fan, both of which could indicate attack by a predator. Although crayfish have well developed eyes, they often live where visibility

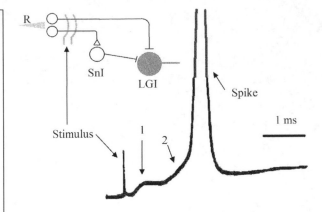

Fig. 4.5 Sensory input to the lateral giant. A nerve that carries axons of sensory neurons was placed on a pair of silver hook electrodes, used to excite the axons with brief electrical shocks. Increasing the voltage of the shock would increase the number of sensory axons that spiked. An intracellular electrode recorded the response from the lateral giant, which consists of two components: 1, directly from sensory neurons; and 2, via sensory interneurons. In the recording shown, sufficient sensory axons were stimulated to excite the lateral giant to its spike threshold. (Recording redrawn from Krasne, 1969.)

in the water is poor even in daytime, so it is an advantage for them to be able to sense a predator's approach in this way. Touch or sudden water movement is detected by stubby hairs that project from the cuticle of the abdomen, augmented by stretch receptors in the tail fan. The abdomen bears about 1000 of these hairs altogether and each is innervated by a pair of sensory neurons that are situated at its base. Deflecting the hair stretches its sensory neuron dendrites, and each neuron responds with a receptor potential and axonal spikes. The hair receptors are well suited to detecting the shock wave in the water produced by the acceleration of a predator towards the crayfish, or to physical contact by the predator. They are sensitive only to touch or to sudden water movements that have a time course of about 10 ms or less. They do not respond to slow water movements and provide a good example of stimulus filtering: because these mechanoreceptors respond only to particular types of disturbance, the lateral giants, which they excite, will respond only to imminent danger and not to water movements caused by waves or movements by the crayfish itself. Other types of hair receptor on the cuticle do respond to slower water movements and they do not excite the lateral giants.

There are 10–20 of the sensory interneurons in each half of an abdominal ganglion and many hair receptors converge onto each one. There is also some divergence, because each receptor axon branches to make contact with several interneurons (Fig. 4.4a). Most of these interneurons have relatively large dendrites and axons, from which synaptic potentials and spikes can be recorded with microelectrodes. Sensory receptor neurons make chemical synapses with the sensory interneurons. Following stimulation of the hair receptors, a compound EPSP with two components can be recorded in the lateral giant (Fig. 4.5). The first component is due to receptors that synapse directly on the lateral giant, and the second is due to input via the sensory interneurons. Because the lateral giant neuron is so much larger than either a sensory neuron or a sensory interneuron, single spikes in any presynaptic neuron cannot generate enough current at the electrical synapses to depolarise the lateral giant above threshold. Instead, single

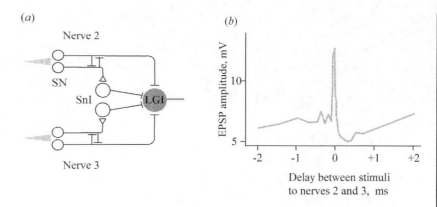

(a)

Nerve 2

SN

SnI

LGI

Nerve 3

(b)

EPSP amplitude, mV

Delay between stimuli
to nerves 2 and 3, ms

Fig. 4.6 Coincidence detection in the crayfish lateral giant interneuron (LGI). (a) Connections made between sensory neurons in two different nerves, 2 and 3, and the LGI. In addition to electrical synapses, different sensory neurons that run in the same nerve make electrical synapses with each other as well as connecting with sensory interneurons and the lateral giant. (b) Results of an experiment in which excitatory postsynaptic potentials (EPSPs) were recorded from the LGI in response to small electrical stimuli to nerves 2 and 3, separated by different intervals. Note how sharply tuned the response is for closely timed stimuli. (Part (b) redrawn from Edwards et al, 1998.)

spikes generate EPSPs which summate in the lateral giant in the usual way (Chapter 2). The first component primes the lateral giant for excitation from the mechanosensory interneurons.

Several mechanisms ensure that the lateral giant spikes only to sudden, intense synaptic inputs. First, the lateral giant is quite hard to excite because it has a high threshold for spike initiation. Second, individual EPSPs die away quickly so that many EPSPs must arrive within a short time of each other if they are to sum and trigger a spike. Third, stimuli to different hairs that are close to each other are much more effective than stimuli to hairs that are further apart. That is because the sensory axons of nearby hairs run alongside each other in nerves and make electrical synapses with each other, providing mutual reinforcement to their excitation (Fig. 4.6a). Fourth, stimuli to different sensory hairs are most effective if the time of stimuli to different hairs coincide in time with each other. This was demonstrated by delivering electrical stimuli to trigger precisely timed spikes in axons of two different nerve branches of the terminal abdominal ganglion. The EPSP in the lateral giant interneuron was largest if the stimuli were simultaneous, and was very much smaller if the stimuli were as little as a fraction of a millisecond apart (Fig. 4.6b). The reason for this is that when the lateral giant is excited, the effectiveness of electrical synapses from other neurons at exciting the lateral giant is reduced (Edwards et al., 1998).

Finally, the chemical synapses between the receptor axons and sensory interneurons decline rapidly in strength with use. Each spike in a receptor axon causes a reduction in the amount of neurotransmitter released when the next spike arrives at the synaptic terminals. A reduction in strength of a specific behaviour evoked by a particular stimulus caused by this type of mechanism is called **habituation**, and it has been most extensively studied in pathways in the sea hare, *Aplysia*, that control protective gill withdrawal when the animal's skin is touched. It helps to adjust an animal's sensitivity to repeated stimuli so that it does not respond inappropriately to repeated stimuli, which are much less likely to represent a substantial threat than a novel, unexpected stimulus. When a crayfish is tapped repeatedly on the abdomen, the probability of a tail flip in response diminishes rapidly. Stimulation at the rate of one tap per minute can diminish responsiveness to zero within ten minutes, and then many hours rest are needed for the response to recover. The reduction in response is not due to fatigue of the muscles because directing a tap to another part of the crayfish elicits a vigorous tail flip.

Triggers for later phases of escape

Following a tail flip, the crayfish must straighten its abdomen again before it can generate another burst of thrust by moving its tail fan. Re-extension is a reflex initiated by proprioceptors stimulated by the flexion movement, rather than being triggered by a lateral giant spike (Reichert *et al.*, 1981). This was shown by using implanted electrodes to record activity in fast flexor and extensor muscles of freely moving crayfish. When the abdomen was restrained so that it could not flex, or else the motor nerves to the flexor muscles were cut, no activity was recorded from the extensors following a stimulus that is adequate to excite the lateral giants. Two classes of proprioceptor make appropriate synaptic connections with the fast extensor motor neurons: the hair receptors on the dorsal surface of the abdomen; and the stretch-sensitive muscle receptor organs, described in Chapter 2. As shown earlier, in Fig. 2.2, each muscle receptor organ spike causes an EPSP in an extensor motor neuron; if the spikes occur at a high enough rate, the EPSPs sum sufficiently to bring the motor neuron to its spike threshold. The hair receptors also have a substantial excitatory input to the fast extensor motor neurons, probably through a pathway that includes interneurons. So, during a normal tail flip muscle receptor organs will first start excitation of the fast extensor muscles, followed by excitation from the hair receptors that are excited by rapid water movement generated by abdominal flexion.

The initial tail flip is normally followed by escape swimming, which consists of a series of tail flips that are not mediated by the giant axons. Swimming is neither initiated by the lateral giants (Reichert and Wine, 1983) nor by sensory feedback from the first movements of flexion or extension. Instead, stimuli to the sensory hairs on the abdomen activate a different pathway that acts in parallel with, and more slowly than, the lateral giant pathway. Evidence for this was obtained by bypassing the normal sensory input to the lateral giants and stimulating them directly with current delivered by electrodes implanted into resting, unrestrained crayfish. Direct electrical stimulation elicited a rapid flexion followed by a re-extension, but fewer than 1% of the tail flips elicited were followed by swimming. So neither activity in the lateral giants nor sensory feedback from the actual tail flip is sufficient to elicit escape swimming, and swimming must be activated by an independent and parallel pathway that leads from the mechanoreceptive hairs.

Taps on the abdomen that are just below threshold for firing the lateral giants will often trigger escape swimming in well rested crayfish. When this happens, the delay between the stimulus and the onset of the first flexor contraction is about 240 ms, slightly longer than the interval between the stimulus and the first flexion in a swimming bout that follows lateral giant spikes and a tail flip. This means that swimming can be triggered through a much slower pathway than that responsible for triggering a tail flip. It also suggests that swimming is not simply inhibited during a lateral giant-initiated tail flip because then one would expect swimming to begin with a shorter delay when it occurs without a preceding lateral giant response. In fact the average delay for swimming that is preceded by a giant-mediated tail flip is somewhat less than the delay for swimming that occurs without an initial tail flip (184 ms compared with 240 ms), which indicates that the lateral giants may facilitate the onset of swimming although they do not trigger it. Therefore, the smooth transition from a tail flip to swimming in normal escape behaviour is largely due to the activation of two different pathways that activate the two types of movement in sequence.

Escape swimming actually begins with a burst of activity in the abdominal extensor muscles, preceding the first flexor contraction by about 50 ms. This extensor activity is a distinct event from the reflex re-extension of the abdomen following a giant-mediated flip, but the two events can overlap. The rate of abdomen movements declines during a swim, and the crayfish coasts with the abdomen in the flexed position before rapidly extending and flexing it at the start of the next cycle. The alternating extension and flexion during swimming is produced by a pattern generator consisting of networks of interneurons that work in a similar way to those described later on for other rhythmically repeating activities (Chapter 7). The proprioceptive reflexes that re-extend the abdomen after a tail flip are not responsible for flexion during swimming.

Executive functions of the lateral giant neuron

A lateral giant not only initiates a tail flip, but it also extensively co-ordinates the sequence of events involved. This executive function is achieved by an extensive, widely distributed array of inhibitory effects that follow a spike in the giant axon. Pathways lead away from the lateral giant to exert inhibition at almost every point in the neuronal network generating a tail flip. The IPSPs produced at these points in the network differ from one another in their delay and duration, in a way that ensures that each part of the response begins and ends at the right time.

The extensor motor system is the first place where inhibition is seen following a spike in a lateral giant axon. This inhibition is accomplished by parallel actions at three points in the motor

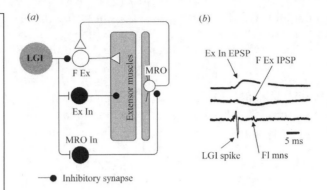

Fig. 4.7 Inhibition of the abdominal extensor muscles by the lateral giant. (*a*) Neuronal pathway generating inhibition of the extensors, showing representative neurons: LGI, the lateral giant; F Ex, fast extensor motor neuron; Ex In, extensor inhibitor; MRO, muscle receptor organ; MRO In, inhibitor of the MRO. Inhibitory neurons are shown in solid black; electrical and chemical excitatory synapses indicated as before. (*b*) Typical recording used to build up the interpretation given in (*a*). Lower trace is an extracellular record from the motor nerve to the flexors, showing the lateral giant spike and subsequent compound spike from flexor motor neurons (Fl mns). Middle and upper traces are intracellular records showing, respectively, an excitatory postsynaptic potential (EPSP) in the extensor inhibitor and an inhibitory postsynaptic potential (IPSP) in the fast extensor motor neuron. Note the short delay of the postsynaptic potentials after the lateral giant spike. The amplitude of the EPSP was about 3 mV, and that of the IPSP was about 1.5 mV. (Part (*b*) redrawn from Wine and Mistick, 1977.)

pathway to the extensor muscles (Fig. 4.7*a*). First, the lateral giant decreases the likelihood that the flexor muscles will be activated during a tail flip by making inhibitory synapses with the excitatory motor neurons of the extensor muscles. Second, the fast extensor muscles are inhibited by a motor neuron, called the extensor inhibitor motor neuron, that causes IPSPs in the muscle fibres. It is quite common in arthropods for muscle fibres to be innervated by such inhibitory motor neurons in addition to those that excite the fibres, although this arrangement is not found in vertebrates. Third, the muscle receptor organ is inhibited by a special accessory cell associated with its sensory neuron. The inhibitory actions begin within a few milliseconds of a lateral giant spike; in fact, onset of the IPSPs in the extensor motor neurons is synchronous with the onset of the EPSPs in the fast flexor motor neurons (Fig. 4.7*b*). The duration of the IPSPs produced at all three points in the extensor pathway is relatively short, having an average value of 30 ms.

These connections clearly function both to reduce the likelihood that extension will interfere with flexion during a tail flip and to ensure that the reflex extension following giant-mediated flexion is properly timed. Inhibiting the extensor pathway at three separate locations makes quite certain that extension cannot occur while the inhibition lasts. The average duration of the IPSPs at the three locations is about the same as the average duration of the giant-mediated flexion, and so the extensor system is released from inhibition just as flexion is completed. The inhibitory action of the lateral giant thus co-ordinates flexion and re-extension effectively, even though the extensor system does not receive any additional excitation from the lateral giant.

Co-ordination of the response is continued by inhibition of the flexor motor system just as re-extension takes place. Two inhibitory neurons have been identified that are important in shutting down the flexor system: the inhibitory motor neuron, which innervates every fast flexor muscle fibre; and the motor giant inhibitor, an interneuron which prevents the motor giant from firing. Both of these inhibitory neurons are excited indirectly by the lateral giant (Fig. 4.8). The motor giant inhibitor is strongly excited by the fast flexor motor neurons and so brings about inhibition of the

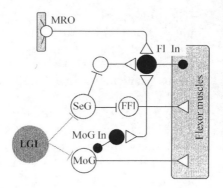

Fig. 4.8 Delayed inhibition of the fast flexor muscles by the lateral giant. (*a*) Neuronal circuit generating delayed inhibition of the flexor system showing representative neurons as in Fig. 4.4 plus: MoG In, the motor giant (MoG) inhibitor; and Fl In, flexor inhibitor. MRO, muscle receptor organ. Synapses indicated as before.

motor giant very soon after the motor giant has fired. The flexor inhibitor is more weakly excited by a variety of sources, including by the fast flexor motor neurons and indirectly by the segmental giants.

This roundabout pathway means that a spike in the flexor inhibitor is not triggered until about 15 ms after the spike in the lateral giant, by which time the motor giant is already inhibited. At this time, the movement of flexion has just begun; but it must be borne in mind that the electrical signals in excitatory motor neurons have finished several milliseconds before the movements they cause are under way because it takes a few milliseconds for the muscles to develop tension and then for the tension to move the exoskeleton. Hence, the delayed inhibition from the lateral giant prevents any additional flexor activity and prepares the way for re-extension. In addition, the flexor inhibitor receives a synapse directly from the muscle receptor organ and this input becomes active as soon as the muscle receptor organ is released from inhibition. This input sums with that from the lateral giant, thereby prolonging activity of the flexor inhibitor for the duration of re-extension.

Inhibition also acts on the giant interneurons themselves and on the sensory inputs to them. A lateral giant spike is followed, after a short delay, by strong inhibition of all of the giant inter-neurons, imposing a veto on startle responses immediately following. This veto imposes a kind of stranglehold on the excitability of the giant neuron because the inhibitory synapses involved are located close to the origin of the axon, where spikes are initiated (Vu and Krasne, 1993). Inhibition is also directed to the presynaptic terminals where the hair receptor neurons make chemical synapses with the sensory interneurons. **Presynaptic inhibition** of this type is common in mechanosensory systems, and its action is to reduce or prevent the release of neurotransmitter. Both post-synaptic and presynaptic inhibition are delayed about 15 ms after the lateral giant spike on account of the indirect pathway that mediates them (Fig. 4.8) and they last about 50 ms. Those hair receptors that provide input to the extensor motor neurons are not inhibited in this way, allowing them to contribute to the re-extension reflex.

The presynaptic inhibition of the first input synapse plays an important part in co-ordinating the startle reflex because, were it not inhibited, this input synapse could cause a perpetual cycle of repeated tail flips by positive feedback. Abdominal flexion is a vigorous movement that stimulates the sensory hairs, which normally trigger a lateral giant spike and so a tail flip. As it is, the onset of inhibition coincides with the onset of movement of the abdomen and so prevents this feedback effect. The long duration of the inhibition makes sure that a second tail flip is not triggered before the first one is completed. The occurrence of presynaptic inhibition at the first synapse is well correlated with the fact that this synapse is the site of habituation of the response. If postsynaptic inhibition alone were present, then synaptic transmission could still habituate during the repetitive stimulation caused by abdominal flexions, and this could render the whole system unresponsive for several hours. Because presynaptic inhibition prevents the presynaptic terminals from being fully activated, presynaptic inhibition has the effect of protecting the terminals from habituation.

Summary of pathways in crayfish startle behaviour

The startle response of crayfish is a simple behaviour, which is initiated with the least possible delay. Nevertheless, the neuronal network underlying this behaviour is quite sophisticated and involves complexities that might not be predicted from the behaviour itself. The main pathways involved in linking the response to an adequate stimulus are shown in Fig. 4.9. The lateral giant gathers sensory information, and distributes it onward, initially to excite flexor muscles. It also delivers inhibition in sequence to the extensor motor system and then to the flexor motor system and to itself. Through these pathways, the first tail flip and re-extension is completed by about 110 ms from the initial stimulus.

Sensory information that is adequate to trigger the giant-mediated tail flip also often triggers escape swimming by an independent, longer neuronal pathway. Swimming is more complex than the initial tail flip as it includes more precisely directed turning to direct the crayfish away from a source of danger. The delay in triggering swimming is such that the first movement of escape swimming, an extension, overlaps with or immediately follows the re-extension of the first tail flip.

The crayfish lateral giant interneuron fulfils the criterion of a 'command neuron' in that it is both necessary and sufficient for a tail flip in response to attack from the rear. All of the sensory information that initiates a tail flip is channelled through it, and it distributes strong excitation by short pathways to the flexor system.

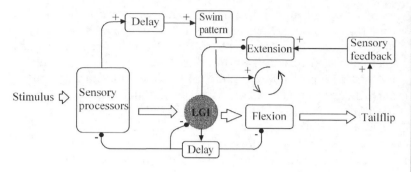

Fig. 4.9 Flow diagram summarising the functional relations between the major components of the startle behaviour in crayfish. Except in the case of the lateral giant, the labelled boxes do not represent individual neurons. Open arrows indicate the fast pathway by which the lateral giant initiates a tail flip. Lines with arrowheads indicate excitation and filled circles indicate inhibition in pathways that initiate, first, re-extension, and then swimming. (Redrawn from Wine, 1984.)

Plasticity in the lateral giant and tail flip response

The way that lateral giants respond to imminent attack, initiate a startle response and ensure it is executed smoothly is fairly straightforward, but that is not the whole story. Some additional elements are involved, for the limbs and claws move during a startle response. Also, crayfish are subject to changes in mood so the startle response is not an automatic response, always elicited with the same reliability whenever the abdomen is touched or disturbed by a sudden water movement. The ease with which it can be elicited depends on what the crayfish is doing, or has been doing recently. The likelihood that a sensory stimulus will elicit a tail flip is reduced, for example if a crayfish is feeding (Krasne and Lee, 1988), or if it is restrained. The mechanism for this includes neurons that descend from the brain and which exert inhibition directly on the lateral giants. This inhibition is called 'tonic' to distinguish it from the brief, vetoing inhibition that follows a giant interneuron spike. Unlike the giant-activated inhibition, the tonic inhibition is made onto relatively remote parts of the dendritic trees of the lateral giants, where it can be integrated with other synaptic inputs. Because the lateral giant acts as a command neuron, tonic inhibition that is directed at it will specifically affect the tail flip response without affecting other behaviours that the flexor and extensor muscles are involved in, such as swimming.

The ease with which startle tail flip responses can be elicited is also modulated by the social status of a crayfish. In aquaria, crayfish are territorial animals. They will fight each other and establish clear dominance hierarchies (Issa *et al.*, 1999), although little is known of the frequency and effectiveness of fights in natural conditions. When two crayfish are initially put together, they grapple each other, and frequently produce tail flips that are not preceded by giant fibre activation (Herberholz *et al.*, 2001). After several minutes, one crayfish moves away, and subsequent encounters reinforce the subordinate status of one and dominant status of the other. Early after the hierarchy has been established, stimuli that would not previously have excited the giant interneurons tend to activate them in both crayfish, indicating that giant interneuron excitability is elevated. After several days, the hierarchy is fully established and the excitability of the giant interneurons in the two crayfish differs: the dominant crayfish shows increased excitability of its giant interneurons, and the subordinate one shows decreased excitability. This is opposite to the effect that might be expected. However, it emphasises that the prime function of the giant interneurons is to signal unexpected attacks. The subordinate crayfish does readily produce tail flips and escape swimming when the dominant one comes near, but these are not initiated by giant interneuron spikes. The subordinate crayfish might be thought of as being in a constant state of vigilance, expecting attacks. The non-giant pathways will allow it greater flexibility in taking an initial escape route away from the direction of a potential attack. The dominant crayfish, on the other hand, does not expect an attack, and elevation of giant interneuron excitability is an insurance measure, protecting it against the decreased likelihood of attack.

The amine serotonin (5-hydroxytryptamine) is thought to be involved in regulation of dominance in crayfish. Serotonin is also involved in regulation of mood and aggression in other animals, including humans, so there is some interest in revealing the cellular basis of its action. When serotonin is injected into a lobster, it adopts an aggressive posture, whereas octopamine, an amine that plays a role in arthropods that is similar to that played by adrenaline in mammals, causes a lobster to adopt a submissive posture (Livingstone *et al.*, 1980). Serotonin also affects the responsiveness of the crayfish lateral giant to stimulation of sensory nerve roots. The effects depend on both the concentration and time course of application. When applied slowly or for long periods serotonin tends to enhance responsiveness, which is partly due to an increase in the strengths of electrical synapses between sensory neurons and the lateral giant (Antonsen and Edwards, 2007). But this effect can be altered by social experience. In particular, in a crayfish that has become subordinate by sharing a tank with a dominant individual, serotonin depresses the lateral giant responsiveness (Yeh *et al.*, 1997). In these kinds of ways, even a straightforward behaviour such as a startle response is subject to modification in ways that integrate it within the overall behavioural repertoire of the animal.

Evolution of the lateral giant and tail flip pathway

There are no fossilised remains of neuronal pathways, but some of the features of the giant interneuron pathways in crayfish and of neurons in other, related crustacea give clues about the possible evolutionary origin of the crayfish startle response. The class of crustacea to which crayfish belong, the malacostraca, diverged from other ancestral crustacean forms nearly 400 million years ago. Prawns and lobsters as well as freshwater crayfish all produce rapid tail flips when threatened, and all have fast-conducting giant interneurons in their nerve cords. The Tasmanian mountain shrimp (*Anaspides tasmaniae*), which is thought to be similar to the ancestor of all of these, has both medial and lateral giants and motor giant neurons (Silvey and Wilson, 1979), indicating that the two tail flip responses that crayfish produce is a primitive character within the malacostraca. A number of groups within this class have secondarily lost one or both giant interneurons. These animals include crabs, squat lobsters and hermit crabs, which perhaps rely more on armour or an ability to burrow than speed for escape. Strangely, spiny lobsters produce tail flips but lack giant interneurons (Espinoza *et al.*, 2006).

One clue about the origin of the tail flip response comes from a study on mantis shrimps (Heitler *et al.*, 2000), which may have diverged from the main malacostracan line earlier during evolution than *Anaspides*. Mantis shrimps produce startle responses when prodded at the front, but not at the rear. They shoot backwards, a movement caused by simultaneous forward-directed movement of their thoracic and abdominal appendages. Unlike the crayfish tail flip, this response varies in strength with the intensity of the stimulus. In response to the most vigorous stimuli, the appendage movement is augmented by abdominal flexion of the abdomen. Mantis shrimps have a single pair of giant axons in the dorsal part of their nerve cord. Stimulating these giant interneurons electrically suggests that they make fairly direct excitatory connections with some of the appendage motor neurons, and less direct connections with abdominal flexor motor neurons.

Together with other data, the observations on mantis shrimps suggest that the most primitive startle response involved simultaneous forward movement of the limbs, and that the neurons responsible for triggering and co-ordinating this became unusually large. But more rapid and effective movements became possible when the tail fan and abdominal musculature that moves it developed. How did the giant interneuron take control of these structures? The segmental giant interneuron that is interposed between the giant interneurons and fast flexor motor neurons in each half segment provides a clue. It has the unusual anatomical feature of a vestigial blind-ending axon in one of the segmental nerves, and this suggests the segmental giant may have originally been a motor neuron innervating a muscle that was lost during evolution, or else is now innervated by a different motor neuron.

Two events during evolution may have linked giant interneurons with flexor muscles. First, the segmental giant lost its function as a motor neuron, and instead distributed excitation to motor neurons of the trunk flexor muscles. Second, one of the flexor motor neurons became larger than the others, and its control was extended to all of the flexor muscles in its half segment. This became the motor giant neuron. This order of events is suggested by the discovery that in hermit crabs the segmental giant excites the motor giant, whereas in crayfish this connection is absent and the motor giant is directly excited by the lateral and median giants. By examining adaptations amongst modern animals, we can derive feasible routes for the evolution of neuronal networks. Although it is hard to establish that neurons with similar properties in different animals are homologous, this is a fruitful field for investigation and speculation. One of the key outstanding issues is how the lateral giant originated.

Mauthner neurons and the teleost fast start

When a sharp tap is delivered to the side of an aquarium, a fish inside exhibits a characteristic startle response consisting of a brisk swivelling movement. In natural circumstances, this is an effective escape movement that enables the animal to dodge the strike of a predator. The key neuron in this startle response is called the **Mauthner neuron**, named after Ludwig Mauthner who, as a 19-year-old medical student, first discovered in 1859 that fish spinal cord contains 'collosal' nerve fibres. Many studies on these neurons have been made in goldfish and zebra fish. There is just one Mauthner neuron on each side of the brain, and it occurs in most species of teleost (bony) fish and of amphibians. Because of the exceptionally large size of Mauthner neurons, it has been possible to study them in both dissected preparations and intact animals; these studies provide one of the few cases in vertebrates where a clear causal relationship has been established between activity in a particular neuron and performance of a specific behaviour pattern. The way the Mauthner neuron operates shows a number of instructive parallels with a crayfish lateral giant neuron, but the Mauthner neuron has a lesser claim to the status of a command neuron.

In teleost fish, the startle movement initiated by a Mauthner neuron is known as a fast start, and consists of a stereotyped sequence of movements that occur in three stages. In the initial stage, the trunk muscles contract all along one side of the body so that the animal assumes a C-like shape with the head and tail bent to the same side. A number of other actions are also initiated, such as closing the mouth, drawing in the eyes and extending the fins. During the second stage, muscle contractions proceed down the other side of the body so that the tail straightens. These first two stages result in a sudden acceleration that propels the animal in a direction that is determined by the extent of the body bend in the first stage and displaces the fish by about one body length. The third stage consists of normal swimming movements, or sometimes coasting, which carries the fish further away.

A startle response by a zebra fish larva (*Brachydanio*) when touched near its tail (asterisk). The Mauthner neuron is involved in triggering this response. Frames in the high-speed video are 4 ms apart. (Photograph supplied by David McLean, Northwestern University, Illinois.)

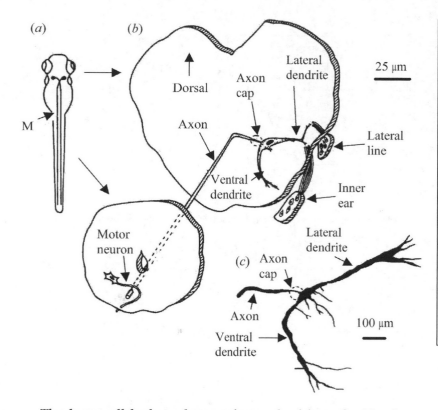

(a)

(b)

Dorsal

Axon cap

Lateral dendrite

25 μm

M

Axon

Lateral line

Ventral dendrite

Inner ear

Motor neuron

Lateral dendrite

(c) Axon cap

Axon

100 μm

Ventral dendrite

Fig. 4.10 The Mauthner neuron in teleost fish. (a) The general location of the bilateral pair of Mauthner neurons (M) shown schematically for a larval zebra fish (*Brachydanio*). (b) The right Mauthner neuron is shown in a thick, transverse section of the hindbrain at the level of the eighth cranial nerve, together with the inputs from the lateral line and inner ear. The output of the Mauthner axon to the spinal motor neurons is shown in a thick transverse section of the spinal cord. (c) The Mauthner neuron of an adult goldfish (*Carausius*), reconstructed from transverse sections of a cell injected with intracellular dye. (Part (a) redrawn from Prugh et al., 1982; (b) redrawn from Kimmel and Eaton, 1976; (c) redrawn from Zottoli, 1978.)

The large cell body and two primary dendrites of a Mauthner neuron are located in the hindbrain (Fig. 4.10). The axon crosses to the opposite side of the brain before descending the nerve cord to contact motor neurons of trunk muscles. This basic anatomy was described by Bartelmez (1915), who first suggested that it was involved in startle behaviour, although for several decades after that it was thought that the Mauthner neuron was involved in normal swimming movements. The role of the neuron in startle behaviour was firmly established when recordings from the neuron were correlated with movements in freely moving animals – spikes in a Mauthner neuron can be identified unambiguously in extracellular recordings because their waveform provides a characteristic signature. When a spike is recorded from a Mauthner neuron in response to a sudden sound, it is always immediately followed by a large muscle potential in the trunk muscles on the side of the body opposite to the Mauthner neuron cell body, and the fish performs a fast start (Fig. 4.11a).

The precise timing of the Mauthner spike in relation to the stages of the fast start is clarified by using an implanted electrode in conjunction with high-speed filming of the response (Fig. 4.11b). This method shows that the average delay from a stimulus to a Mauthner spike is about 6 ms. There is a further delay of about 2 ms until the muscle potential starts and then a delay of 6–10 ms until the muscle develops sufficient tension for actual movement to begin. The speed of this startle response is comparable to that of

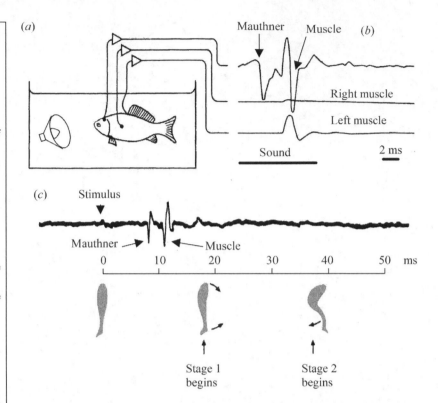

Fig. 4.11 Activity of the Mauthner neuron during startle behaviour. (*a*) Goldfish in an aquarium with electrodes on the skin to record a Mauthner neuron and from both left and right trunk muscles. The stimulus is delivered by a loudspeaker to the front and right of the fish. (*b*) Typical recording of a startle response to a sound stimulus. The electrode on the head (top trace) picks up the Mauthner spike (M) and also the prominent electrical signal that spreads from the trunk musculature. The two muscle electrodes show that this muscle potential originates from the left muscles rather than the right muscles, which are on the same side as the speaker. (*c*) Simultaneous recording with electrodes on the brain and high-speed filming in dorsal view (representative silhouettes at bottom), establishing the precise timing of the Mauthner spike in startle behaviour. (Parts (*a*) and (*b*) modified from Zottoli, 1977; (*c*) redrawn from Eaton *et al.*, 1981.)

the crayfish tail flip, which is not surprising as both are natural responses for evading the strike of a predator. Another parallel with the crayfish startle system is that only a single spike occurs in the Mauthner neuron, and this precedes the initial C-like bend. Mauthner spikes do not accompany the second stage or subsequent swimming, which must therefore be due to the activity of other interneurons acting in parallel with the Mauthner neuron.

The Mauthner neuron on the side closest to the stimulus is the only one that produces a spike and, since this excites the con-tralateral ('opposite side') musculature, the initial, C-like turn is made to the side away from the stimulus. The size of this initial turn is not constant, but it varies with the angle of the threat-ening stimulus with respect to the fish. This was shown by drop-ping a 4.25-cm diameter ball on to the water near a goldfish and analysing high-speed films of the resulting startle response. In this experiment, the angle through which a fish had turned by the end of stage 1 was inversely proportional to the direction of the impact of the ball (Eaton and Emberley, 1991). When the startle response was filmed in goldfish with electrodes implanted in the trunk muscles (as in Fig. 4.11*a*), the results showed that progressively larger turns in stage 1 were correlated with progres-sively longer and more complex muscle potentials. The fish evi-dently controls how far it turns by varying the activity of motor neurons supplying the trunk muscles. However, direct electrical stimulation of the Mauthner neuron consistently results in a short and simple muscle potential. Hence the size of the C-like turn must be controlled by other interneurons that are active in parallel with the Mauthner neuron (Eaton *et al.*, 1981).

Excitation and inhibition in the Mauthner neuron

The Mauthner neuron is excited by wide range of different sensory modalities. As a result, a large number of sensory neurons and interneurons converge on this neuron. This is reflected in the great size of the two main dendrites (Figs. 4.10 and 4.12*a*). The cell body and the dendrites sum the synaptic potentials which are generated by inputs from the sensory neurons, and this summed signal is conducted passively to the initial part of the axon. The Mauthner neuron dendrites have large space constants (Chapter 2) and conduct the passive signals that originate at synapses onto them several hundred microns towards the axon. The initial part of the axon plays a crucial part in the integrative mechanism of the Mauthner neuron because it is where spikes are initiated when the input delivered from the dendrites exceeds threshold. Both the dendrites and the cell body are incapable of supporting spikes.

The input that has been studied in most detail is that coming from the auditory system. The underwater sounds and pressure wave that accompany the onrush of an attacking predator can be detected by the sensory receptors in the inner ear of the threatened fish. Using a loudspeaker or a high-pressure jet of water as a stimulus shows that auditory stimuli are certainly capable of initiating fast starts. Visual stimuli and input to the lateral line may augment the auditory stimulus but do not appear to be capable of triggering a spike in the Mauthner neuron on their own. For example, in the above experiment where a ball was dropped on the water, the goldfish could see the ball coming but a fast start was not initiated until several milliseconds after the ball hit the water. This suggests that the triggering stimulus for the Mauthner neuron was the auditory one from the impact of the ball on the water.

The sensory neurons from the inner ear provide the major input to the lateral dendrite, where they make direct contact in the form of large club-shaped endings. When postsynaptic potentials are recorded by a microelectrode in the lateral dendrite, they clearly contain two components. The first is a sharply rising EPSP after a short latency of about 0.1 ms; and the second is a more gradually rising EPSP with a longer latency, of about 1 ms (Fig. 4.12*b*). The latencies of these two components suggest that the first is mediated by electrical synapses, and the second by chemical synapses, possibly from sensory interneurons. The first component can be elicited by relatively weak shocks to the eighth cranial nerve, which are strong enough to excite the largest sensory axons but not to excite the smaller sensory axons that are responsible for the chemical EPSP. Electron microscopy of the

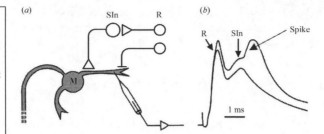

Fig. 4.12 Sensory input to the Mauthner neuron. (*a*) Schematic representation of input to the lateral dendrite from the eighth cranial nerve, showing receptors with direct electrical synapses onto the Mauthner neuron (M) and those with indirect chemical synapses by way of sensory interneurons (SIn). (*b*) Intracellular recording from the end of the lateral dendrite, as indicated in (*a*), showing two superimposed records. Following electrical stimulation of the receptor axons, a compound excitatory postsynaptic potential (EPSP) is recorded, with an early component due to the electrical synapse directly from receptors and a later component due to the chemical synapses from sensory interneurons. In one record, the EPSPs sum to trigger a spike, which is small in size because it has spread passively from the spike initiating zone to the recording site. (Part (*b*) redrawn from Diamond, 1968.)

distal part of the lateral dendrites confirms that the club endings connect to the Mauthner neuron by way of tight junctions between the membranes of the two cells, which are characteristic of electrical synapses.

As stimulus strength is increased further, both the components grow in size until their EPSPs are sufficiently large to trigger a spike. In the recording shown in Fig. 4.12*b*, the spike appears small and rounded, and the chemical EPSP that triggered it appears smaller than the preceding electrical EPSP. This is because the site of the recording was towards the end of the lateral dendrite, in the region where it receives the electrical synapses. Both the spike and the chemical EPSP, which is caused by synapses closer to the cell body, were conducted passively along the cell back to the recording site, and became smaller and more rounded in form as a result of this passive conduction.

Thus, the electrical EPSP has a priming role. In natural conditions, it promotes early excitation of the spike initiation zone, so that a following, large, chemically mediated EPSP rapidly triggers a spike. As in the input network to the crayfish lateral giant, both electrical and chemical synapses play roles in conveying sensory information to the Mauthner neuron. In addition, the electrical synapses that auditory neurons make with the Mauthner neuron play a role in enhancing the response to near simultaneous excitation in several presynaptic neurons (Pereda *et al.*, 1995) as happens at the electrical synapses between hair sensory neurons and the crayfish lateral giant. Both the large diameter of the Mauthner neuron axon and the axon's extensive insulation with myelin ensure that, once initiated, a spike travels rapidly down the spinal cord. In 10-cm long goldfish, the conduction velocity for this spike is 85 m/s, which enables the signals to travel the whole length of the spinal cord within 1 ms. Consequently, all the motor neurons to the trunk muscles on that side of the body are excited virtually simultaneously.

Outputs and executive functions of the Mauthner neuron

The Mauthner neuron makes direct synaptic connections with many of the spinal motor neurons that it excites. A short collateral (sub-branch) of the Mauthner axon makes contact with a special region of the motor axon a little distance from the cell body (Fig. 4.10b). A spike in the Mauthner axon produces an EPSP in the motor neuron with a synaptic delay of about 0.6 ms, which indicates the transmission is chemical. The excitation of a substantial number of other motor neurons is relayed by way of a pre-motor interneuron and so involves a slightly greater delay. The motor axons travel a relatively short distance to the trunk muscles, where they have normal chemically transmitting synaptic endings on the muscle surface. This short and direct pathway is responsible for the short delay of about 2 ms between a Mauthner neuron spike and the onset of a spike in the trunk muscles.

A spike in the Mauthner axon inhibits the spinal motor neurons contralateral to the axon at the same time as it excites those ipsilateral to (on the same side as) the axon. The inhibition appears to be mediated by inhibitory interneurons that cross the midline and are activated by electrical synapses from the Mauthner axon (CI in Fig. 4.13a). This crossed inhibitory pathway makes sure that, if only a short interval in time separates spikes in the left and right Mauthner neurons, only the earlier of the two Mauthner spikes is able to fire its motor neurons. If the two Mauthner neurons spike simultaneously, the crossed inhibition prevents any motor output at all.

Within the brain, branches from the axon of the Mauthner neuron excite a number of cranial relay neurons, which carry out some important functions in the startle response (Fig. 4.13b). The cranial relay neurons excite motor neurons of muscles that close the jaw and draw in the eyes, which helps to streamline the fish for its escape. In addition, the cranial relay neurons excite a group of interneurons that inhibit both the Mauthner neurons.

These Mauthner inhibitor neurons have cell bodies clustered around the ventral dendrite and they act to inhibit the Mauthner neuron in several ways. Some of them exert an unusual form of

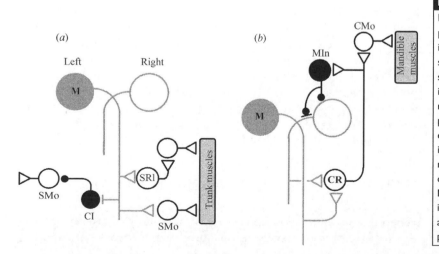

Fig. 4.13 Output connections of the Mauthner neuron. (a) Neuronal pathways generating excitation and inhibition of spinal motor neurons, showing: the Mauthner neuron (M); spinal relay neurons (SRl); crossed inhibitory interneuron (CI); and spinal motor neurons (SMo). (b) Neuronal pathways generating excitation of the cranial motor neurons and self-inhibition, showing the Mauthner neuron (M); cranial relay neuron (CR); cranial motor neurons (CMo); and the Mauthner inhibitor (MIn). Electrical inhibition of the Mauthner neuron axon cap is indicated by the filled circle plus short line.

electrical inhibition: they have axons coiled tightly round the origin of the Mauthner axon, contributing to the axon cap (Fig. 4.10*c*). When a spike is generated in one of these axons wrapped around the Mauthner neuron, it makes the outside of the Mauthner neuron locally more positive, which has the same effect as driving the Mauthner intracellular voltage more negative. Spikes from the axons of the inhibitor neurons thus exert a kind of brief vetoing stranglehold on the Mauthner neuron, right at the zone where its spikes are initiated. The electrical inhibition follows a Mauthner spike with a delay of about 1 ms and it helps to prevent the Mauthner neuron from firing twice in response to a stimulus. Furthermore, the Mauthner inhibitor neurons are excited by branches of both left and right Mauthner axons, with the result that the contralateral Mauthner neuron also receives electrical inhibition with the same short delay after a spike in the ipsilateral axon.

The electrical inhibition of both Mauthner neurons is a brief event, lasting no more than about 2 ms, but it is followed by inhibition through chemical synapses that lasts much longer. The Mauthner inhibitors produce this inhibition through their axon terminals on the cell body of the Mauthner neuron, which make conventional chemical synapses. The chemical inhibition begins about 1 ms after the electrical inhibition, and lasts for about 45 ms, which takes it well into the second stage of the startle response. This direct inhibition is backed up by presynaptic inhibitory synapses made by the Mauthner inhibitors onto the synaptic terminals of the sensory neurons on the Mauthner neuron dendrites, and also by separate inhibitory pathways that are activated directly by the sensory neurons. The result of these inhibitory mechanisms is not only that the active Mauthner neuron and its contralateral partner are shut down as soon as a spike has been generated, but also that they are kept shut down long enough for the startle response to go to completion.

Although the involvement of Mauthner neurons in triggering startle responses in teleost fish is clear, this does not mean that this is the only function that these neurons serve. Sudden lunges by goldfish to capture prey on the water surface involve similar movements to those of a startle response, and follow a Mauthner neuron spike (Canfield and Rose, 1993). Besides integrating information about a source of potential danger, a Mauthner neuron must be able to collect information that triggers an attack at the appropriate time. The input and output networks of a Mauthner neuron are arranged so that it can initiate and co-ordinate a rapid, directed movement.

Is the Mauthner neuron a command neuron?

When it is excited, a Mauthner neuron inevitably causes a fast tail flip in a fish, and it collects and processes sensory information that elicits such a response. However, a Mauthner neuron does not necessarily fulfil the strict criteria for a command neuron for startle because it is neither necessary nor does it often act on its own. It was initially a surprise when in 1982 Robert Eaton and colleagues found that goldfish in which a Mauthner neuron had been knocked out produced fast C-starts in response to acoustic stimuli (Eaton *et al.*, 1982). Although electrical stimulation of a Mauthner neuron consistently elicits the same stereotyped movement, showing it can on its own elicit a fast flip of the tail, the startle responses by free-swimming fish are more variable (Nissanov *et al.*, 1990), reflecting the ability of the fish to steer itself away from a source of danger rather than just to the left or right. This all points to the involvement of other neurons in a natural startle response which add a capability for steering that could not be derived from a single Mauthner neuron spike.

The Mauthner neuron belongs to a class of neurons called the reticulospinal neurons. There is one cluster of reticulospinal neurons on each side of each of the seven segments of the hindbrain, and several neurons in each cluster are unique individuals. The Mauthner neuron is in segment 4, and in segments 5 and 6 are homologous neurons called MiD2 cm and MiD3 cm, which share characteristics with the Mauthner neuron including a contralateral axon, a lateral and medial dendrite as well as developmental history (Fig. 4.14*a,b*). Their cell bodies and axons are smaller than those of the Mauthner neuron, but like it they are uniquely identifiable neurons. All three receive auditory information, but MiD2 cm and MiD3 cm generate trains of spikes when excited rather than the single spikes characteristic of the Mauthner neuron and they respond to lower sound intensities than the Mauther neuron (Nakayama and Oda, 2004). The smaller neurons also have axon collaterals that branch to both sides of the spinal cord, unlike the Mauthner neuron whose axon and its branches are restricted to the same side.

There is good evidence that MiD2 cm and MiD3 cm are involved in initiating and steering startle responses. Direct evidence is harder to obtain than with the Mauthner neuron because these interneurons are smaller and so it is much more difficult to make electrical recordings from them or to stimulate them. But larval zebra fish are both small and sufficiently transparent to allow individual brain cells, when suitably stained, to be seen under a microscope through the dorsal surface of the head. A stain that has been used is calcium green dextran, which fluoresces with a particular colour in the presence of calcium ions. Calcium levels in cells are normally very low, but rise dramatically in neurons when they are electrically excited. So, when the calcium green dye is introduced into a neuron, changes in the intensity of its fluorescence can be used to indicate changes in intracellular calcium concentration which, in turn, are an indirect indication of changes in electrical excitation. In experiments, a

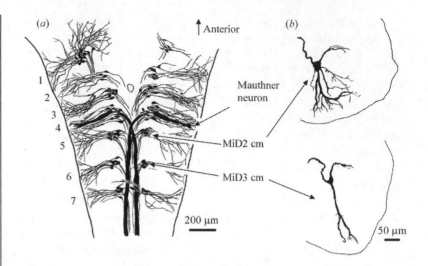

Fig. 4.14 Mauthner neurons and their segmental homologues. (a) dorsal view of the brain of a goldfish in which reticulospinal neurons were revealed by injecting an enzyme, horseradish peroxidase, into the spinal cord. Neuronal axons take up this substance and it is transported throughout each neuron. Later, it is visualised by reacting it with a labelled antibody. There are seven distinct groups of reticulospinal neurons, corresponding with segments of the hindbrain, which are numbered. The Mauthner neuron, in segment 4, is the largest, and shares certain characteristics such as an axon that crosses to the contralateral side, and a dorsal and ventral dendrite, with neurons such as MiD2 cm and MiD3 cm in other segments. (b) Transverse sections through the brain of an adult zebra fish, showing MiD2 cm and MiD3 cm. Note similarities with the goldfish Mauthner neuron in Fig. 4.11c. (Part (a) redrawn from Lee et al., 1993; (b) redrawn from Lee and Eaton, 1991.)

small amount of the dye was injected into the ventral part of the spinal cord in the larval fish tail (O'Malley *et al.*, 1996). Stimuli were presented by a small vibrating piezoelectric device positioned either behind or in front of the little zebra fish. Stimuli from behind excited only the Mauthner neuron, but stimuli from in front excited all three neurons. In other experiments with slightly different techniques, either MiD2 cm plus MiD3 cm or the Mauthner neuron were killed by directing a laser light spot for several minutes at the neurons that were stained with this technique (Liu and Fetcho, 1999). When the Mauthner neuron was killed, no difference was found in startle responses to stimuli from the front of the fish, but the fish no longer produced startle responses to stimuli from behind.

Thus, it is possible that the Mauthner neuron is a real command neuron for startle responses to danger from behind a fish, acting on its own to trigger a quick bending movement. But when a fish receives a startling stimulus from another direction, other reticulospinal neurons will join in. Exactly which neuron responds and how vigorously it responds will presumably depend on the direction of the threat. The Mauthner neuron will be excited earlier than its cousins because of its greater size. But it only produces one spike. For stimuli in front of the fish, this single spike might offer a small time advantage in initiating the startle movement, but may not be very significant in comparison with excitation from trains of spikes from MiD2 cm, MiD3 cm and similar neurons. In a larval zebra fish, just over 200 neurons send axons from the brain into the spinal cord, and at least 80% of these are probably activated during startle (Gahtan *et al.*, 2002), with the Mauthner neuron, MiD2 cm and MiD3 cm being the first to be excited.

We can expect these neurons will differ between each other in the way they respond to stimuli and in the effects they have on behaviour, allowing a fish to react appropriately to different types of stimuli from different directions and to ensure its startle reactions are tailored to other activities it is engaged in.

Conclusions

Startle behaviour provides simple and dramatic examples of how behaviour is initiated as a result of a decision-making process in the central nervous system. It also illustrates the way in which it is possible to trace the orderly flow of information through pathways of interconnected neurons. Both of the examples in this chapter enable us to understand how a specific behaviour pattern is triggered and executed, and so provide some general lessons about the role of nerve cells in behaviour. They also have specialisations: both have clear adaptations for speed, including several electrical synapses, and fast spike conduction velocities provided by giant axons; and the crayfish lateral giant and the fish Mauthner neuron operate with single spikes rather than trains of spikes.

Startle behaviour begins with the stimulation of specific groups of sensory neurons, and these in turn excite one or more types of interneuron. Sensory filtering ensures that the animal responds only to significant stimuli. The filtered information is integrated by interneurons that act as a switch or trigger, determining whether or not that particular behaviour is initiated. In the lateral giant in crayfish, this executive function is concentrated in a single giant neuron, which yields an extremely rapid response and acts as a command neuron. Like the crayfish lateral giant, a Mauthner neuron in a fish can trigger a startle response, but it normally acts in concert with others in directing the fish's movement. The cockroach giant interneurons described in the previous chapter also work in concert with each other to direct a startle response, and the direction of turn made by the animal is made on the basis of weighing up the spike pattern each giant produces in response to wind puffs.

The neuronal pathways in which the crayfish lateral giant and fish Mauthner neuron are involved are known in considerable detail. One important aspect of the way these neuronal networks work is that both inhibition as well as excitation plays a vital role. The pattern of inhibitory connections is essential for ensuring that the behaviour is executed efficiently. This is seen clearly in the way that motor systems that would generate movements incompatible with a startle response are inhibited, such as the inhibition of the extensor system during flexion of the crayfish abdomen and the mutual inhibition between left and right Mauthner neurons. A second role for inhibition is in switching off sensory systems that are likely to be stimulated by movements originating from the startle response itself. This prevents undesirable consequences such as triggering fresh startle responses and habituation in sensory pathways. Finally, it is common for an excitatory network that triggers a startle response to inhibit itself, after a delay, which prevents fresh activity from being triggered until the initial movement has gone to completion. The effectiveness of this inhibitory control is often ensured by having inhibitory input at several separate points in a particular neural pathway.

Questions

If there was a predator that specialised in catching crayfish, what behavioural adaptations would you expect it to show?

How might one particular reticulospinal neuron in fish have evolved into the Mauthner neuron?

Summary

- Startle behaviours are often triggered by networks that are simple and fast acting.
- They often include giant neurons through which sensory information is channelled for distribution to motor control networks.
- Networks including giant neurons also ensure the startle behaviour is executed smoothly.
- Some giant neurons act as command neurons, defined as being necessary and sufficient to trigger a behaviour.

Crayfish lateral giant interneurons
- Startle in a crayfish is a tail flip, involving rapid contraction of abdominal flexor muscles followed by re-extension.
- In each half ganglion are eight motor neurons that each innervate a group of flexor muscle fibres, plus a motor giant and an inhibitory motor neuron that each innervate all flexor fibres in their half segment.
- The lateral giant receives input from sensory hairs in the abdomen and triggers a somersault-like tail flip in which only the three most anterior abdominal segments flex.
- The lateral giant is present in each segment and strongly electrically coupled with its neighbours.
- Sensory hairs that are sensitive to sudden water movements make electrical synapses with a lateral giant and chemical synapses with sensory interneurons that also make electrical synapses with the lateral giant.
- Several mechanisms underlie sensory filtering that ensures a lateral giant responds only to sudden, intense sensory stimuli.
- A single lateral giant spike causes flexor muscle contraction. The lateral giant excites the motor giant, and also the segmental giant which then excites all fast flexor motor neurons in its half segment.
- Abdomen re-extension is a reflex triggered by proprioceptors.

- A lateral giant ensures flexion is executed smoothly by inhibiting the extensor motor system at different locations.
- After a delay, it ensures that re-extension occurs smoothly by inhibiting the flexor motor system and also itself.
- Escape swimming often follows a tail flip and is triggered by a slower pathway not involving the lateral giant.
- Lateral giant excitability is influenced by several factors including dominance hierarchies.
- Lateral giant interneurons probably evolved from neurons that triggered forward movement of the limbs, and later became associated with the tail fan.

Fish Mauthner neuron

- Bony fish have a single, large Mauthner neuron on each side of the hindbrain and an axon that descends the contralateral spinal cord.
- It is excited by auditory and water movement stimuli through sensory neurons and interneurons that make electrical and chemical synapses with the lateral dendrite.
- A Mauthner neuron axon directly excites motor neurons of trunk and tail muscles on its side of the spinal cord, and inhibits motor neurons on the other side.
- It also excites Mauthner inhibitor neurons in the brain which inhibit left and right motor neurons and their networks at several locations, ensuring their actions do not interfere with each other and are brief.
- A Mauthner neuron plays a role in initiating a rapid bend away from a source of startle. It normally acts with other reticulospinal interneurons to initiate a response that points the fish accurately away from the stimulus, and the initial bend is followed by swimming.

Further reading

Edwards, D. H., Heitler, W. J. and Krasne, F. B. (1999). Fifty years of a command neuron: the neurobiology of escape behavior in the crayfish. *Trends Neurosci.* **22**, 153–161. A clear review of research on the crayfish lateral giant interneuron, particularly concentrating on coincidence detection, modulation of responses and evolution.

Korn, H. and Faber D. S. (2005). The Mauthner cell half a century later: a neurobiological model for decision-making? *Neuron*, **47**, 13–28. The Mauthner neuron is an important model subject for investigations of many branches of neuroscience, including neuroethology, as shown in this review.

5

Eyes and vision: sensory filtering and course control in insects

Diptera (the true, or two-winged, flies) clearly rely on their eyes. They usually escape capture with apparent ease, and during fast flight they turn often, rarely colliding with their surroundings. Good eyesight goes along with aerobatic manoeuvrability: some dipteran behaviours include the most rapid reactions made by animals. An airborne hover fly, blow fly or fruit fly can turn a right angle in less than 50 ms, and can make 10 turns per second (Schilstra and van Hateren, 1999; Fry *et al.*, 2003). Flies rarely crash into other objects while airborne, and are able to land precisely on appropriate perches, including cup edges. The spectacle of house flies chasing each other is familiar, and high-speed filming reveals that the chasing fly, usually a male, tracks every turn made by the target fly, often a female but sometimes a rival male (Fig. 5.1*a*). Robber flies also chase house flies, in this case to prey on them. Although the neuronal pathways involved in visual behaviours of flies and other insects are more complex and involve much larger numbers of neurons than the startle responses of crayfish or fish described in the previous chapter, several visual interneurons of insects have been studied in some detail. Like the T5(2) neurons of toads (Chapter 3), the interneurons respond to particular defined stimulus configurations and so act as feature detectors, and they have been shown to play specific roles in guiding behaviour. Studying these neurons has provided information on the ways in which networks of neurons work to ensure that particular stimulus features are filtered out and recognised. For many of these insect visual interneurons, the feature of interest is the direction of movement by a stimulus over the eye.

The eyes and visual system of insects and vertebrates have different architectures, but the problems they need to solve in capturing and processing images are the same, and there are striking parallels in the kinds of operation that neurons perform in the two different types of animal. In the first operation, an eye collects light to form an image, which is sampled by an array of photoreceptor cells. The optics of the eye need to ensure each photoreceptor collects

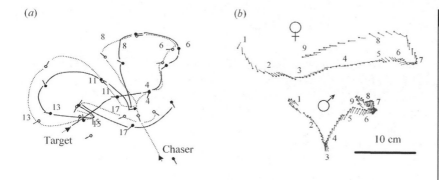

(a)

(b)

Target

Chaser

10 cm

Fig. 5.1 Visual behaviour by flies, shown by flight paths recorded from below with a high-speed camera. (*a*) Extract from a recording in which one male house fly, *Musca domestica*, chased another. The track of the target is a solid line and that of the chaser a dotted line. Location of each fly at the same time is indicated by the numbers, which are multiples of 100 ms from the start of the whole recording sequence. At numbers 4 and 15 the chasing fly was immediately below the target. Head and body positions are indicated. (*b*) A 360-ms excerpt recording head and body positions of a male and a female hover fly, *Syritta pipiens*. Numbers indicate intervals of 400 ms. (Part (*a*) redrawn from Wagner, 1986; (*b*) redrawn from Collett and Land, 1975.)

sufficient light to enable it to produce a clear receptor potential and that adjacent photoreceptors sample neighbouring segments of the image so that that detail in the image can be distinguished. Second, photoreceptors must adapt so that their photoreceptor potentials are appropriate to the ambient light intensity, whether a brightly lit summer day, or during a gloomy night. Responses to changes in light are further enhanced in the neurons that are post-synaptic to the photoreceptors. Third, features such as edges in the image are enhanced by lateral interactions between visual neurons, which compare the signals provided to them by neighbouring photoreceptors.

Both in vertebrates and in arthropods, neurons that pick out movements of objects over the eye occur at quite early stages in the visual system, and this kind of neuron in insects provides the focus of this chapter. Movement detecting neurons, or **movement detectors**, usually have responses that are selective for particular types of movement, for example responding only to a small object moving in a particular direction, which is a characteristic of some of the retinal ganglion cells in frogs and toads that respond only when small dots move within their receptive fields, as described in Chapter 3. Insects possess many types of motion selective neurons, and their response properties tune them to different aspects, which can be combined in different ways. For example, some are tuned to movement in a particular direction such as from the front of the animal towards the back, and some neurons with a particular pre-ferred movement direction will respond best to small objects moving and other neurons will respond best if the whole visual field moves. This kind of segregation into separate pathways that deal with par-ticular aspects of a stimulus occurs on a broader scale, too: both in arthropods and vertebrates colour and motion are processed by dif-ferent neuronal pathways. A similar kind of segregation occurs in other sensory modalities, too; for example, in the auditory system of owls the time and loudness of sounds are dealt with in different pathways, as described in Chapter 6. Of course, the visual or auditory world is not perceived in a fragmented way, so the different process-ing streams must be recombined in an appropriate way to guide behaviour.

Visual behaviour of flies

A high speed film of two hover flies, *Syritta*, made by Tom Collett and Mike Land (1975) provides an excellent introduction to the visual behaviour of flies. An excerpt from the film (Fig. 5.1*b*) traces the courses of a male and a female hover fly over a period of 3.6 s. The female was cruising slowly, probably inspecting flowers that she might visit to drink nectar. Although the female was continually moving, her direction of gaze tended to remain in a particular direction for periods of a few 100 ms or more, during which time she would either hover in a stationary position, or drift forwards, backwards or sideways. Many insects that hover move in ways that hold images on their eyes in constant positions, and the movements that achieve this stability are called **optomotor responses**. They are a means to maintain a constant position, or a constant heading while moving through the environment, and are elicited when all or most of the visual field moves in a coherent way. Most animals with good form vision exhibit optomotor responses. The female fly in Fig. 5.1*b* changed her direction of gaze abruptly between periods in which her gaze direction was fixed. Such abrupt changes in gaze direction are called **saccades**, and are a feature of vision in many animals. Humans usually move their eyes in saccades, even while reading text: it is hard to read while moving the eyes at a constant speed across the page. One reason is that moving images are blurred across the retina, and another is that detailed examination of an image is achieved by a small part of the eye, the fovea, which is excellent at making fine distinctions, but not at processing moving images.

For much of the time, male hover flies behave in the same way as females. But if another fly comes into view, the male's behaviour changes, as in Fig. 5.1*b*. While moving gently backwards and then forwards, this male closely tracked the female, keeping her within 5° of his body axis. He was keeping her within his fovea for close inspection, and probably to plan an approach flight.

Many dipteran flies rely on eyesight to find mates, to navigate, to find perching places and in some cases to find prey. This photograph shows a robber fly (*Asilus*), a fly that waits until prey such as this house fly (*Musca*) has come into view, and then pounces on it. (Photograph by Peter Simmons.)

Overview of the insect visual system

All insect visual systems share common structural features, although there are many variations in detail, especially in the Diptera. The diagram in Fig. 5.2 outlines the structure that most insect orders share. The main eye of most adult insects is called a **compound eye**, a name that means it consists of many lenses each associated with its own set of photoreceptor cells, usually eight in number (in contrast, a 'simple eye' is one that has a single lens). Information from each compound eye is processed in a series of distinct neuronal regions in part of the brain called the **optic lobe**. The basic compound eye unit of a lens plus its photoreceptors and some other associated structures is called an **ommatidium**. Most of the photoreceptors of an ommatidium have axons that run together into the first area of neuropile, where synapses are made, the **lamina**. In the lamina, neurons are arranged in column-like elements called lamina cartridges, each of which in most insects corresponds with and receives signals from an ommatidium. Each cartridge contains about a dozen neurons of several distinct anatomical types, a few of which interconnect nearby cartridges.

The pattern in which ommatidia correspond with individual cartridges and preserve the spatial layout of the image is called **retinotopic**. The axons of most photoreceptors end in the lamina, but a few that carry information about light colour or plane of polarisation bypass it. The next layer of neuropile is called the **medulla**. Its neurons are also arranged in repeated units, each of which corresponds with a particular lamina cartridge, preserving the retinotopic

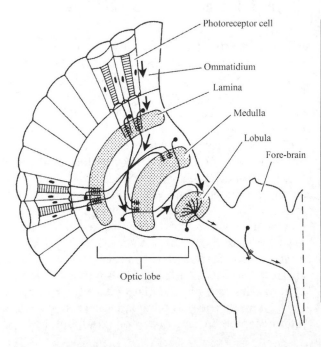

Photoreceptor cell

Ommatidium

Lamina

Medulla

Lobula

Fore-brain

Optic lobe

Fig. 5.2 The visual system of a diurnal insect, such as a dragonfly or a locust, showing the main areas involved in movement detection. The three optic neuropiles (lamina, medulla and lobula) are indicated by light stippling. The small arrows show the pathway followed by information originating in the uppermost ommatidium drawn. The general arrangement is similar in flies, except that the photoreceptors of one ommatidium are not fused, and the lobula contains a distinct neuropile called the lobula plate, which would lie behind the lobula in this drawing and contains the large fan-shaped neurons.

projection. However, the axons from the front of the lamina connect to the back part of the medulla, crossing over the axons from the back part of the lamina, which project to the front part of the medulla. An arrangement where axons cross each other like this is called a **chiasm**. There is a greater diversity of neuron types in the medulla compared with the lamina, and the medulla's neurons are small, which makes them a challenge to study.

The retinotopic arrangement is preserved into the next neuropile area, the **lobula**, and there is a further chiasm between the medulla and the lobula. The lobula includes column-like modules of small cells, an arrangement similar to that in the medulla. But the lobula also includes some large neurons with extensive dendritic fans that cover large areas of the visual field. Instead of being concerned with events in one small part of the image, the large, fan-shaped neurons in the lobula collect information over a large part of the visual field. Many of them are sensitive to particular movements; for example, some respond to movements of the image in a back-to-front direction over the eye, and others respond to objects that are approaching towards the eye. Their responses illustrate very well the way in which the image is processed to extract features of significance to the insect's behaviour. Most of the fan-shaped neurons have axons that project into the brain where their targets include some of the brain neurons that are involved with controlling movements.

In Diptera, the large fan-shaped neurons are found in a region that is distinct from the neighbouring main part of the lobula, and is called the **lobula plate**. There are about 60 of them, collectively called **lobula plate tangential** cells, or LPTCs.

Optical design of compound eyes

In most insects, light that enters an ommatidium is caught by a central, rod-like **rhabdom** that acts as a light guide. Light from the environment is caught by a hexagonal corneal lens and then passes through a second lens, the crystalline cone, to be focused onto the tip of a rhabdom (Fig. 5.3a). In insects that are active in bright light, pigment in special cells around each ommatidium shields them from each other by reducing light scatter between ommatidia. Individual ommatidia do not function if they are too small, for two main reasons. First, a light-catching structure such as a rhabdom must be at least about 1 μm wide, or light would bounce off the outside instead of being caught within it. Second, lenses with narrow apertures do not perform very well. They catch less light than wider lenses, so the images they form are less bright. In addition, light rays coming from different directions interfere with each other when they pass through narrow apertures, so a lens with a small diameter will focus pinprick spots of light or dark into blurred blobs rather than sharp points. Figure 5.3b shows how a scene of three small black spots would be projected onto photoreceptors in a small part of a compound eye.

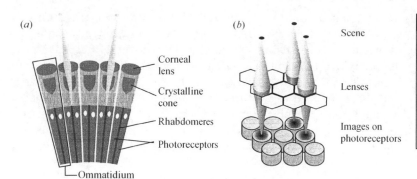

(a)

Corneal lens
Crystalline cone
Rhabdomeres
Photoreceptors
Ommatidium

(b)

Scene
Lenses
Images on photoreceptors

Fig. 5.3 Introduction to optics of compound eyes. (*a*) Five ommatidia are shown. For two of them, the way that light is focused onto the rhabdom is indicated. (*b*) An array of nine ommatidia viewing a simple scene of three dark dots. Each dot is focused as a blurred blob onto the photoreceptors of an ommatidium.

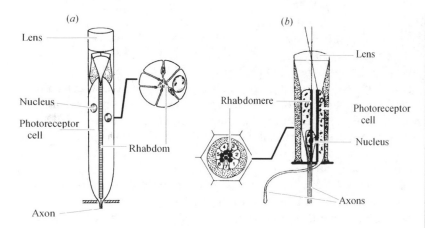

(a)

Lens
Nucleus
Photoreceptor cell
Rhabdom
Axon

(b)

Rhabdomere
Lens
Photoreceptor cell
Nucleus
Axons

Fig. 5.4 The arrangement of cells in the ommatidia of two insects. (*a*) A locust, *Locusta*, ommatidium. A transverse section of the ommatidium at the level indicated (on the right) shows how the rhabdomeres of individual photoreceptor cells form a single, central light collecting rod, the rhabdom. (*b*) A blow fly (*Calliphora*) ommatidium. Here, the rhabdomeres of different photoreceptors within an ommatidium remain separate. Each looks out at a different direction as indicated. Seven of the photoreceptors are numbered in the cross section (on the left). Also shown are pigment cells that surround each ommatidium. (Part (*a*) redrawn from Wilson *et al.*, 1978; (*b*) redrawn from Hardie, 1986.)

The process of transducing light energy into an electrical signal starts in the rhabdomere of a photoreceptor cell, the contribution each photoreceptor makes to the rhabdom. In a rhabdomere, the cell membrane is folded into very regular finger-like projections, or microvilli, that contain the photopigment, which absorbs light and initiates transduction. In most insects, including bees, locusts and dragonflies, the rhabdomeres of different photoreceptors of the same ommatidium abut each other to form a central, rod-like rhabdom (Fig. 5.4*a*). In this eye design, all the photoreceptor cells of one ommatidium share the same field of view and connect to the same underlying lamina cartridge. However, advanced dipteran flies such as blow flies, hover flies and fruit flies have a different structure called an open rhabdom, where individual rhabdomeres remain separate (Fig. 5.4*b*). Each photoreceptor of an ommatidium looks in a slightly different direction, which it shares with a different photoreceptor in each adjacent ommatidium. The axons of photoreceptors that share the same field of view all come together in the same lamina cartridge, so that as in other insects the lamina contains a retinotopic representation of the visual world. The advantage of this arrangement is thought to be that flies retain very good responsiveness to moving visual stimuli over a wide range of light intensities.

Compound eyes give the animal a very wide field of view; a fly can see in almost every direction from its head. But the size of eye that an insect can carry around is obviously limited and so is the number of ommatidia. Large dragonflies have the largest number of ommatidia of any insect eye, about 30 000. By contrast, blow flies have about 4000, and fruit flies about 750. In honey bees, which are typical of insects that fly on bright days, most facets are about 25 μm across, and collect light from a cone with an angle of a little over 1° (roughly the angle occupied on a human eye by a finger tip at arm's length). To resolve two small light spots in a visual scene as separate rather than one larger spot, a bee would have to use three ommatidia – one for each spot, with an unilluminated ommatidium between them. Therefore, bees can distinguish between two objects that are about 3° apart. The best resolution that a compound eye achieves, in the most acute region of a dragonfly eye, is 0.25°.

Compound eyes are not constructed homogeneously, but the density with which ommatidia are packed together as well as the widths of their lenses usually varies in different parts of the eye. An excellent example of this is the eye of the mantis *Tenodera*, which depends on vision to detect, locate and then capture its prey. In order to achieve good resolution for this task, there is a need to dedicate as many ommatidia as possible to sample the visual environment. But each ommatidium must have a crisp image of a small part of the environment, which means that each must have a lens that is as large as possible. Only a small portion of each mantis eye, the fovea, has the high resolution required to examine potential prey. The fovea of a mantis eye is a small region at the front of the eye (Fig. 5.5*a*). Individual ommatidia here have lenses that are 50 μm in diameter,

Fig. 5.5 The fovea of the mantis (*Tenodera*) eye. (*a*) Anterior view of the head; the broken circle within each eye indicates the fovea. (*b*) Facet diameters and interommatidial angles plotted against position in the eye for the row of facets indicated in (*a*). Note how facet diameters increase as interommatidial angles decrease. The acceptance angles of individual ommatidia follow the curve for interommatidial angles very closely. (Redrawn from Rossel, 1979.)

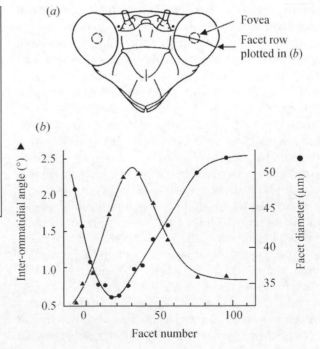

larger than elsewhere in the eye, and rhabdoms that are 1.5 to 2 μm across, narrower than elsewhere in the eye. Also the angle between adjacent ommatidia, 0.6°, is narrower in the fovea than elsewhere in the eye (Fig. 5.5b), so the surface of the eye is less curved here than elsewhere. When a prey-like object appears anywhere in its field of view, a mantis swivels its head to direct its fovea at the object, and the mantis tracks further movements of the prey by moving its head so that the prey's image remains on the fovea. The foveae of the two eyes overlap and the mantis is unusual among insects in that it uses binocular cues to determine the distance to the prey and then to strike accurately at the prey with its grasping front legs.

In fly eyes, too, the width and packing of ommatidial lenses varies across eye. Flies tend to have relatively narrow lenses and large angles between ommatidia in the parts of the eyes that look sideways, but larger lenses and smaller angles between ommatidia in forward-facing parts. A reason for this is that the images of objects side-on to the fly move relatively fast across the eye during forward flight. There is little time for photoreceptors in this region of the eye to respond as the image moves across them, so there is little point in constructing a fovea here compared with the front-facing part of the eye, in which the image generally moves more slowly and photoreceptors have a longer time to sample it. In many fly species, part of the eye of the male that faces forwards and upwards has wider ommatidia than the equivalent part of the eye of the female. This eye region of the male is called the 'love spot' (Franceschini et al., 1981), as it is the part of the eye that a male keeps directed at a female he is chasing. It is particularly well developed in the hover fly Syritta, in which the angle between ommatidia is 0.6°, about three times smaller than the angle elsewhere in the male's eye or in the same region of the female eye.

In absolute resolving power, insect foveae are about an order of magnitude poorer than human peripheral vision and two orders of magnitude poorer than human foveal vision. It has been calculated that a compound eye would need to be a metre across to give a similar performance to the human eye. However, both in terms of numbers of individuals and of species, animals that use compound eyes to see things outnumber those that use simple eyes. A small insect will not need such good resolution as a monkey: the fly needs to react to objects that are only a few centimetres away, whereas the appropriate reaction distance for the monkey may be several metres. So compound eyes provide their small owners with resolving power that is as good as that of vertebrates when considered in terms of biological needs rather than physical ideals.

Photoreceptors and the receptor potential

If an eye has been left in darkness for some time to maximise its sensitivity, the photoreceptors are capable of responding to the

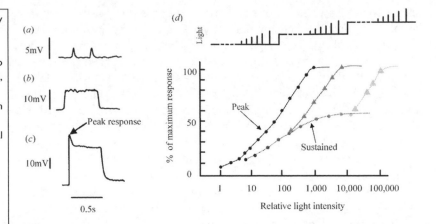

Fig. 5.6 Coding of light intensity by an insect photoreceptor cell. (a–c) Recordings from a dragonfly (*Hemicordulia*) photoreceptor cell to three different levels of light stimuli, delivered in darkness. Extremely dim light (a) elicits two single photon bumps; moderately dim light (b) elicits a steady receptor potential in which small fluctuations, caused by the random nature of photon arrival, can still be discerned; and bright light (c) elicits a receptor potential with a sharp, initial peak, followed by a more sustained level. Notice how the voltage calibration changes in (a) to (c). (d) Graphs of response against light intensity in which the amplitude of the receptor potential is expressed as percentage of the maximum response recorded, and light intensity is expressed on a logarithmic scale. Above the graphs is a schematic diagram to show how light stimuli were organised. The dark curve on the left shows the size of the peak response to flashes of increasing intensity in a dark adapted eye. The dotted curve (bottom) indicates the level of sustained response to different levels of background illumination. The two curves to the right show responses to flashes of light superimposed on two different levels of background illumination. (Redrawn from Laughlin, 1981.)

smallest units of light, single photons. In an insect photoreceptor, each photon gives rise to a discrete, depolarising potential, often called a 'bump' from its shape. Bumps are usually 1–2 mV high in intracellular recordings (Fig. 5.6a). In absolute darkness, bumps are very rare, and the evidence that each bump represents the absorption of a photon comes from a comparison with the statistical properties of the arrival of photons from a very dim light source (Lillywhite, 1977). As light intensity increases, the bumps fuse together, and the photoreceptor response to a change in light intensity becomes much smoother (Fig. 5.6b). In daylight, an insect photoreceptor cell receives millions of photons a second. If each bump generated a signal of 1 mV, this would lead to a signal of about a thousand volts in the photoreceptor, which it is clearly incapable of generating. To cope with this, the photoreceptor decreases its sensitivity when the overall light level increases, so that each additional photon generates a much smaller response. This is called **light adaptation**; conversely, adaptation to decreasing levels of ambient illumination, which causes sensitivity to increase, is called **dark adaptation**. A number of different mechanisms are involved, including changes in pigment-containing cells that regulate the arrival of light at a rhabdom (analogous to the pupil of vertebrate eyes), and changes in the enzymes involved in transduction.

It is important for photoreceptors to be able to adjust their sensitivity to ambient lighting conditions, because these can vary enormously. In sunlight, each photoreceptor receives about 40 million photons per second, and this number decreases to 40 thousand inside a room and 40 in moonlight. Because natural light intensity varies so much, it is convenient to use a logarithmic scale to express it, in which each unit represents a tenfold change in light intensity. A change in intensity from 10 to 100 occupies the same space on such a scale as a change from 100 to 1000 (10 is 10^1 so the logarithm of 10 is 1; 100 is 10^2 so the logarithm of 100 is 2; and so on).

One effect of light adaptation is that, at quite moderate light intensity, the response to a step increase in light consists of a sharp, initial depolarising peak, which quickly drops to a more sustained or plateau receptor potential (Fig. 5.6c). If the increase in light is maintained, the photoreceptor potential declines very gradually, but stabilises to a sustained level over a timescale of a few minutes. Using the terminology introduced in Chapter 2, the sustained receptor potential in response to a change in overall illumination is a tonic characteristic, and the initial peak response, which emphasises a change in light intensity, is a phasic characteristic.

The second effect of adaptation is apparent in the way that information about light intensity is encoded as a particular value of receptor potential. This pattern of adaptation is usually described by constructing graphs of light intensity against response (Fig. 5.6d). The way the experiment is conducted is indicated schematically at the top of Fig. 5.6d (note that light intensity is not represented accurately). At the start of an experiment, the eye is kept in the fully dark adapted state and the sizes of the peak responses to steadily brighter flashes of light are recorded. The time intervals between successive flashes are kept long enough to ensure that the eye remains fully dark adapted throughout the experiment. For low intensities of stimulus, there is a gradual rise in the size of the peak receptor potential as stimulus intensity increases (left curve). The maximum intensity that the receptor is capable of signalling is about 10 000 (four log units) on the horizontal axis of the graph. At this intensity, the response has reached its maximum value, or **saturated**. However, over a range of intensities of three log units (a thousandfold change in absolute intensity) the size of the response is proportional to the logarithm of light intensity.

The next stage in the experiment is to superimpose the stimulus light on a steady, background level of illumination, which is more like the natural situation. The responses to superimposed test flashes are then recorded in the same way as before. Fig. 5.6d shows that increasing background illumination shifts the intensity-response curve to the right along the intensity axis, and the size of the shift depends on the intensity of the background light. At low intensities, where adaptation in the response waveform is small, the shift is also small; but at higher intensities the shift produced becomes larger. Hence, the range of intensities to which the cell responds is shifted to match the level of background illumination. The right-ward shift in the curves represents a loss in sensitivity: with an increase in background light, a brighter stimulus is needed for the photoreceptor to generate the same voltage response. The curves keep approximately the same shape, indicating that the relation between test flash relative intensity and peak response remains much the same, and response amplitude changes in proportion to the logarithm of stimulus intensity.

It is a common pattern in sensory receptors for the amplitude of the response to be proportional to the logarithm of stimulus intensity, and this makes excellent functional sense. First of all, the large range of light intensities to which an eye is exposed in the day-to-day life of an animal is compressed into a manageable scale.

Second, responding in this way has the effect of making equal *relative* changes in light intensity generate equal *absolute* changes in the size of the receptor potential. One log unit on the light intensity scale (a tenfold change in absolute light intensity) causes a change in receptor potential of about 35% of its maximum amplitude. A tenfold change in light intensity is an enormous change in Nature, and a photoreceptor would normally experience much smaller changes in intensity as it scans a natural image. Coding light intensity in this way enables the eye to recognise different objects because they are distinguishable by differences in the proportion of light that they reflect – their **contrast**. Contrast of different objects does not vary when ambient illumination changes – for instance, the contrast between the print and the white paper on this page will be the same whether you are reading the book in a dimly lit room or outside on a beach in bright sunlight. On the beach, the black print will actually reflect more light than the white paper when the book is read in the room. The signal that a photoreceptor generates, therefore, consists of a small, steady depolarising potential, which depends on the mean or background level of light, plus fluctuations of a few millivolts caused by viewing objects of different contrasts.

The photoreceptor potential is conducted by passive spread along the axon. Sometimes, the axon can be as long as 2 mm, but the cable properties of these axons are such that little of the signal is lost between the cell body and the synaptic terminals. Photoreceptor cells do not usually generate trains of spikes, and communication with the next layer of neurons is achieved by graded changes in the receptor potential.

Comparative aspects of photoreceptor potentials

The extent of adaptation varies between types of eyes, and a very good illustration of this comes from research on the eyes of different species of dipteran flies. *Tipula*, a slow-flying, nocturnal crane fly, and *Sarcophaga*, a fast-flying, diurnal flesh fly, are two of the 20 species of dipteran fly surveyed by Simon Laughlin and Matti Weckström (1993). The receptor potential in *Tipula* rises relatively slowly, and does not adapt during the light pulse, whereas that of *Sarcophaga* rises rapidly and, for the brighter pulses, adapts by decaying quickly from an initial peak to a more sustained level (Fig. 5.7). *Tipula* is more sensitive to light than *Sarcophaga*, as shown by the intensity versus response curves in Fig. 5.7 and by the larger bump responses in dim light. Photoreceptors of *Sarcophaga* generate fast responses, and they are able to respond quickly to fluctuations in light as the insect flies around among twigs and leaves. Those of *Tipula* are too slow to help a fast-flying animal collide with objects, but are good at detecting light and dark at night-time. *Sarcophaga*'s photoreceptors are fast because

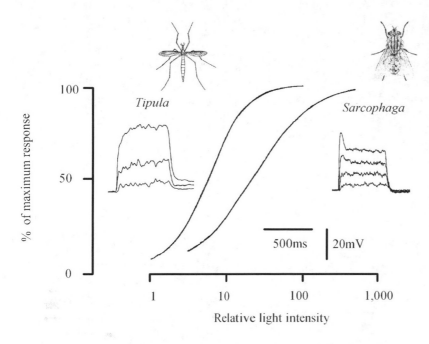

Fig. 5.7 Fast and slow photoreceptors in two dipterans, *Tipula*, a slow-flying, nocturnal crane fly, and *Sarcophaga*, a fast-flying, diurnal flesh fly. The intracellular recordings in the figure are responses to half second long pulses of light delivered to dark adapted eyes. The graphs of light intensity against photoreceptor response show that *Tipula* is more sensitive to light than *Sarcophaga*. (Redrawn from Laughlin and Weckström, 1993.)

they contain a special potassium channel. The lifestyle of *Tipula* means it does not need fast photoreceptors – in fact, to have them would be a significant metabolic cost to the animal. Even within the same eye, physiological properties of photoreceptors vary, too. For example, photoreceptors in the love spot area of male flies adapt more quickly than those elsewhere in the eye or in the female eye, reflecting their task of detecting fast-moving, small blobs (Burton *et al.*, 2001).

Neurons in the lamina

The principal cells of each lamina cartridge are called lamina monopolar cells (LMCs). As the term 'monopolar' suggests, the axon of an LMC gives rise to just one, long process (Fig. 5.8*a*). A tuft of short brush-like dendrites arise from it and, in blow flies, these dendrites make almost exactly 200 discrete anatomical synapses with each connecting photoreceptor (Nicol and Meinertzhagen, 1982). The LMC process then becomes a smooth axon which, along with the axons of smaller neurons, conveys information from the lamina to the medulla. In flies, five LMCs collect synaptic input from six of the photoreceptors that terminate in the lamina and deliver it to different layers in the medulla. Within the lamina, amacrine cells make lateral connections between adjacent cartridges, and a neuron called a tangential cell also connects the medulla with the lamina. Most of the recordings of electrical responses of lamina neurons have been made from the LMCs with the widest axons, numbers 1 and 2 in blow flies.

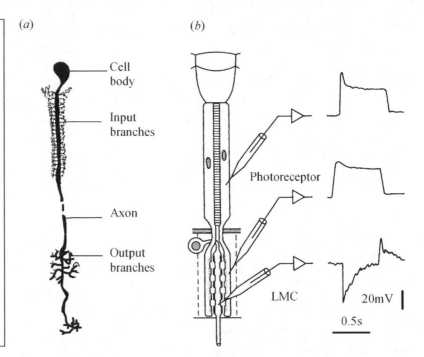

Fig. 5.8 A lamina monopolar cell (LMC) and the transfer of signals from a photoreceptor to an LMC across the first synapse in an insect visual pathway. (*a*) Structure of one of the large lamina monopolar cells, showing the regions where it receives signals from photoreceptors and where it delivers signals to the medulla, to two different layers in the case of the cell shown. (*b*) Drawing on the left shows how the photoreceptor cells of an ommatidium send short axons down to the layer of second-order neurons in the lamina. Intracellular recordings of responses to a half second long flash of light were recorded separately from the receptor cell body and axon, and from an LMC. (Recordings in (*b*) redrawn from Laughlin, 1981.)

Neuronal coding in the insect lamina

The LMCs are specialised to respond to contrasts, or changes in the visual signal. Like the photoreceptor cells, they convey signals as small, graded changes in membrane potential and do not generate trains of spikes. They are particularly sensitive to small fluctuations in light intensity about an average value. Their responses also depend on the contrast between light received by their own photoreceptors and those of neighbouring cartridges.

Characteristic intracellular responses from a photoreceptor and an LMC to a pulse of light are shown in Fig. 5.8*b*. The photoreceptors make inhibitory synapses with the LMCs, so an LMC responds to an increase in light with a hyperpolarising signal, and to a decrease in light with a depolarising signal. The response by an LMC is not, however, a mirror image of the response in a photoreceptor. The LMC response to a change in light is more phasic than that of a photoreceptor, as shown by the waveforms of the responses to pulses of light in Fig. 5.8*b*. The photoreceptor response shows an initial peak depolarising potential, followed by a sustained, smaller depolarisation that lasts until the end of the stimulus. The LMC response starts with a large, transient hyperpolarising signal, and ends with a transient depolarising signal, but during the light stimulus, the LMC membrane potential repolarises back almost all the way to its level in darkness. The result is that in the LMC response information about mean levels of illumination is lost but information about changes in illumination is enhanced. The loss of information about the mean level of

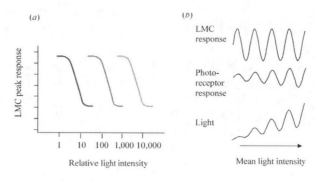

(a)

LMC peak response

1   10   100 1,000 10,000

Relative light intensity

(b)

LMC response

Photo-receptor response

Light

Mean light intensity

Fig. 5.9 Coding of light intensity by insect lamina monopolar cells (LMCs) (a dragonfly). (a) Peak-response amplitudes plotted as a function of the intensity of illumination delivered from a small source of light, centred on the receptive field of an LMC. Curves are drawn to illustrate responses recorded at three different mean light intensities. (b) A schematic diagram to show coding of light intensity in photoreceptors and LMCs. The light signal varies sinusoidally and gradually increases in mean intensity, representing the envelope of light intensity variations that might be experienced by scanning a visual scene at different intensities of ambient light. Because of adaptation, when the mean light level increases the photoreceptor cell response does not grow as dramatically as the light intensity signal. The LMC response is proportional to the contrast in the light signal: most of the signal about ambient light intensity is removed, and the remaining signal is amplified. (Redrawn from Laughlin and Hardie, 1978.)

illumination is a good example of filtering in a sensory system – useful information is more likely to be obtained from changes in the light signal than from its absolute strength.

The way in which the LMCs code visual stimuli has been studied by Simon Laughlin and colleagues in the same way as has been done for photoreceptors. First, in the dark adapted state, responses are measured to brief light stimuli of different intensities delivered from a dark background. Then a series of graphs to plot light stimulus intensity against LMC response amplitude are plotted, each one recording responses to increases and decreases in light from a particular background intensity (Fig. 5.9a). Because adaptation quickly returns the membrane potential of an LMC to near its dark resting potential following a change in light, there is no steady state response to background illumination. Consequently, the LMC response to an increase or decrease in light always starts from the same value of membrane potential and the intensity-response curve moves horizontally by an amount equal to the change in background light intensity. Small increases and decreases in intensity produce, respectively, hyperpolarising and depolarising departures from resting potential. These graphs have steeper gradients for LMCs than for photoreceptors, which shows that LMCs have greater sensitivity to small fluctuations in light than photoreceptors.

In summary, the synapses that connect a photoreceptor with an LMC process the visual signal in two important ways. They filter it, so that information about background intensity is subtracted; and they amplify it, so that the signal about small changes in intensity is transmitted. Amplification, to make the signal as large as possible at an early stage in the visual pathway, is important because every time the signal passes across a synapse from one neuron to another it can become contaminated with unwanted, accidental electrical signals or noise (Laughlin *et al.*, 1987; de Ruyter van Steveninck and Laughlin, 1996). The amplitude of responses by an LMC is determined by the relative change in light intensity, irrespective of what the mean level of light actually is, as indicated in Fig. 5.9b. Consequently, a fly's LMCs will produce the same size of signals as the fly's eyes sample a particular visual scene whether it is the middle of a bright day or during a dull, overcast evening.

Lateral inhibition and edge detection

Another kind of transformation that occurs involves mutual inhibition between nearby cartridges. This kind of inhibition between neighbouring units in a sensory system is widespread, and is known as **lateral inhibition**. It is a mechanism for sharpening the receptive field of a neuron so that it responds well to small stimuli centred on its own receptive field, but not to stimuli that fall just outside it.

Lateral inhibition was first discovered in a compound eye of *Limulus* (a 'living fossil' that is related to scorpions and spiders despite being commonly called the horseshoe crab). It has a complement of a number of different eyes, and the compound eyes have been the focus of most attention. Neurons that are postsynaptic to photoreceptors in this *Limulus* eye are called eccentric cells because their axons are displaced to one side of the ommatidium. Their axons carry trains of spikes, unlike the axons of fly LMCs. The optic nerve is easily accessible, and Hartline devised a method of carefully dissecting it to lift individual eccentric cell axons onto a pair of extracellular recording electrodes (Hartline *et al.*, 1956; Fig. 5.10*a*). He found that although individual eccentric cells were briskly excited by a small spot of light that covered their own ommatidium, they were not much impressed by larger light patches, including turning the room lights on or off. On further investigation, Hartline found that illuminating just one extra ommatidium would greatly reduce responses by an eccentric cell to a spot of light on its own ommatidium. This suppressive effect works in both directions, and it is strongest for immediately neighbouring ommatidia, dying off gradually as the distance between two ommatidia increases.

The explanation is that eccentric cells inhibit each other: inhibition between nearest neighbours is stronger than that between cells from more distant ommatidia. The inhibition has the effect of sharpening the difference in responses by eccentric cells whose ommatidia sample either side of a contrasting border, such as an edge in the image (Fig. 5.10*b*). This is a basic operation in image processing – enhancing edges in the image helps subsequent neurons to analyse the image to determine features of significance, such as the identity or movement direction of objects. In the case of *Limulus*, the objects it uses its eyes to detect include other *Limulus* during the mating season, which has been shown by a variety of techniques including fitting cameras onto the carapaces of individuals to record a horseshoe crab's view of the world (Barlow *et al.*, 2001).

Lateral inhibition is important in the vertebrate retina (Fig. 3.3*b*), where lateral interactions are mediated by horizontal cells and enable bipolar cells and ganglion cells to respond to bars, edges or dots rather than report the pattern of light and dark provided by photoreceptors. It occurs in insects, too, and individual LMCs provide greater responses to small spots of light directed into their own visual

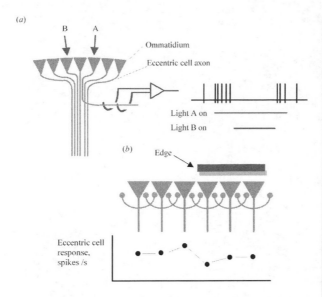

Fig. 5.10 Lateral inhibition in the eye of *Limulus*. (*a*) Each ommatidium sends the axon of an eccentric nerve to the brain in the optic nerve. In an experiment by Hartline, a single axon was dissected from the nerve and placed onto a pair of silver hook electrodes. Spikes picked up by the electrodes were amplified and displayed. A small light spot directed at the axon's own ommatidium excited it to spike briskly, but when a second light that illuminated a nearby ommatidium was added, the eccentric cell was inhibited. (*b*) Network of lateral inhibition between neighbouring ommatidia, indicated by the usual symbols for inhibitory synapses. Only the inhibition between immediate neighbours is indicated here. The location of an edge of an object (which throws a shadow over the eye) is enhanced as a difference in spike rate between the two ommatidia that border the edge on either side.

fields than to larger spots that drive their neighbours (Srinivasan *et al.*, 1982). Lateral inhibition also occurs in other sensory systems, and generally has the function of enhancing the detection of contrasts in responses provided by neighbouring neurons in an array.

An insect's LMCs, therefore, provide signals that have been conditioned to enhance some aspects of the neuronal image at the expense of others. Adaptation, both in the photoreceptors and in the LMCs, removes much of the signal about sustained illumination while enhancing signals about changes in the local stimulus. Adaptation also enables the neurons to generate reasonably large voltage signals irrespective of the mean level of illumination, which varies considerably with changes in time of day, season or weather. Lateral inhibition sharpens the contrast in signal between LMCs that sample neighbouring areas of the image, improving the detectability of borders in the image.

Optomotor neurons in flies

Unlike the large monopolar cells, the group of roughly 60 lobula plate tangential cells (LPTCs) of flies are not concerned with stimuli that affect small areas of the retina, but they respond to particular movements that occur anywhere within a large receptive field. Many of them are thought to be concerned with optomotor responses, which flies exhibit reliably while walking or flying. Each LPTC is a unique individual neuron and generates its largest responses to particular types of stimulus movement. It filters out particular stimulus features and ignores others. Often the direction of movement is the most significant feature for these neurons: they show **direction selectivity**. Some respond best to coherent movement of an extended visual stimulus, as would occur when the fly moves

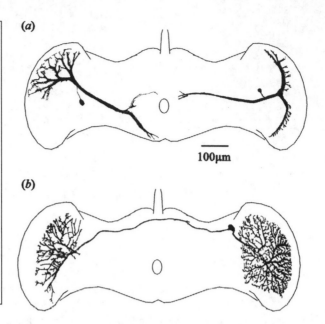

Fig. 5.11 Lobula plate tangential cells (LPTCs) in a blow fly (*Calliphora*) viewed looking from behind the brain. (*a*) On the left is one of the three HS neurons that responds to anticlockwise movements around the yaw axis; and on the right is one of the VS neurons, which responds to downward motion over its eye. (*b*) Neuron H1, which is excited by movements in the back to front direction over its own eye, the right eye for the H1 drawn here. Its axon carries information from one lobula plate to the other. (Part (*a*) redrawn from Dvorak *et al.*, 1975; (*b*) redrawn from Franceschini *et al.*, 1989.)

through the world, and others respond best to small objects, such as other flies or possible targets to land on. For example, the three HS neurons respond most strongly to images that rotate around the animal in the horizontal plane and the ten VS neurons respond to movements upwards or downwards (Fig. 5.11*a*). Each HS neuron in the right lobula plate is most strongly excited by movements occurring in the clockwise direction around the animal (backwards over their own eye or forwards over the other eye), and those in the left lobula plate are most excited by anticlockwise movements. The most effective direction at exciting a neuron is called the **preferred** direction. Often, neurons are inhibited by movement in the opposite direction, which is called the **null** direction.

The way that the HS and VS neurons respond to moving stimuli suggests that they could be important in controlling optomotor responses and there is strong, indirect evidence that they play such a role (Hausen and Egelhaaf, 1989; Borst and Haag, 2002). The output processes of the HS and VS neurons intermingle with dendrites of a small group of neurons that carry information from the brain to motor neurons that control flight and neck muscles. Optomotor responses are impaired in flies after minute cuts are made into the brain to sever the axons of these neurons, or if the HS neurons are killed by irradiating them with a laser. Also, in *Drosophila*, some mutants that lack the HS neurons do not produce normal optomotor responses. It is, therefore, likely that these neurons are an important part of the cockpit of the fly, helping it to maintain a particular flight course.

An LPTC that is often studied is H1 (Fig. 5.11*b*) because it is relatively easy to use an extracellular electrode to record spikes from its axon. It carries signals across the brain from its own optic lobe to the other, which provides a pathway that enables the fly to compare stimuli that affect the two eyes, so it can distinguish whether it is rotating or moving forwards. During rotation, movement flows forwards over one eye and backwards over the other, whereas when the fly moves forwards, both eyes see backwards movement. H1 is most excited by movements of the visual scene in a forwards direction over its own eye, the right eye in Fig. 5.11*b*.

A mechanism for directional selectivity

H1 is a good example of a neuron that acts as feature detector or filter, responding most vigorously to one particular stimulus configuration, in this case movements of the visual surroundings in a particular direction. The selectivity of these neurons for their preferred stimulus must arise from the way in which their presynaptic neurons are arranged, and study of H1 and other LPTCs has provided insights into the way in which one particular kind of selectivity – directional selectivity – arises. Each LPTC is considered to be driven by an array of local motion detecting units, each of which detects the movement of images in a particular direction over a small part of the eye. These units are commonly referred to as **elementary motion detectors** (EMDs). Their properties are known in some detail, although the neurons of which they are composed have not been identified. They undoubtedly include some of the tiny neurons in the medulla.

Many of the properties of EMDs were first discovered in the 1950s from experiments by Reichardt and Hassenstein on optomotor behaviour of beetles and flies, well before methods had been developed for recording signals from single neurons in the fly brain. More recently, their properties have been inferred by recording the responses by lobula plate neurons to moving stimuli that are seen by small regions of the eye, when the number of local motion detectors stimulated would be very small.

When the image of a moving object travels across an eye, different photoreceptors are stimulated in a particular order. Information about the direction of movement could, therefore, be obtained from the sequence in which the photoreceptors are stimulated. The open rhabdom design of the fly eye has enabled Nicholas Franceschini and colleagues (1989) to perform a remarkable experiment in which directionally selective responses from H1 were elicited by stimulating just two photoreceptors in sequence (Fig. 5.12*a*). A special optical instrument enabled the experimenters to view a single ommatidium and to direct tiny spots of light onto each of two individual photoreceptors which view adjacent points in space. H1 is excited when the posterior receptor is stimulated just before the anterior receptor, and it is inhibited when the two receptors are stimulated in the reverse order (Fig. 5.12*b*). This corresponds with the directional selectivity by H1 for moving objects, which is forwards over its own eye.

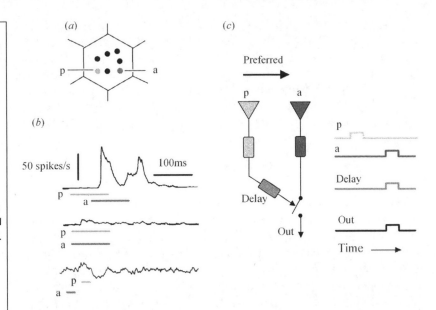

Fig. 5.12 Directional selectivity in the house fly (*Musca*) H1 neuron to stimulation of two individual photoreceptors. (*a*) A fly ommatidium viewed through its lens, showing the seven individual rhabdomeres that are visible. The anterior (a) and posterior (p) photoreceptors that were stimulated in the experiment are indicated. (*b*) Responses by H1 to different sequences of stimulating the two photoreceptors. The times when a pinpoint of light was directed at each photoreceptor are indicated. There was a vigorous response when stimulus to the posterior receptor preceded that to the anterior, but almost no response when the two receptors were stimulated simultaneously. Note that the light-on and light-off stimuli generated separate peaks in the response. In another experiment, shown in the bottom trace, stimulation of the anterior receptor before the posterior one (the null direction) caused inhibition of H1. Each recording was an average from 100 stimulus repetitions, and shows the response of H1 as instantaneous spike frequency. (*c*) Diagram of an elementary motion detector. Open triangles represent the two photoreceptors. To the right are responses at different parts of the elementary motion detector to an object moving over p and then a. (Part (*b*) redrawn from Franceschini *et al.*, 1989.)

Illumination of either receptor alone, or both at the same time, does not cause any response in H1; there must be a delay between stimulation of one and stimulation of the other. Illumination of the posterior receptor **facilitates** a response by H1 to stimulation of the anterior receptor. This facilitation does not start immediately when the posterior receptor is stimulated, but after a delay of at least 10 ms. The delay in facilitation ensures that H1 is not excited when overall illumination of the eyes alters or if light intensity flickers, which would occur when the insect flies under dappled shade. One of the limitations in this experiment is that the lights used to stimulate the photoreceptors switch on or off very rapidly, whereas the rate of change in light caught by a photoreceptor when a real object moves over it is rather more slow. In the experiments by Franceschini and colleagues, H1 was most sensitive to movements that stimulate the anterior receptor 50 ms after the posterior receptor, equivalent to images travelling over the surface of the eye at a speed of 72°/s. In fact, H1 responds very well to movements that are faster and slower than this.

The diagram in Fig. 5.12*c* summarises how an artificial directionally selective motion detector with a preferred direction from posterior (p) to anterior (a) might be organised. The diagram summarises how information is processed in different elements, which represent the operations that occur within the EMD and do not correspond with particular identified neurons. The inputs are elements that respond to light-on or light-off stimuli, and in a fly each would probably include a photoreceptor and various neurons in the lamina and medulla. The signal from the posterior photoreceptor is delayed before it is combined with that from the anterior photoreceptor. The combination is indicated as a simple switch that enables one photoreceptor to regulate the output of the other, although in reality the way the outputs from the two photoreceptors are combined is more sophisticated than a simple switch. The two signals are thought to be multiplied together rather than added, which ensures that the output is very small when only one of the two receptors is stimulated (because multiplying by a small number gives a result that is also a small number). The result is that a strong output from the motion detector is only produced when stimulation of the second photoreceptor follows stimulation of the first photoreceptor with a particular delay.

It would be reasonable to assume the optic lobe contains an array of very many EMDs, perhaps one per column of neurons in the medulla. Because many EMDs converge onto H1, it is excited by movement in its preferred direction over any part of its eye and the response by H1 is greater when large areas of the eye rather than small areas are stimulated. In fact, there are probably at least four similar copies of each EMD. The first duplication is necessary to account for inhibition of H1 by stimuli moving backwards over the eye, the null direction. Inhibition is apparent in experiments where H1 shows quite a high rate of spontaneous spike discharge in the absence of any movement stimulation, and sequential stimulation of the two photoreceptors in the null direction briefly inhibits H1, reducing its spike rate. This inhibition shows the same time-dependent properties of facilitation as excitation by movement in the preferred direction. To account for this, there must be two mirror images of an EMD, one exciting H1 and the other inhibiting it. The second duplication, in which there are two of each excitatory and inhibitory EMD, is necessary because light-on stimuli are processed separately from light-off stimuli. Although photoreceptors are excited by light-on stimuli and inhibited by light-off stimuli, neurons such as H1 respond to moving light or dark objects, so signals from photoreceptors must be channelled through different, parallel pathways before they reach the lobula plate. Microstimulation of single pairs of receptors shows that light on to the first receptor facilitates responses to light on, but not to light off, delivered to the second. This duplication ensures that a local motion detector will respond only when there is correspondence in the stimulus that the two photoreceptors see.

In another type of experiment, Martin Egelhaaf and Alexander Borst recorded intracellular responses from an HS neuron (Egelhaaf and Borst, 1993). They stimulated the eye with a pattern of vertically oriented stripes that drifted backwards or forwards over the eye (Fig. 5.13a). The darkness of each stripe varied sinusoidally, so there were no sharp borders between light and dark in the pattern. A mask with a narrow slit cut into it was placed between the moving pattern and the eye so that only a narrow, vertically oriented band a few ommatidia wide was stimulated. When the stimulus moved, the photoreceptors in this band experienced sinusoidal changes in light intensity, rather than the abrupt switches in light that occurred during the microstimulation experiments on H1. When the stimulus

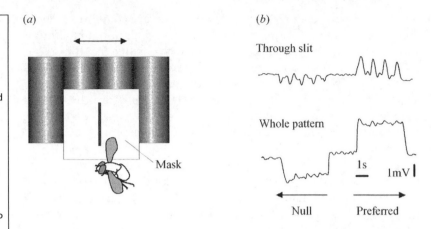

Fig. 5.13 Responses by elementary motion detectors revealed in recordings from an HS neuron. (*a*) When a fly viewed a moving pattern through a vertical slit cut in a mask, only a narrow band of ommatidia and elementary motion detectors was stimulated. (*b*) A wide field HS neuron responded to each lightening or darkening of this band, as shown in the intracellular recording. When the stimulus moved in the null direction, each response by the strip of motion detectors caused hyperpolarising potentials; and when the stimulus moved in the preferred direction another set of motion detectors responded, causing depolarising potentials. When the mask was removed so the eye saw the whole moving pattern, responses from different strips of motion detectors summed so that sustained excitatory or inhibitory responses were recorded from the HS neuron. (Part (*b*) redrawn from Egelhaaf and Borst, 1993.)

moved in the preferred direction, the response by the HS neuron consisted of a series of discrete depolarising potentials in which a large potential alternated with a smaller one (Fig. 5.13*b*). Each large potential corresponded with a darkening of the photoreceptors in the band of stimulated ommatidia, and each small potential corresponded with a lightening of this band. When the stimulus moved in the null direction, the neuron responded with a series of alternating large and small hyperpolarising potentials. When the mask was removed so that a large part of the eye saw the moving pattern, the intracellular recording showed a sustained depolarisation of the neuron, indicating excitation, or a sustained hyperpolarisation, indicating inhibition, depending on which direction the stripes moved. These results are consistent with the hypothesis that motion is detected by local networks such as the one summarised in Fig. 5.12*c*. When the eye sees a large, extended pattern rather than a narrow slit, an HS neuron will receive signals from many vertical bands of EMDs. Because the outputs from the different bands are out of phase with each other, the summed signal in the HS neuron will be a steady depolarising or hyperpolarising potential in which the fluctuations in output from individual EMDs are ironed out.

One approach to discovering the physical identities of the neurons from which a real EMD is composed is to make intracellular recordings from neurons of the lamina, medulla and lobula, and then to inject them with intracellular stains so that they can be identified anatomically (Douglass and Strausfeld, 2003; Higgins *et al.*, 2004). But the small size and high density with which these neurons are packed makes this task extraordinarily difficult. An alternative approach is to use genetic engineering to examine the effects of removing particular neurons in the optic lobes of *Drosophila* (Rister *et al.*, 2007). The experimental technique allowed the output synapses of particular types of neuron in the lamina to be selectively inactivated. Furthermore, the allele responsible for inactivating the synapses is temperature sensitive, so particular neurons could be turned on or off by heating or cooling a fly. The optomotor responses of *Drosophila* to moving backgrounds were tested, and the experiments indicated that two types of LMC, numbers 1 and 2, are needed for a fly to produce optomotor responses. In a parallel set of experiments, a strain of fly was used in which none of the lamina neurons were able to respond to the neurotransmitter that photoreceptors release, histamine, which is not used as a neurotransmitter by other neurons of the optic lobe. In this fly strain, the responses of either LMC1 or LMC2 to photoreceptor signals could be restored, and the results of these experiments showed that the function of these two particular LMCs is sufficient for the ability to generate optomotor behaviours. A complicating factor is that the way in which LMC1 and LMC2 work depends on light intensity. At high light intensities, either alone is sufficient for movement detection; at low light intensities, both are needed; and at intermediate intensities, LMC1 is needed to detect movements in one direction and LMC2 is needed to detect movements in the opposite direction. Other types of lamina neuron, including the remaining LMCs, are not needed for a fruit fly to express its optomotor behaviour. But the amacrine and tangential neurons, rather than the LMCs, are involved in another visual response, orientation towards objects in the visual field. This indicates that particular types of lamina neuron feed information into particular types of feature-detecting networks in the central part of the optic lobe.

Speed and pattern in motion detection

Neurons like H1 and the HS neurons are excited most strongly by movement of large field stimuli in a particular direction. The amount of excitation also depends on the speed and repeat pattern of the stimulus. These two stimulus features, speed and repeat pattern, cannot be distinguished from each other by an EMD because the response to a narrow series of stripes moving quite slowly over the eye is the same as the response to stripes that are twice as broad but moving twice as rapidly (Fig. 5.14). The strength of excitation, therefore, depends on the frequency with which the detector is stimulated with changes in the contrast of light: the **contrast frequency**. Slowly moving, closely spaced stripes will stimulate an EMD with the same contrast frequency as more rapidly moving, broader stripes. The fact that the responses of H1 and the HS neurons also depend on contrast frequency indicates strongly that they are driven by this type of EMD. The strengths of optomotor turning responses by a tethered fly also depend on contrast frequency of stimuli, rather than speed or spatial pattern, and this lends support to the hypothesis that the HS neurons are responsible for controlling these behaviours.

Some movements by insects, however, are controlled in a way that shows they can measure stimulus speed independently of image structure. Bees can be trained to fly along tunnels between a food source and their hive, and they normally fly along the middle of the

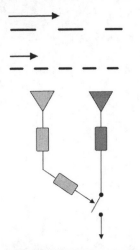

Fig. 5.14 An elementary motion detector circuit cannot distinguish between narrow stripes moving slowly and wider stripes moving more quickly. In this circuit, each receptor is excited whenever a light–dark edge moves over it. The amount by which the detector is excited depends on the delay between excitation of the two receptors, which is short if either the speed of movement is fast or if the stripes are spaced close together.

tunnel, an equal distance from left and right walls. They do this even if one wall has closely spaced, narrow vertical stripes and the other wall has broad, widely spaced stripes (Kirchner and Srinivasan, 1989; Srinivasan, 1992). This might not be expected if the bees were comparing the signals made by EMDs in their left and right optic lobes, and experiments in which one of the walls had stripes that moved demonstrated that bees can measure the speed of visual stimuli independently of the repeat pattern of the stripes. If one wall of stripes moves in the same direction as the bee, it interprets that as if it was further away from that wall because the stripes on that wall move more slowly over the eye on that side than the stripes on the non-moving wall seen by the other eye. Instead of flying down the centre of the tunnel, the bee flies close to the wall with moving stripes. Conversely, if a wall of stripes moves in the opposite direction to the bee's flight, it adjusts its path close to the stationary wall. So the bee optic lobes must be capable of detecting movement using devices that are different from the EMDs described above, perhaps by comparing the rates of change in light caught by photoreceptors in the same neighbourhood of the retina (Borst, 2007). It is likely that bees, flies and other insects make use of more than one means to detect movements of their surroundings. Therefore, visual systems channel information about movements through a number of independent pathways, each focusing on particular aspects of the stimulus. Ambiguities, such as that between stimulus velocity and pattern, can be removed at later stages by combining the outputs of different pathways.

What do a fly's LPTCs tell it?

As a fly moves through the air, some of the movements it makes are intentional and others result from disturbances such as gusts of wind. Its visual system can provide it with information about the way it is moving, so allowing it to correct for unwanted changes in course and helping it move towards objects such as other flies or perches to land on. Particular movements result in a specific kind of **flow field** of the image, which is a term used to describe a pattern of coherent movement over an eye. For example, if a fly is moving straight forwards the images of the visual field move backwards over the two eyes in a similar way, a front-to-back flow field (Fig. 5.15a). Movements in straight lines, such as forwards, upwards, downwards or sideways are called 'translational'. The normal motion of an airborne fly includes combinations of translational and rotational movements. Flow fields that result from rotations about the three main axes are illustrated in Fig. 5.15b. If a fly yaws as it is hovering, the flow field over one eye will move forwards and the flow field over the other eye will move backwards. During a roll, the flow field on one eye moves upwards while the flow field on the other eye moves downward; and during a pitch rotation, the flow fields on

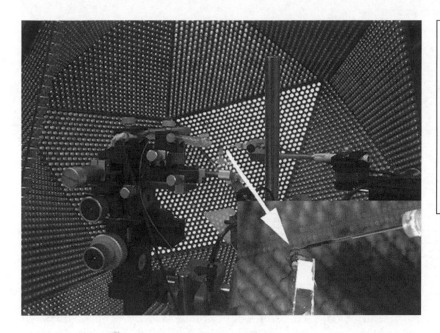

A blow fly (*Calliphora*) is placed in a 3-D cinema, called 'FliMax', in which moving images are created on an array of light-emitting diodes. The inset detail shows the fly waxed to a clear plastic holder and a glass microelectrode, used to make intracellular recordings, which is partly coated with black ink to help an investigator see its tip. (Photographs supplied by Department of Neurobiology, Bielefeld University, Germany.)

the two eyes will move in the same direction, up or down, but at different speeds over different parts of each eye. Usually, rotations will occur while the fly is also translating so the flow fields will change appropriately. For example if it yaws slowly while flying forwards at a reasonable speed, flow fields on both eyes will move backwards but the field on the eye on the side outside of the turn will move fastest. Many species of fly, including hover flies, often move sideways during flight (Fig. 5.1*b*). This may enable them to gauge relative distances to different objects because the images of nearer objects move further over the retina than the images of more distant objects (Fig. 5.15*c*).

Ideally, we would like to be able to implant electrodes into the LPTCs of unrestrained flies to record how these neurons respond in a freely behaving fly, and to deliver stimuli to see how individual neurons change the way the fly is moving. Although the electronic and mechanical apparatus associated with electrodes are much too bulky to achieve this, it is possible to study response by LPTCs to nearly natural stimuli. For example, recordings have been made from H1 while flies are mounted on a motor that twists them rapidly to left and right while viewing woodland outside, so their eyes view images of natural brightness and contrast. Under these conditions, H1 codes rotational speeds up to 3000°/s in its spike trains (Lewen *et al.*, 2001).

Another approach has been to make high-speed videos of free-flying flies either outside or moving around a small arena patterned with images of leaves, stalks and twigs. Then the fly's own view of its flight is recreated in a special device that is essentially a 3-D cinema for flies, called 'FliMax', consisting of a dome-shaped array of light-emitting diodes. A fly sitting in the middle of 'FliMax' would experience visual stimuli very similar to those seen by the moving fly and,

Fig. 5.15 Flow fields created by movements of images of the environment over the eyes as a fly moves in different ways. (*a*) During the fly's forward flight or walking, the flow fields are directed backwards over each eye. (*b*) While the fly is hovering, the three axes of rotation generate different combinations of flow field over the two eyes. (*c*) When a fly moves sideways, the image of an object that is near moves further and more rapidly over the eyes than the image of an object that is further away. Three scenes are shown, in which a stick with a dark top is further from the fly than the stick with the light top. As the fly moves to one side or the other, the locations of the two sticks in the image as seen by the fly change as shown.

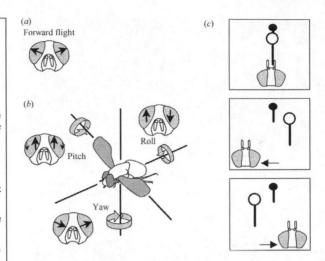

(a) Forward flight

(b) Pitch Roll Yaw

(c)

because it is stationary, it is possible to record from individual neurons. In one study, recordings were made from an HS neuron while a fly viewed a 3-D video as if made by another moving airborne fly (Kern *et al.*, 2005). As is usual with flies, the flights that were recorded consisted of saccades, in which the fly rotated rapidly, separated by longer periods in which its movements were almost entirely translational. A blow fly makes about 10 saccades a second. The H1 neuron or HS neurons spike very briskly in response to the fast rotations during saccades, but also respond well to the slower movements of the scene that occur between saccades. This was confirmed in experiments in which the video recording was manipulated to remove the translational movements, leaving only the fast saccades. The ability to report very fast rotational movements (during saccades), but also to report much slower movements of the environment (during sideways movements) is a result of fast adaptations by the visual system. In a similar way to the way that adaptation enables photoreceptors and LMCs to adapt their response to the mean ambient level of illumination, the responses by LPTCs adapt very quickly to particular movement speeds.

Quite remarkably, flies make very quick movements of their heads using their neck muscles so that the time the head spends in saccades is much briefer than the rest of the body (Kern *et al.*, 2006). This was determined by using minute conducting coils mounted on free-flying flies which were developed to measure body–head angle (Schilstra and van Hateren, 1998), which is beyond the resolution of the video images. The body of a fly has much more inertia than the head, so its direction cannot be altered so rapidly. The flexible neck and its muscles not only allow the head to yaw rapidly, but also ensure that the head remains level while the body rolls and pitches.

There are about 60 LPTCs in each lobula plate, a number that greatly exceeds the complement needed to describe whether the fly is moving forwards, upwards or rotating about the three different principal axes. At least in the case of the VS neurons, an individual

LPTC is well tuned to respond best to a particular combination of rotation and pitch. Each VS neuron responds to movements in a downwards direction over its own eye, and responds to a different axis of rotation from the other VS neurons. A fly would rarely roll precisely about its longitudinal axis, and most VS neurons respond best to rotations around axes offset from vertical. Holger Krapp and colleagues (1998) demonstrated this in a series of experiments in which they made recordings from individual VS neurons using dye-containing microelectrodes so the neurons could be later inden-tified. As a visual stimulus, they used an array of dark spots that each travelled around a small, circular path driven by a motor (Fig. 5.16a). The image of a spot was 7.6° wide at the fly's eye, and it moved over a circular path 10.4° wide at 2 cycles per second, either in a clockwise or anticlockwise direction. Each spot would, therefore, test a local area of the eye with movements in different directions, and for each VS neuron, the direction of movement over a small part of the eye that produced the best response was measured.

By using an array of these stimuli, most of the visual field of the eye could be tested, allowing a map to be plotted showing how each VS neuron responds to particular directions of movement in each part of the eye (Fig. 5.16b). The arrows indicate which direction of movement at each location over the eye excited the particular VS neurons; the strengths of response were also recorded, but are not shown in this figure. Each VS neuron responds best to movements within a vertical strip-like receptive field and there is an orderly progression in the angle of rotation that best excites each VS neuron. VS6 responds best to motion over the central part of the eye, so this neuron is the one that is most excited by a pure roll towards its eye. At the ends of the array of VS neurons, VS1 and VS10 are excited by pitching rather than rolling movements, VS1 by a head-up pitching movement that causes downward motion over the eye, and VS10 by a head-down pitching movement. Other VS neurons, such as VS3 and VS9 (Fig. 5.16b), respond best to a particular mixture of rolling plus pitching movement.

Each VS neuron, therefore, behaves as if it is strongly excited by EMDs that detect downward motion in one strip-like region of the eye. But from regions just outside that strip, a VS neuron responds to different but appropriate directions of movement. For example, the arrows in the diagram of the receptive field for VS6 (Fig. 5.16b) show that it responds to backward movements over the front and top of its eye and to forward movements over the front and back of its eye. It would be reasonable to suppose that each retinotopic location is served by a number of EMDs, each detecting motion in a particular direction. Each VS neuron collects signals from EMDs in particular regions of the eye, and the relative strengths of the signals vary for each VS neuron in a systematic way within different eye regions.

The position of the receptive field of a VS neuron correlates with the part of the lobula plate its dendrites connect with (Fig. 5.16c), providing an anatomical basis for the idea that each collects signals

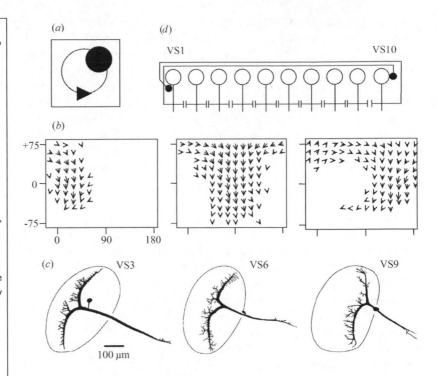

from EMDs in particular eye regions. VS3's main dendrite is located towards the outside part of the lobula plate, so collects signals from neurons connected with a strip of eye towards the front. VS9's dendrite is located further towards the brain, corresponding with a strip of eye further towards the back; and VS6's dendrite is centrally located, corresponding with a sideways-looking strip of the eye. But the area of visual field over which each VS neuron responds to movements is considerably wider than would be expected from the width of the strip of lobula plate each connects with. This arises because the VS neurons are connected with their neighbours by electrical synapses (Fig. 5.16*d*; Haag and Borst, 2004). These synapses have the effect of broadening each neuron's receptive field, with each neuron receiving a degree of excitation from its nearest neighbours, and weaker excitation from other VS neurons further away. The responses to upward motion over part of the eye by the two neurons at the ends of the chain, VS1 and VS10, are due to the chemical, inhibitory synapses they make with each other. Coupling neighbouring VS neurons with electrical synapses compensates for natural variations in the structure of natural scenes. Some parts of a scene will contain a lot of contrasting features, such as twigs and leaves, whereas other parts, such as the sky, will have little structure and so provide relatively weak signals to movement detectors such as VS neurons (Elyada *et al.*, 2009).

The HS neurons, VS neurons and other LPTCs form a highly interconnected network, enabling a fly to respond quickly and reliably, for example enabling it to escape capture or correct its flight course when deflected by a gust of wind (Borst and Haag, 2007). These adjustments are not made by direct communication between visual interneurons and flight motor neurons. The first response a fly makes to a visual stimulus is to move its head, and motor neurons of the neck muscles are driven by LPTCs, so individual neck motor neurons respond in a similar way to VS neurons to visual stimuli, although motor neurons have a receptive field over both eyes (Huston and Krapp, 2008). In the thoracic motor centres, the visual information is channelled to motor neurons that control vibrations of the fly's balancing organs, the halteres, which are highly modified hindwings. A haltere is a rod-like structure that beats up and down in time with and out of phase with the wings. Various sense organs monitor stresses and strains at its base and it acts like a gyroscope to detect changes in flight course, a more rapidly acting sensor than is provided though the optic lobes (Sherman and Dickinson, 2003). The way in which visual information is fed into the flight control system, therefore, is that the fast-acting mechanical haltere system is adjusted, and the fly alters its wing movements accordingly. If the visual system were to affect wing movements more directly, then the visual and haltere systems might work against rather than with each other to steer the fly.

Detecting and tracking moving objects

Optomotor responses help to stabilise an animal when it is stationary or proceeding along a straight course, and it does this by referring to movements that occur in the background. Many animals also need to be able to respond to individual objects, such as potential predators, mates or perching sites. In order for a male fly to detect and chase another fly it needs to have information about individual objects rather than the whole visual scene. Tethered, suspended flies turn strongly towards a small object that is moving backwards relative to the fly, and this response is enhanced if the object is moving over a background pattern that is also moving, rather than remaining stationary (Kimmerle *et al.*, 1997). Likely candidate neurons for recognising a small object and initiating a turn by a blow fly towards it are a group of four neurons called **figure-ground neurons** in each lobula plate (Egelhaaf, 1985). They select a figure, or object, that is distinct from the general background. Like the HS neurons, they are excited by movements backwards over the eye. They also have large receptive fields but, unlike the HS neurons, they respond much more briskly to movements of small targets than to large parts of the visual field. This is because they receive a balance of excitatory and inhibitory synaptic inputs in response to moving stimuli (Warzecha *et al.*, 1993). A figure-ground neuron responds strongly to a stimulus that extends over about 5° of the visual field, but progressively less strongly to stimuli that are seen by larger regions of the eye. A particularly rich collection of neurons that respond to small, moving objects has been found in the lobula of a hover fly (Nordström *et al.*, 2006; Barnett *et al.*, 2007). Some of them respond to images smaller than one ommatidium.

Collision-warning neurons in the locust

Fly LPTCs behave as if they collect signals from EMDs that each serve small segments of the visual field, but the experiments with bees flying in tunnels described above hint that insects have other ways of detecting movements. One of these has been studied in a large fan-shaped neuron found in the lobula of locusts such as *Locusta migratoria* and *Schistocerca americana*, which acts as a collision-warning neuron by responding specifically to objects approaching directly towards the eye. It is called the 'lobula giant motion detector', or LGMD (Fig. 5.17), and it has a postsynaptic follower neuron with a large axon in the nerve cord, the 'descending contralateral motion detector' or DCMD. Because of its axon's size, it is relatively easy to recognise DCMD spikes in extracellular recordings made with wire hook electrodes under the connective nerve, and for this reason the DCMD is often studied in undergraduate practical classes. It was one of the first identified neurons to be studied (Rowell, 1971). The LGMD and the DCMD respond vigorously to the images of approaching objects and can trigger sudden dives during flight that may enable escape from bird predators or collision with other locusts flying in a swarm (Chapter 7; Santer *et al.*, 2006).

The LGMD and DCMD are extremely sensitive to any sudden movements, but in order to determine the visual image features that they are really specialised to detect, Claire Rind made recordings from the DCMD axon while a locust viewed a video of the movie *Star Wars* (Rind and Simmons, 1992). This was an effective way of showing a locust a wide variety of moving stimuli and results indicated that

An electron micrograph showing micronetworks of neurons in the locust brain that are involved in detecting rapidly approaching stimuli. The membrane that surrounds each neuron stains darkly, and sections through several different neuronal processes can be seen. The largest belongs to the lobula giant motion detector (LGMD) neuron. It receives synapses from an array of smaller processes (some indicated with asterisks). Each synapse includes a darkly stained bar, and small, membrane-bound vesicles that contain neurotransmitter. The synapses occur where two presynaptic processes come together, so each neuron that is presynaptic to the LGMD also makes a synapse with its neighbours. (Photograph supplied by Claire Rind, Newcastle University.)

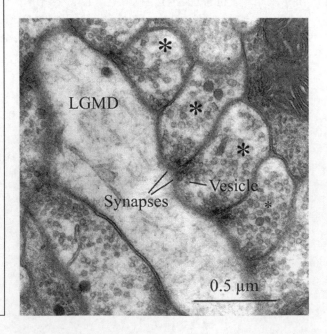

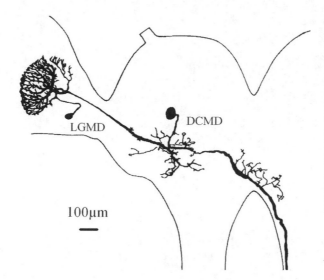

LGMD

DCMD

100µm

Fig. 5.17 Two feature-detecting neurons in the visual system of a locust (*Locusta*), the lobula giant motion detector (LGMD) and the descending contralateral motion detector (DCMD). The drawings were made from neurons that had been stained by injecting them with cobalt ions. The LGMD receives excitatory input from lower order interneurons on the fan-like array of dendrites on the left. (Redrawn from Rind, 1984.)

the LGMD responded more vigorously to images of approaching objects than to other types of movement. For example, it was strongly excited whenever a spaceship approached the screen but much less excited when the spaceship crossed the screen. Responses were then analysed in more detail using carefully controlled computer-generated moving images which showed that two different features of the image are important for the LGMD: the edges of an approaching object's image grow in length; and they move faster and faster over the retina during approach (Simmons and Rind, 1992). The LGMD only responds weakly to overall decreases in light intensity, to a box moving sideways, or to a receding object. The frequency of LGMD spikes increases throughout the approach movement, as if the neuron has locked on to the approach movement (Fig. 5.18a). Deviations as small as 2–3° from a direct collision course result in a reduction in response in the LGMD by a half, so it is remarkably tightly tuned to respond to objects that will collide with the animal (Judge and Rind, 1997). Wide field movements, like those that cause optomotor responses, inhibit responses by the LGMD.

As with the lobula plate neurons in the fly, selectivity by the LGMD for a particular kind of moving stimulus is established by networks among small neurons in the medulla. A locust eye has about 8000 ommatidia, so an LGMD probably receives inputs from many thousands of individual neurons connecting the medulla with the lobula. Electrophysiological recordings, made before responses to colliding stimuli had been studied, demonstrated that lateral inhibition operates among the neurons that are presynaptic to the dendrites of the LGMD – columnar neurons in the medulla (O'Shea and Rowell, 1976). When one area of the retina is stimulated by a moving image, the response to stimulation of other parts of the retina is depressed but no inhibitory postsynaptic potentials are recorded from the LGMD. This means that the inhibition must occur at an earlier stage, presynaptically to the LGMD. Using electron

Fig. 5.18 Detection of approaching objects by the locust lobula giant motion detector (LGMD) neuron. (*a*) Response by the LGMD to the image of a 3-cm diameter disk approaching and then receding at 3.5 m/s. The top trace is an intracellular recording from the LGMD, and the lower trace is a monitor of the width of the object's image, which was displayed on a screen. PSPs, postsynaptic potentials. (*b*) Diagram to illustrate mechanisms for feature detection by the LGMD neuron. The LGMD is shown with its dendrites in the lobula receiving excitatory synapses from an array of small neurons in the medulla. The medullary neurons form a lateral inhibitory network with each other. The LGMD is excited if the rate at which the image over the eye grows outstrips the spread of lateral inhibition between neurons pre-synaptic to the LGMD, and also adaptation in the membrane of the LGMD's dendrites. The LGMD is simplified in this diagram; it has two additional fields of dendrites (one is shown at the top in Fig. 5.17), which are thought to suppress its responses to wide-field movements. (*a*) Redrawn from Rind, 1996.

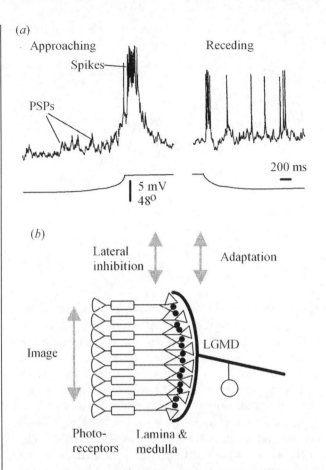

(*a*) Approaching   Receding

Spikes

PSPs

200 ms

5 mV
48°

(*b*)

Lateral inhibition   Adaptation

Image   LGMD

Photo-receptors   Lamina & medulla

microscopy, Claire Rind and Peter Simmons (1998) showed that the dendrites of the LGMD are covered with input synapses from neurons that originate in the medulla. The synapses are arranged in a remarkable manner: each neuron that synapses with the LGMD also synapses with its neighbours, so that the neurons that drive the LGMD also synapse with their neighbours. These microanatomical networks provide a possible route for local lateral inhibition among the elements that excite the LGMD. Lateral inhibition, therefore, can serve different functions. In the lamina, the lateral inhibition between cartridges sharpens the detection of edges in the image, but in the input network to the LGMD, it provides selectivity for objects that are approaching rather than moving in other directions.

The way in which the LGMD filters out the image of an approaching image can be envisaged as a kind of race between excitation by the growing image and two different kinds of suppression in the optic lobe (Fig. 5.18*b*). The excitation comes from photoreceptors that are stimulated sequentially by the edges of the image as it travels over the eye. Early on, when the image is small, a relatively small number of photoreceptors are stimulated as the edge of the growing image crosses them. As approach continues, an increasing number of photoreceptors are stimulated because both the size of the image and the rate of travel of its edges over the eye increase. One mechanism for suppression is by lateral inhibition among elements presynaptic to the LGMD as outlined in the previous paragraph, and the race is over the LGMD's dendritic fan between the travel of excitation caused by the moving image edges and the spread of lateral inhibition. For excitation to win the race, the extent of the edges in the image or the speed with which they move must be increasing. A computer model of a lateral inhibitory network like this responds as predicted, being excited most strongly by the images of approaching objects (Rind and Bramwell, 1996), and it works when fitted to an artificial eye to warn of impending collisions by cars. The second mechanism for suppression is that LGMD dendrites adapt strongly when excited. Injecting a pharmacological agent that blocks this adaptation into the LGMD increases its responsiveness to images of small bars that move across the eye and so reduces its selectivity for approaching over translating objects (Peron and Gabbiani, 2009). So properties of both the network of small neurons that excite the LGMD and of the LGMD's own dendrites play significant roles in shaping its properties as a feature detector, responding best to rapidly approaching objects.

The LGMD is not simply a passive collector of signals, but the way it works depends on what the locust is doing and on its state of arousal (Rind *et al.*, 2008). In a resting locust, repeated presentation of an approaching stimulus leads to a dramatic decline in response by the LGMD, a neuronal analogue of habituation (Chapter 4) that enables an animal to ignore a repeated stimulus, which is less likely to be a real threat than a novel one. During a series of repeated approaching stimuli the LGMD response declines, but is boosted if the leg is stroked between visual stimuli (Fig. 5.19*a*). A locust that is flying is in a constant state of arousal, and little or no decline in response to repeated approaching stimuli was found (Fig. 5.19*b*). In these experiments, locusts were tethered to a rod (see Fig. 7.8*a*) and wire electrodes implanted near the nerve cord were used to pick up spikes from the DCMD axon. Even at repeat intervals for the approaching stimulus as short as 10 s no decline in LGMD response is seen in a locust that is flying, in comparison with responses by locusts that are resting on the ground or tethered in a wind stream but not flying. The arousal of the LGMD during flight may be through the action of octopamine, a neurohormone with an action similar to noradrenaline in mammals. A drug that blocks the action of octopamine counteracts the effect that flight has on boosting LGMD responses; blood levels of octopamine rise during the early part of a flight; and there are some octopamine-containing neurons in the locust's brain that could release the neurohormone over the LGMD dendritic fan. A reason why flying requires an increase in responsiveness by the LGMD is likely to be to enhance its function in warning of collisions, which is vital for an animal that, like locusts, sometimes flies in dense swarms.

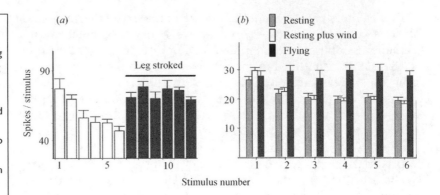

Fig. 5.19 Arousal in the locust lobula giant motion detector (LGMD). Responses to approaching disks were measured as descending contralateral motion detector (DCMD) spikes recorded in the nerve cord, with mean and standard error from several stimulus repetitions shown. (a) Responses to successive approaches separated by 40 s. During the first six stimuli in a series, the locust was resting and the response declined. Just before the sixth stimulus and then just before each stimulus up to number 12 a hindleg was stroked, which has the effect of reversing the previous decline in LGMD response. (b) Spikes per stimulus in response to an approaching disk. Six stimuli were presented 10 s apart in three different situations. While the locust was flying, no reduction in response with stimulus repetitions was recorded. (More spikes were recorded in a compared with b due to a difference in stimulus speed.) (Redrawn from Rind et al., 2008.)

Conclusions

Many animals, including flies, devote large proportions of their brains to vision. Each part of the image that falls on the retina needs to be processed by networks that filter out information about significant features, including movements of the visual world and of objects within it, as well as identifying objects such as potential mates or prey. In the first stage of vision, each photoreceptor collects light from a small receptive field, and adaptation adjusts its sensitivity so that it generates signals that are large enough for neurons downstream in the optic lobe to work with whatever the level of ambient illumination. The electrical signal is conditioned further in the lamina, where signals about contrasts are enhanced in two main ways. First, in transmission from photoreceptors to lamina monopolar cells, almost all of the standing signal about absolute light level is discarded, so that an LMC's signal range is fitted appropriately to the level of ambient illumination, whether it is bright daylight or much dimmer. Second, lateral inhibitory interactions sharpen responses so that the neurons either side of borders in an image give signals that stand out as significantly greater or smaller than the signals produced about featureless areas of the image. In the medulla, further comparisons are made between changes in the light signals in time and in space, extracting particular image features. Detection of image motion in particular directions is a good example of one of the operations that occurs in the lobula and medulla. Almost all of the neurons of the lamina, medulla and then the lobula are confined in columns, each of which is dedicated to processing signals from a small part of the image. The fly lobula plate tangential cells, or locust lobula giant motion detector, are the first neurons of the visual system that combine signals from many different retinotopically arranged units. Each of these large, fan-shaped neurons responds to image motion over a wide area. Each acts as a feature detector, responding most vigorously to a particular type of motion such as a front-to-back flow field or a rapidly approaching object. The selectivity each has for a particular kind of image feature can be explained by the network of connections among its presynaptic neurons, combined with its own physiological properties. In the fly, an assembly of LPTCs differ amongst each other in their receptive fields and their preferred directions of movement. The whole moving visual scene, therefore, is fragmented into the responses of about 120 individual neurons. The connections these neurons make with premotor interneurons and directly with motor neurons must work as a coherent network, so that the movements the animal makes cause it to behave appropriately to particular stimuli.

An insect's compound eye is constructed in such a way that it gains an almost all-round view of the world, much more coarsely defined but more panoramic than the view a mammal has. Despite the very different optical arrangement of the eyes and anatomy of the brain, many of the neuronal processes involved in vision are similar between insects and mammals. In both, signals are conditioned in the early parts of the visual pathway to enhance contrasts between neighbouring parts of the image, or in the same part of the image as light changes as a result of image movement. Some of the changes in the signals as they pass from photoreceptors to bipolar cells and then ganglion cells in the vertebrate retina (Fig. 3.3b) are similar to those occurring in the early stages of the insect visual pathway. Inhibitory interactions ensure that bipolar cells and ganglion cells respond well to edges or small spots; and some amacrine and ganglion cells respond to particular directions of image motion. In the mammalian brain, different tasks in dealing with particular aspects of the image are assigned to different regions. In the early stages, individual neurons respond best to simple features, such as the orientation of an edge in their receptive field. Later on, neurons can have sophisticated requirements, such as the face of another individual or the sight of another animal performing a particular act ('mirror neurons', described in Chapter 9). Many insects are very good at flying, able to change course rapidly several times a second. Although their feature recognition capabilities fall far short of those of mammals, the requirements of their optic lobes are to distinguish rapidly between different moving stimuli, and to organise appropriate steering movements.

Questions

In an experiment in which you record responses from single neurons to various visual stimuli, what tests could you employ to check whether a particular stimulus feature is the one that a neuron is really tuned to detect?

It is common for particular image features, such as shape, direction of movement or location of a moving object, to be segregated for processing in different neuronal pathways. How can a nervous system recombine the information from the different pathways, and ensure that information provided by the different pathways is balanced appropriately?

Summary

- This chapter focuses on movement-detecting neurons in insects, which are good examples of feature detectors. Some are involved in optomotor responses, and others in tracking movements of objects.

Insect visual system
- In the optic lobes, there is a retinotopic projection from photo-receptors (in ommatidia) to lamina to medulla to lobula. In the lobula or lobula plate are wide-field, motion-sensitive neurons.
- In most insects, light is focused onto a central rhabdom, to which all photoreceptors of the ommatidium contribute. Diptera have an open rhabdom design in which individual rhabdomeres remain separated.
- Compound eyes give an animal a wide field of view, but their resolution is limited. Many compound eyes have foveas.

Coding by photoreceptors and lamina neurons
- Photoreceptors and lamina monopolar cells (LMCs) signal with graded potentials.
- In a dark adapted eye, light photons cause depolarising bumps in photoreceptors. As ambient light increases, bump size decreases and bumps fuse to give a smooth photoreceptor potential.
- Light and dark adaptation enable photoreceptors to signal changes in light intensity, or contrasts, in a wide variety of ambient lighting conditions. Receptor potential is proportional to the logarithm of stimulus intensity.
- The extent of adaptation varies between species and between photoreceptor types.
- Insect photoreceptors make inhibitory synapses with LMCs.
- Compared with photoreceptors, LMCs are more sensitive to changes in light and generate more phasic responses. This enhances responses to stimulus contrasts in time.
- Stimulus contrasts in space are enhanced by lateral inhibition, which helps to detect borders in the visual image.

Optomotor neurons in flies
- Many lobula plate tangential cells are directionally selective, for example H1 responds to back-to-front movements over its own eye.
- H1 can be excited by sequential stimulation of two photoreceptors in the same ommatidium, which look in different directions.
- H1 is excited by an array of elementary motion detectors, each of which detects movement in a particular direction over a small region of the eye. HS neurons also collect input signals from arrays of small-field motion detectors that each serve small parts of the eye.
- In one model for elementary motion detectors, a signal from a posterior photoreceptor is delayed before it is combined with a signal from an anterior photoreceptor.

- This type of motion-detection mechanism involves a correlation between signals from two nearby photoreceptors. Motion detectors like this respond to contrast frequency of a moving image – when stimulated with a moving grating pattern, they cannot distinguish speed from repeat pattern.
- The strength of a fly's optomotor movements also depend on contrast frequency, suggesting that elementary motion detectors drive optomotor behaviours.
- In some behaviours, insects are clearly able to measure the speed of movement of an image, so must have additional motion-detecting mechanisms.
- Many fly lobula plate neurons are tuned to respond to particular flow fields and can inform the fly about its direction of movement as it flies or walks. Each of the ten VS neurons responds to a different, particular combination of roll and pitch movements.
- In flight a fly makes rapid optomotor movements of its head that stabilise the position of the visual field on the eyes. The visual system causes steering movements of the wings by first changing the ways the halteres are moved.

Object movements

- Fly figure-ground neurons respond well to movements of small objects rather than of the whole scene. They might enable flies to track each other.
- Locusts possess large, fan-shaped lobula neurons that respond selectively to objects approaching the eye on a collision course.

Further reading

Frye, M. A. and Dickinson, M. H. (2001). Fly flight: a model for the neural control of complex behavior. *Neuron* **32**, 385–388. A review of how a fly translates information about its visual environment into mechanical movements to steer its flight.

Land, M. F. and Nilson, D-E. (2002). *Animal Eyes*. Oxford, UK: Oxford University Press. A very readable book that describes the many types of eye found throughout the animal kingdom, including their optical systems, function and evolution.

Sensory maps: hunting by owls and bats

The interaction between a predator and its prey represents a dramatic example of animal behaviour, in which an evolutionary arms race has greatly stretched the capabilities of nervous systems. This can be seen clearly in the battle for survival that takes place when a toad attempts to catch a cockroach (Chapter 3). In such an encounter, the hunting animal faces the fundamental problems of detecting and localising the prey, and it must solve them on the basis of purely passive information given out inadvertently by the prey. This is a formidable task and it has led to the evolution of some remarkably sophisticated neuronal systems in species that are adapted for hunting.

Predatory birds and mammals do, indeed, possess central nervous systems with the necessary sophistication to handle the complex task of tracking prey, but this sophistication makes most of them unsuitable as subjects for neuroethological research. However, the difficulty can be overcome by looking at species with a highly specialised method of hunting, based on a sensory system that is dedicated to the specialised method of prey detection and localisation. It then becomes easier to correlate the properties of particular neurons in that system with the particular behavioural task, as has been achieved to great effect with the specialised nose of the star-nosed mole (Chapter 1).

Such dedicated systems are found in two groups of animals – owls and bats – that use specialised auditory systems to hunt at night when visually guided predators are at a disadvantage. Like T5(2) visual neurons of the toad (Chapter 3) or lobula and lobula plate neurons of insects (Chapter 5), the auditory systems of owls and bats include neurons that act as feature detectors for relevant stimuli. This is possible because the responses of auditory interneurons become progressively more complex and selective as one travels along the auditory pathway. The result is that biologically important features of the environment are extracted from the basic parameters of sound received by the ears and are mapped into specific regions of the brain.

A series of flash photographs showing a barn owl, *Tyto alba*, swooping onto its prey, a mouse, in darkness. Owls have an auditory system that is adapted to allow them to locate the source of a rustling sound in darkness, so can hunt at night. (Photograph supplied by Masakazu Konishi, California Institute of Technology.)

Owls are able to locate small animals on the ground by listening for the tiny rustling noises made by an animal moving among fallen leaves and twigs. On most nights, an owl's hearing is used in conjunction with its excellent eyesight, but on very dark nights some species can hunt by sound alone. Insectivorous bats both hunt and find their way around by sound, using a method that is akin to human sonar and is known as **echolocation**. This method involves the bat emitting loud pulses of sound and then analysing the returning echoes in order to find out what lies ahead. Using echolocation, a hunting bat can detect, identify and then capture a flying insect in less than a second.

Neuroethological research on owls has concentrated on their ability to localise a sound source in space by using sound pressure level and frequency as cues. Most research on bat echolocation has paid particular attention to the bat's ability to determine the distance from which an echo has returned and to the suitability of different bat sounds for different sonar techniques. Taken together, these studies on bats and owls are providing valuable insight into the neuronal basis of some sophisticated behaviour patterns.

Sound

The basic properties of sound and their variation over time determine what is possible by way of detection and localisation for a hunting owl or bat. Sound is created when air molecules are set in motion by a vibrating structure such as a loudspeaker. The vibrations of the speaker generate alternating waves of compression and rarefaction of the air, which propagate out from the speaker at the speed of sound. The molecules involved in propagating the sound move back and forth from regions of high pressure into regions of low pressure, which thereby become regions of high pressure and so on.

The greater the pressure changes, then the louder the sound is. The pressure generated by a sound wave, or its loudness, is expressed as sound pressure level, most often using a unit called the **decibel** (dB), which is equivalent to an increase or decrease of about 12.2% in relative sound pressure (a special kind of logarithmic scale). Absolute pressure levels are described relative to a reference sound pressure, usually 20 μPa (= 2×10^{-5} N/m^2), which is roughly the threshold of human hearing to sound with a frequency of 1 kHz. The interval from a given point on one sound wave to the equivalent point on the next sound wave is the wavelength, and its reciprocal is its frequency, or number of waves passing a fixed point each second. Frequency is expressed as cycles per second (or hertz). Most sounds that animals produce or listen to have frequencies of thousands of cycles per second or kilohertz (kHz).

Prey localisation by hearing in owls

Adult owls hunt within a well-defined territory, which they know well and patrol regularly at night. During its patrol, an owl visits a number of observation perches, from which it can survey the ground round about. If it hears potential prey, the owl swiftly turns its head so that it directly faces the object of interest. Then, after adequate scrutiny, it flies down to capture the prey in its outspread talons (Fig. 6.1a). An owl listening from a perch or in low level flight must be able to pinpoint the sound source both in the horizontal plane (azimuth) and in the vertical plane (elevation). In fact, unless an owl is looking down from directly above its prey, its orientation is more critical in the vertical than in the horizontal plane (Fig. 6.1b). The use of hearing in prey capture has been studied mostly in the ubiquitous barn owl, *Tyto alba*. Early behavioural studies were carried out using tame individuals, which proved able to locate prey even in total darkness, and a number of simple experiments demonstrated that hearing is used to accomplish this (Payne, 1971). For example, it was found that the birds could strike accurately at a concealed loudspeaker quietly broadcasting a recording of leaf rustling noises. This result not only showed the barn owl's ability to locate prey by hearing alone but also opened the way for testing which features of the sound are important in localisation.

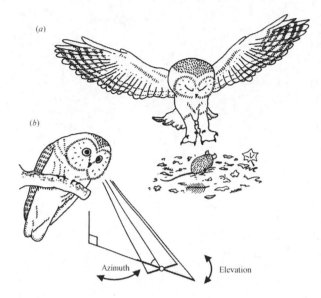

Fig. 6.1 The hunting technique of an owl, drawn from photographs of Tengmalm's owl (*Aegolius funereus*) in its natural habitat. (*a*) The owl about to strike prey with its talons, after flying down from an observation perch. (*b*) The owl on its perch immediately before striking, with a diagram showing the errors involved in localising prey by hearing. The prey (o) is observed at a shallow angle in elevation with the result that a given angle of error converts into a greater distance along the ground for a vertical (elevation) error than for a horizontal (azimuth) error. (Redrawn from Norberg, 1970, 1977.)

Using this technique, the accuracy of localisation was found to vary with sound frequency. The greatest accuracy was achieved with a sound containing frequencies from 6 kHz to 9 kHz. Typically, birds are not sensitive to frequencies above 5 kHz but the barn owl can hear up to 10 kHz, and 6–9 kHz is the range to which its ear is most sensitive. In addition, more than half the auditory neurons in a barn owl's ear are devoted to a range of sound frequencies between 5 kHz and 10 kHz. The usual pattern is for approximately equal numbers of auditory neurons to be devoted to each doubling of frequency (octave) but the barn owl devotes a disproportionate number to the higher frequencies (Koppl *et al.*, 1993). Hence the owl can analyse sounds with these important frequencies in most detail.

The accuracy of sound localisation has been examined closely by exploiting the natural response in which an owl turns to face a novel sound. The owl is trained to remain on its perch and the angle through which its head turns in response to a sound is measured using an electromagnetic angle detector (Fig. 6.2*a*). In each test, the head is first aligned by attracting the owl's attention with a sound from the zeroing speaker, and then the owl is stimulated with a sound from the target speaker. Mazuo Konishi and collaborators (Konishi, 1993) have isolated the sensory cues that owls use to locate the source of a sound, and have then extended the investigation to reveal how the neuronal mechanisms that are responsible work and develop.

When the target speaker is placed in front of its face, the barn owl's localisation is exceptionally accurate, with an error of less than 2° in both azimuth and elevation. But the owl's accuracy deteriorates in both planes as the angle between the source and the axis of the head increases (Fig. 6.2*b*). The rapid flick of the head, with which the owl responds, is initiated about 100 ms after

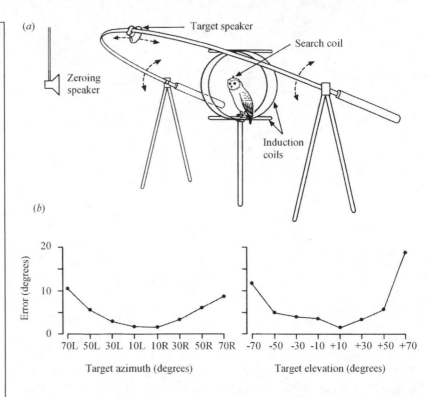

Fig. 6.2 Orientation of the head to sounds by the barn owl (*Tyto alba*). (*a*) The method used to measure the accuracy with which the owl locates sounds coming from different positions in space. Sound stimuli originate from either a fixed source (the zeroing speaker) or a movable source (the target speaker). The search coil on top of the owl's head lies at the intersection of horizontal and vertical magnetic fields generated by the induction coils. Movement of the head in response to sound from a speaker induces a current in the search coil, which is analysed by computer to give horizontal and vertical angles of movement. (*b*) Localisation accuracy as a function of the position of the target speaker, showing the mean degree of error in judging target position in the horizontal plane (left) and in the vertical plane (right) for an individual owl. (Redrawn from Knudsen and Konishi, 1979.)

the onset of the sound. However, maximum accuracy can be achieved even with brief sounds (75 ms duration) that end before movement of the head begins. This shows that the owl can determine the sound's precise location in space immediately without successively improving the accuracy of its movement by constantly referring to the sound as its head moves.

An indication of just what cues are involved in locating a sound is provided by partially blocking one ear, which effectively reduces the sound pressure level at that ear without altering the sound's time of arrival. A plug in one ear leads to significant errors in elevation but only slight errors in azimuth (Fig. 6.3*a*). A plug in the left ear causes the owl to direct its head above and a little to the right of the target, and a plug in the right ear results in the owl facing below and a little to the left. The tighter the ear plug, the greater is the degree of error. This result indicates that the intensity difference between the ears is the principal cue for locating a sound in elevation.

The owl is able to use a comparison of sound intensity between the left and right ears to locate a source of sound accurately in elevation due to asymmetries in the ears and in the arrangement of feathers on its face. The ear openings and the protective, preaural flaps are vertically displaced, the left flap being above the mid-point of the eye and the right one below it (Fig. 6.3*b*). There is also a slight asymmetry in the facial ruff, which is composed of dense, tightly packed feathers and forms a vertical trough behind each ear

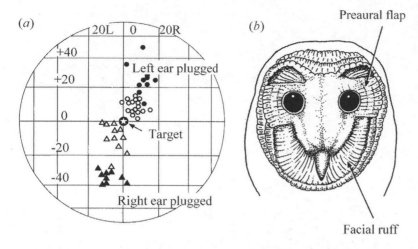

(a)

(b)

Preaural flap

Facial ruff

Fig. 6.3 The barn owl's ability to locate sounds in elevation. (*a*) Plot of auditory space in front of the owl in degrees of azimuth (L and R) and of elevation (+ and −). The symbols show individual errors in open-loop localisation of a sound source in the centre of the plot (Target) produced by partly blocking one ear. A tighter ear plug (closed circles and triangles) produces a greater localisation error than a looser ear plug (open circles and triangles). (*b*) Facial structures of the barn owl that contribute to localisation of sound in elevation: the facial ruff is formed from tightly packed feathers projecting from the relatively narrow skull, and the ear openings are located behind the preaural flaps. These structures are revealed by removing the sound-transparent feathers of the facial disc, which give the owl's face a flat appearance. (Redrawn from Knudsen and Konishi, 1979.)

opening: the left trough is orientated downwards and the right one upwards. Because of its dense feathers, the facial ruff acts as an effective sound collector at frequencies above 4 kHz. Consequently, the asymmetries in the ruff and in the ear openings give rise to a vertical asymmetry in the directionality of the two ears: the left ear is more sensitive to high-frequency sounds from below and the right ear from above the horizontal plane. If the ruff feathers are removed, the owl is unable to locate sounds in elevation and always faces horizontally regardless of the true elevation of the source, but it can still locate in azimuth.

When a source of sound is not directly in front of or behind the head, sound will reach the two ears at slightly different times. In order to test the importance of these time differences, sound stimuli are delivered using miniature earphones installed in the owl's ear canals rather than from a distant speaker, which would inevitably produce differences both in time and intensity. With the earphones, stimuli can be made equal in intensity but different in time of arrival at the two ears. An owl responds to such a stimulus by turning its head horizontally in a direction that the time difference would represent if it were an external source of sound. The owl turns its head to the side that receives the stimulus earlier, and the angle of turning is positively correlated with the magnitude of the time difference between the ears (Moiseff and Konishi, 1981). These experiments were then elaborated to show that an owl uses the continual disparity in the time at which the same sound wave arrives at each ear (the alternative, that the owl uses the time at which the sound starts, was eliminated experimentally).

Auditory interneurons and sound localisation

The fact that a barn owl can localise brief sounds accurately without access to the sound as it moves its head implies that its auditory system can recognise each point in space from a unique combination of the

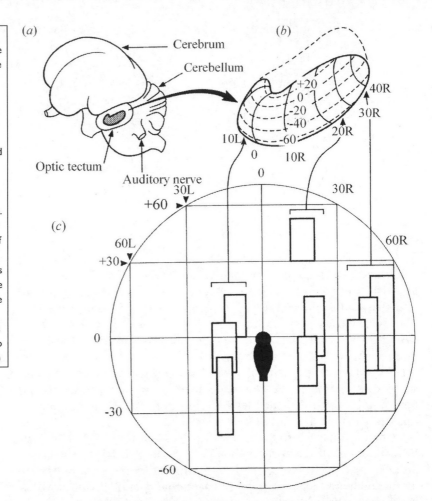

Fig. 6.4 The neuronal map of auditory space in the midbrain of the barn owl (*Tyto alba*). (*a*) The left side of the owl's brain, showing the external nucleus of the auditory midbrain or inferior colliculus (shaded and in bold outline) on the inner side of the optic tectum.
(*b*) The left external nucleus enlarged from (*a*), with the co-ordinates of the neuronal map indicated in degrees of azimuth (L and R) and of elevation (+ and −) of auditory space.
(*c*) Plot of auditory space in front of the owl in degrees of azimuth and of elevation, showing the receptive fields (bold rectangles) of 10 neurons recorded in three separate electrode penetrations. The penetrations were made with the electrode parallel to the transverse plane at the positions indicated by the arrows linking (*c*) to (*b*). (Redrawn from Knudsen, 1981.)

cues that specify azimuth and elevation. Interneurons that respond to such complex combinations of stimulus features are found among higher-order neurons in the midbrain. The main auditory region of the midbrain (often called the inferior colliculus) is situated on each side of the brain on the inner edge of the left or right optic tectum (Fig. 6.4*a*) and is divided functionally into a central nucleus and an external nucleus. In the central nucleus, the interneurons are tuned to particular frequencies and are arranged according to their best frequency. The great majority of these neurons respond to their best frequency regardless of where the sound source is located in space.

However, in the external nucleus of the auditory midbrain there are interneurons that respond quite differently: all have similar best frequencies near the upper end of the owl's range (6–8 kHz) and respond only when the sound originates from a specific region of space. These are therefore called space-specific neurons. When recording from a space-specific neuron with a microelectrode, the size and shape of the specific region to which it responds (its receptive field, a term that we introduced for visual neurons in Chapters 3 and 5) can be determined with the movable speaker (Fig. 6.2*a*). Typically, the neuron is excited by sounds coming from a region of space shaped like a vertically elongated ellipse. Surrounding this central, excitatory part of the receptive field is a region where sound sources inhibit any spontaneous spike discharge by the neuron, which has the effect of sharpening the neuron's ability to pinpoint a sound source. Unlike most auditory neurons, the space-specific neurons are insensitive to changes in intensity, and even a 20 dB increase in intensity has little effect on the size of the receptive field.

As the recording electrode is advanced through the external nucleus of the auditory midbrain, it samples neighbouring interneurons, which are found to have receptive fields representing neighbouring regions of space. In fact, the space-specific neurons are arranged systematically according to the azimuth and elevation of their receptive fields, so that they form a neuronal map of auditory space (Fig. 6.4b, c). Two-dimensional space is mapped on to the external nucleus of the auditory midbrain with azimuth being arrayed longitudinally and elevation being arrayed dorsoventrally. On each side of the brain, the map extends from 60° on the opposite side to 15° on the same side, but a disproportionately large number of neurons are devoted to the region within 15° of either side. This arrangement means that the 30° of space directly in front of the owl is analysed by an especially large population of neurons on both sides of the brain. In elevation, the map extends from 40° upward to 80° downward but the majority of neurons are devoted to the region below the horizontal (Fig. 6.4c).

Partially blocking one ear, which changes the normal intensity differences between the two ears, causes a significant vertical displacement of the receptive field in the space-specific neurons but makes little difference in azimuth. When changes in time differences between the ears are delivered with the miniature earphones, it is found that individual neurons respond only to a narrow range of time differences, which correspond with the horizontal location of the neurons' receptive field. Neurons that respond to small time differences have receptive fields in front of the face and those that respond to larger time differences have receptive fields at greater angles to the face. This shows that the azimuthal position of the receptive field is determined largely by time differences between the ears (Moiseff and Konishi, 1981). The space-specific neurons are exclusively binaural – they never respond to sounds delivered to just one ear. Their response properties are built by comparing signals in the left and right auditory pathways.

Thus the acoustic cues that are used to create the receptive fields of the space-specific auditory interneurons are exactly the same cues as the intact owl depends on for sound localisation. This fact makes it almost certain that the neuronal map of auditory space in the midbrain underlies the barn owl's skill in rapid sound localisation. Certainly, the greater density of space-specific neurons devoted to the 30° arc in front of the face would account for the owl's greater accuracy in locating sounds in this region. The large proportion of the neuronal map that is devoted to the region of space somewhat below and in front of the face also makes good functional sense since this is likely to be the prey-containing region for an owl scanning the ground from its perch. Usually, sounds that an owl is most interested in will come from a moving source, such as a running vole. If an owl hears sound from a moving, rather than a stationary, sound source the receptive fields of space-specific auditory neurons shift in a subtle way towards the direction of movement of the sound source, and the size of the shift depends on the speed of movement of the sound source (Witten *et al.*, 2006). This is an excellent example of adaptation of the properties of neurons to perform specific functions, in this case to enable an owl to locate moving prey.

Synthesising a neuronal map of auditory space

Each space-specific neuron in the owl's midbrain responds to stimuli delivered via the earphones only when both the time difference and the intensity difference fall within the range to which it is tuned. It is not excited by either the correct time difference alone or the correct intensity difference alone. Evidently, the receptive fields are formed by tuning of the neurons to specific combinations of time differences and intensity differences, which are coded separately by lower-order neurons. The initial separation takes place at the start of the auditory pathway in the hindbrain, in which each cochlear nucleus receives axons from its ear.

Auditory receptor cells are located in the inner ear, or cochlea, which is a slightly curved, elongated bony box in birds. A functional diagram of the cochlea is drawn in Fig. 6.5a. The receptor cells are called hair cells and they transduce sound energy into receptor potentials. Each hair cell has a tuft of cilia, which give them their name and are where transduction starts. The hair cells are arranged along the length of the basilar membrane, which almost completely divides the inside of the cochlea into two fluid-filled chambers. Sound vibrations are transferred from the auditory canal to the cochlea fluid by way of two lightweight membranes, the tympanum (eardrum) and oval window, which are linked across the cavity of the middle ear by a small bone called the stapes. The basilar membrane varies in stiffness along its length, which means that sounds of low frequencies cause parts of the basilar membrane furthest from the oval window to vibrate most vigorously, whereas sounds of high frequencies cause parts of the basilar membrane that are close to the oval membrane to vibrate most vigorously. The result of this is that each hair cell is sharply tuned to respond to a particular frequency of sound: those nearest the eardrum picking out high frequency sounds and those nearest the brain picking out low frequency sounds. This arrangement is called **tonotopic**, which means that the tone, or frequency of sound, to which a cell responds depends on its location along the length of the basilar membrane. A natural sound stimulus is usually made up of many different frequencies, which are parcelled up into responses by

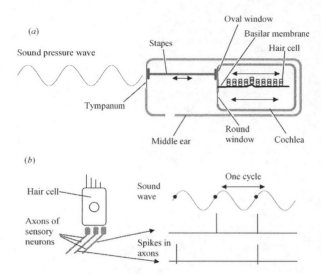

(a)

Sound pressure wave

Tympanum

Middle ear

Stapes

Oval window

Basilar membrane

Hair cell

Round window

Cochlea

(b)

Hair cell

Axons of sensory neurons

Sound wave

Spikes in axons

One cycle

Fig. 6.5 Functional organisation of the ear of an owl. (*a*) A sound wave oscillates the tympanum (2-headed arrows), and is transferred into the cochlea, where it vibrates the basilar membrane on which sit hair cells, the cells that transduce sound into receptor potentials. Each portion of the basilar membrane, and so the hair cells on it, is tuned to respond to a particular sound frequency. (*b*) Each hair cell connects with several sensory neurons that have axons in the cochlear nerve. Spikes in a sensory axon are phase locked to a particular instant within each cycle of the sound wave (dots). Each axon generates a spike during only a proportion of sound wave cycles.

thousands of hair cells along the basilar membrane. Each generates a receptor potential with an amplitude that depends on the relative loudness of that cell's best frequency within the sound stimulus.

Each hair cell excites about half a dozen sensory neurons that have axons in the auditory nerve, connecting the ear with the brain (Fig. 6.5*b*). The hair cell receptor potential has a waveform similar to the tone the cell is tuned to respond to, and the postsynaptic potentials that result in the sensory neurons trigger spikes in the sensory axons. These spikes are **phase locked** to the waveform of the sound stimulus. That means that each spike occurs at the same instant during a sound cycle (Fig. 6.5*b*). But because the sound waves have frequencies of several thousand per second, which is a rate of spikes that axons cannot achieve, only a proportion of the individual cycles trigger spikes. Phase locking of spikes is important for the owl's ability to locate the source of sound – it ensures accuracy of the timing information each ear provides. Precise phase locking is found on all but the highest sound frequencies an owl can hear. For a loud sound, an individual sensory axon will generate more spikes for a larger number of sound wave cycles than a quiet sound, and the number of axons excited by a particular hair cell that spike for each cycle will also be greater. In this way, sensory neurons use spikes to convey information about both the timing and intensity of sounds that arrive at one ear. Within the brain, signals from left and right ears are compared to extract the location of a source of sound.

When it reaches the brain, each sensory axon divides into two branches: one going to the angular nucleus and the other to the magnocellular nucleus. These pathways are present on both sides of the brain, and connections between the left and right sides of the brain allow comparisons of signals that arrive at the two ears. The role played by the angular and magnocellular nuclei was revealed by an ingenious experiment, in which a tiny amount of local anaesthetic was injected into one of them so as to inactivate

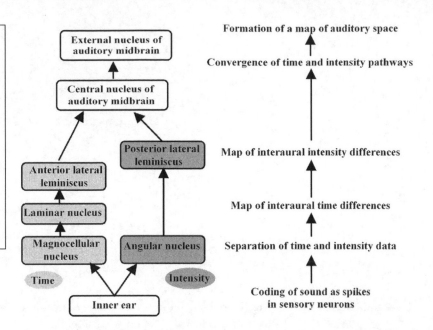

Fig. 6.6 Simplified flow diagram showing how the neuronal map of auditory space is synthesised in the brain of the barn owl (*Tyto alba*). The boxes on the left represent successive regions of the brain, and the process that takes place in each region is shown on the right of the corresponding box. The arrows indicate the flow of information along pathways (left) and along the sequence of computational steps (right). Note the separation of the time and intensity pathways. (Redrawn from Konishi, 1992, 1993.)

most of its neurons (Takahashi *et al.*, 1984). The responses of the space-specific neurons were then re-examined to see what changes occurred. The results were clear cut: inactivation of the angular nucleus altered the responses of the space-specific neurons to interaural ('between ears') intensity differences without affecting their responses to time differences, and inactivation of the magnocellular nucleus altered their responses to time differences without affecting their responses to intensity differences.

When recordings were made from the interneurons in these two nuclei, their properties were found to be consistent with this result. Neurons of the magnocellular nucleus preserve the phase locking shown by the sensory neurons but are insensitive to changes in intensity, while the neurons in the angular nucleus are sensitive to intensity changes but do not show phase locking. Evidently, these two cochlear nuclei serve as neural filters, which pass along information about either time of arrival or intensity but not both. Data about time of arrival and intensity is then processed separately in a series of auditory nuclei, as shown in Fig. 6.6, before information from the two streams is recombined in the auditory midbrain.

In the sound intensity processing pathway, the angular nucleus contains a number of different types of neurons that respond in various ways to sound stimuli. Its neurons send axons to the posterior part of the lateral lemniscus, both on its own side and on the other side of the brain. Axons travel between the two posterior lateral leminsci. Neurons in the posterior lateral lemnisci integrate excitatory and inhibitory signals from the two ears in a way that makes each of the neurons pick out a particular difference in sound intensity between the two ears. Furthermore, the neurons of the posterior lateral lemniscus vary systematically in the intensity difference that causes them to respond maximally, forming a topographical array

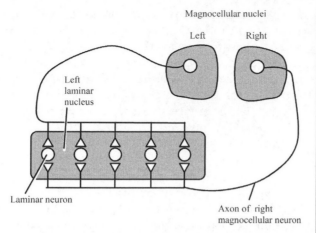

Magnocellular nuclei

Left   Right

Left
laminar
nucleus

Laminar neuron

Axon of right
magnocellular neuron

Fig. 6.7 Neuronal circuit for measuring interaural time differences in the barn owl (*Tyto alba*), shown as a highly diagrammatic section through the brain. The left laminar nucleus receives excitatory input from both the left and right magnocellular nuclei, represented here by a single axon from each. When spikes conducted along the left and right magnocellular axons reach a given laminar neuron simultaneously, that neuron will be strongly excited. This will happen whenever the difference in a sound's arrival time at the two ears compensates for the difference in time taken for spikes to travel to that laminar neuron along the left and right magnocellular axons. (Redrawn from Konishi, 1992.)

along the dorsoventral axis (Manley *et al.*, 1988). In effect, this array constitutes a neuronal map of interaural intensity differences for a given frequency band. This information is passed along by the axons of the lemniscal neurons to the central nucleus of the auditory midbrain, where it converges with the information on interaural time differences (Fig. 6.6). From each frequency band in the central nucleus, intensity difference detecting neurons are connected to the external nucleus of the auditory midbrain so as to generate the elevation selectivity for space-specific neurons. Combining information about different frequencies of sound helps to sharpen sensitivity for a particular intensity difference.

The first place in the auditory pathway at which the arrival time of sound at each ear is compared is the lamina nucleus in the hindbrain. There is one lamina nucleus on either side of the brain, and the neurons in each receive excitatory input from neurons in both the left and right magnocellular nuclei. As its name suggests, a magnocellular nucleus contains particularly large nerve cells, an adaptation that helps transmit information about sound timing faithfully. It seems clear from the anatomy and physiology of the lamina nucleus that this nucleus serves to measure interaural time differences. The laminar neurons are arranged in an array, and the axons of the left magnocellular neurons pass along this array from one end while the right magnocellular axons pass along from the other end (Fig. 6.7). Both left and right axons give off terminal branches that make synaptic contact with the laminar neurons.

This arrangement provides a way for individual laminar neurons to respond to particular interaural time differences on the principle of **delay lines** and coincidence detection in the following way. Each laminar neuron is maximally excited when it receives excitatory input simultaneously from axons of both left and right magnocellular neurons and so it functions as a coincidence detector – a spike in either a right or a left magnocellular neuron alone is insufficient to trigger a spike in a laminar neuron. The magnocellular neuron axons function as delay lines because the time it takes each spike to travel along the axon from one end of the lamina nucleus toward the other end causes a delay in the spike's time of arrival at a given laminar neuron. The delay in spike conduction balances the delay in arrival of sound coming from a particular direction at the two ears. The point in the array of laminar neurons where the left and right magnocellular spikes arrive simultaneously will, therefore, vary systematically as a function of the difference in a sound's time of arrival at the left and right ears plus the conduction time along the magnocellular axons from the two ears.

Consequently, the array of laminar neurons effectively constitutes a neuronal map of interaural time differences in which the relative time of arrival of a sound at the two ears corresponds with a particular location in the array (Carr and Konishi, 1990). In Fig. 6.7, the laminar neuron at the right end of the array will respond best to sounds coming from the left of the owl, the neuron at the left end of the array will respond best to sounds coming from the right side of the owl, and the neuron in the middle will respond best to sounds coming from the front (or the back) of the owl. The array is duplicated several times to respond to many different sound frequencies: as with sound intensity, localisation of a sound source is greatly improved by combining information about different sound frequencies. The axons of the laminar neurons convey this information forward first to the anterior part of the lateral lemniscus, which then carries it to the central nucleus of the auditory midbrain where, as in the sound intensity pathway, information from neurons responding to different sound frequencies is combined. Finally, information about the relative time of sound arrival at the two ears is combined with information about relative intensity in the external nucleus of the auditory midbrain (Fig. 6.6).

By this means, the neuronal map of auditory space is synthesised centrally from sensory cues that are not themselves spatially organised. The inner ear is organised tonotopically, with the topographical array of receptors coding frequency rather than space. For each frequency band, the ear monitors intensity and time of arrival, cues that are separated in the brain and processed in parallel pathways. In both pathways, the information from the left and right ears is brought together to form a topographical array of interaural differences. Finally, the intensity and time pathways are brought together, thereby enabling neurons in the auditory midbrain to encode specific combinations of time and intensity differences.

Alignment of the auditory with the visual map

Barn owls normally use both sight and hearing to track their prey at night. On hearing a novel sound or sight stimulus, the initial response by a perching owl is to swivel its head to look towards the stimulus – owls do not move their eyes much within the eye sockets, so need to swivel the whole head. So that the owl stares in the correct location whether the stimulus is an auditory or a visual one, its auditory and visual maps of space need to coincide with each other. To achieve this, the auditory map of space in the external nucleus of the auditory midbrain is passed on point-by-point to the optic tectum. Thus the optic tectum also contains an auditory map of space and this is precisely aligned with the visual map of space, which is derived by topographical projection from the retina. These aligned maps of auditory and visual space in the optic tectum are used to control networks in the hindbrain that generate the movements responsible for orienting the owl's head to its prey (Masino and Knudsen, 1990).

Given the important role played by these two space maps, the question arises as to how they come to be aligned correctly during development of the brain in a young owl chick. This question has been carefully examined by Eric Knudsen and collaborators, who initially showed with a variety of techniques that sound localisation by barn owls is adjusted in the light of visual experience. One observation that shows that the auditory map is changeable is that although an owl misdirects its gaze towards a sound source when an earplug is inserted into one ear to reduce sound intensity reaching it, over the course of some weeks the accuracy of the owl's gaze returns. But if the earplug is then removed, the owl misdirects its gaze again, this time in the opposite direction.

More detailed examination of the way auditory behaviour can be recalibrated by the visual system took advantage of an owl's response in accurately directing its gaze towards either a small flashing light or a source of sound. If a baby owl is fitted with goggles that deflect its vision to the right by 23°, it misdirects its attention towards the flashing light, and it does not correct for this if the goggles remain in place for many weeks (Fig. 6.8a). Initially, the owl with the goggles still directs its gaze accurately towards a sound source (Fig. 6.8b), but over several days, gazes towards the sound become misdirected and correspond with the misdirected visually directed gazes (Fig. 6.8c). If the goggles are removed, the visual gaze direction is immediately corrected, but the direction of the sound directed gaze is not (Fig. 6.8d). These effects depend on the age of the owl (Knudsen and Knudsen, 1990): maximum errors in gaze direction towards a sound occur in owls up to three weeks old. For older owls, the change in gaze direction becomes progressively smaller. These observations show, first, that the auditory map can change under instructions from the visual map of space; and second, that the ability to change depends on the age of the owl. The first few days of life for the owl are, therefore, a **critical period** in which it can make particular adjustments to the connectivity of its brain neurons and to its behaviour. However, the critical period for adjusting the auditory map in owls does not have a sharply defined ending because if the owl's vision is deflected by small amounts using weaker prisms, then adult owls are able to make some changes to the alignment between their visual and auditory maps.

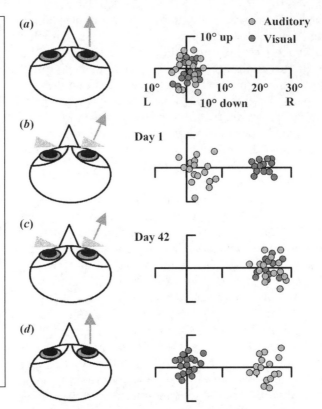

Fig. 6.8 Experiment that shows that auditory behaviour in a young owl is aligned with visually guided behaviour. (*a*) A baby owl directs its gaze accurately towards a small flashing light or a sound source. Circles show gaze direction for individual stimuli, measured as in Fig. 6.2. (*b*) The owl is fitted with prisms that deflect its line of sight by 23°. It initially misdirects gaze direction towards the light but not towards the loudspeaker. Notice that the scatter of gaze directions is not affected by this treatment. (*c*) Over a few weeks, the direction of gaze towards the loudspeaker drifts until it coincides with the misdirected gazes triggered by the light stimulus. (*d*) When the prisms are removed, the owl's ability to gaze accurately towards the light stimulus is restored immediately, but the gazes towards the loudspeaker initially remain misdirected. (Redrawn from Knudsen, 2002.)

When a young owl is fitted with prisms to deflect its line of sight, receptive fields of the auditory space-specific neurons in the external nucleus of the auditory midbrain shift with the same time-course as the change in gaze direction (Brainard and Knudsen, 1993). Auditory receptive fields of cells in the optic tectum also shift, but there are no changes in the properties of neurons in the central nucleus of the auditory midbrain. These and other observations point to the map of auditory space in the external nucleus of the auditory midbrain as a site of plasticity. Under instruction from the optic tectum, individual neurons in the external nucleus of the auditory midbrain adjust their receptive fields. Local lesion experiments show that the tectum calibrates the auditory map in a point-by-point manner, adjusting the fields of individual neurons rather than adjusting the whole field together using a single error signal generated by a disparity between the visual and auditory maps (Hyde and Knudsen, 2002).

The echolocation sounds of bats

Whereas owls locate prey by listening passively to the noises produced by prey, insectivorous bats actively interrogate the environment using the technique of echolocation. For this purpose, a flying bat produces a succession of loud calls, each of which

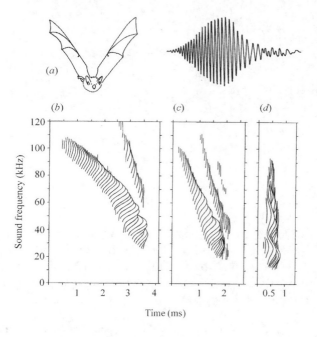

(a)

(b)   (c)   (d)

Sound frequency (kHz)

120

100

80

60

40

20

0

1   2   3   4   1   2   0.5 1

Time (ms)

Fig. 6.9 Echolocation sounds of the mouse-eared bat (*Myotis myotis*). (a) A single frequency-modulated pulse, recorded from a bat in flight (left) and displayed on screen (right). (b) Computer-generated sonagram of a single pulse, showing the downward sweep in frequency in more detail. The computer generates the sonagram by plotting curves of the relative intensity of different frequencies at successive intervals of time during the sound pulse. This pulse was emitted by the bat at a distance of 4 m from the target. Parts (c) and (d) are sonagrams of pulses emitted respectively at 36 cm and at 7 cm from the target. (Part (a) redrawn from Sales and Pye, 1974; (b) to (d) redrawn from Habersetzer and Vogler, 1983.)

consists of a brief pulse of sound. These sounds are often described as ultrasonic because they contain frequencies (20–200 kHz) beyond the range of human hearing. The sound pulses travel out in front of the bat and, when they encounter a target, are reflected back and picked up by the ears. The essence of echolocation lies in the ability of the brain to reconstruct features of the target, most importantly its position in space, by comparing the neural representation of the echo with that of the original signal.

Analysis of the sound pulses used for echolocation reveals what is at first sight a bewildering variety of form as one compares one species with another, but it has become clear that there are basically two kinds of sound signal used by bats. The first kind are broadband signals, which consist of short pulses, less than 5 ms in duration, that are frequency modulated (FM pulses). An example of this kind of signal is found in the mouse-eared bat (*Myotis*) in the widespread family Vespertilionidae: each pulse starts at a high frequency and sweeps downward in frequency during the course of the pulse (Fig. 6.9a, b). At any given instant, the sound within the pulse is a fairly pure tone, corresponding with the fundamental frequency generated by the larynx, with traces of a second harmonic towards the end of the pulse. But the downward sweep results in the pulse having a total bandwidth of 60–70 kHz.

Frequency modulated pulses appear to have evolved in bat echolocation because the wide range of frequencies make them suitable for target description and accurate ranging. Behavioural tests show that vespertilionid bats can estimate target range from the time it takes sound pulses to travel out to the target and return as echoes. This is confirmed by studies of the brain,

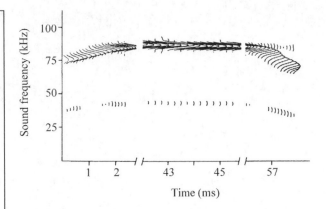

Fig. 6.10 Echolocation sound of the greater horseshoe bat (*Rhinolophus ferrumequinum*). The computer-generated sonagram shows the three components of the long pulse: the initial, upward sweep in frequency, part of the long constant-frequency component (note breaks in time axis) and the final, downward sweep in frequency. The faint traces of the fundamental frequency at around 40 kHz indicate that the loudest sound in the broadcast frequency is the second harmonic. (Redrawn from Neuweiler *et al.*, 1980.)

which show that bats take advantage of FM signals for target ranging by making multiple estimates of pulse–echo delay.

The second basic kind of echolocation sound used by bats consists of narrowband signals, in which the sound has a constant frequency (CF pulse). These signals are generally longer, between 10 ms and 100 ms in duration, and form part of an alternative strategy of echolocation employed by many species. This method for echolocation is particularly well developed in the horseshoe bats, which are members of the specialised family Rhinolophidae, and is often used in hunting in woodland rather than open meadows. The echolocation sound of the greater horseshoe bat (*Rhinolophus ferrumequinum*) consists mainly of a long component (about 60 ms) with a constant frequency of just over 80 kHz, which is followed by a brief downward frequency-modulated sweep and is often preceded by an even briefer upward sweep (Fig. 6.10). The bat's cry is not a single tone; as shown in Fig. 6.10, the loudest sound during the CF part of the cry has a frequency about 80 kHz, but there is a softer tone at about 40 kHz. The 40-kHz tone is the fundamental frequency, or first harmonic, of the emitted sound, and the 80 kHz tone is the second harmonic. Some bats emit cries with third and fourth harmonic components as well, and the FM parts of the cry also contain two or more harmonics. It is common for the first harmonic to be relatively soft, and bats can use this to distinguish their own broadcast sounds from those of neighbouring bats.

A long CF signal is unsuitable for target description but is well suited to measuring the Doppler shift, which is the shift in sound frequency experienced by an observer listening to a moving sound source. That horseshoe bats actually perceive the Doppler shifts generated during echolocation is shown clearly by the way they modify their sounds in flight. If a horseshoe bat is trained to fly down a long room to a landing platform, it is observed to alter the sound frequency of its echolocation pulses so as to keep the Doppler-shifted echoes from the landing platform at a constant, species-specific frequency, which is 83 kHz for the loudest harmonic component in the greater horseshoe bat.

In human affairs, a Doppler shift is experienced when someone listens to a car approaching and then passing. To an observer sitting in the car, the sound made by the engine does not change. But to someone standing by the roadside, the engine's pitch increases as the car comes nearer, and then the pitch decreases as the car goes by. Similarly, bats using a long CF signal are able to determine the relative velocity of their prey by comparing the frequency of the outgoing pulse with that of the returning echo. In addition, it has been found that horseshoe bats are able to perceive the relatively small Doppler shifts in echo frequency produced by the beating wings of a flying insect. They use this as a means of detecting insect prey in the face of the extensive echo clutter produced by dense foliage or other background objects.

CF and FM sound pulses thus represent two different strategies for extracting information from the environment by means of echolocation. It is evident that *Rhinolophus* depends heavily on CF while *Myotis* employs FM exclusively. However, some species employ a mixture of the two strategies and emit pulses in which both CF and FM components are well developed. One example that has been the subject of a detailed neuroethological study is the moustached bat from Central America (*Pteronotus*) in the family Mormoopidae. Research in the field has shown that each species of bat has its own particular set of echolocation sounds, and these are closely related to the ecological situation in which it hunts. A number of proposals have been made for grouping bats according to their habitat and echolocation sounds (e.g. Schnitzler and Kalko, 2001; Jones and Holderied, 2007).

Interception of flying prey by bats

The way in which bats use their echolocation signals to track prey has been studied by combining high-speed photography with audio recordings of the normal sequence of airborne interception. These observations are mostly made under controlled conditions but numerous, and increasingly exact, observations have been made in the field. All the species studied so far follow a fairly standard routine, which can be broadly divided into three stages, known as the search, approach, and terminal stages (Fig. 6.11). Much the same sequence is followed whether the bat is intercepting prey, avoiding an obstacle or landing on a perch.

During the search stage, a bat emits sound pulses of a constant, species-specific form at a low repetition rate of about 10 Hz or less, as described in the previous section. As its name implies, the main function of the search stage is to detect potential prey or obstacles. Behavioural tests with bats using FM signals show that they can detect a small sphere (0.5 cm diameter) at almost 3 m and a larger sphere (2 cm diameter) at over 5 m. Calculation of

A big brown bat (*Eptesicus fuscus*) approaches a moth that it has found by using its echolocation system. The auditory system of bats is specialised to recognise details of potential prey, such as size and distance from the bat. Here the moth was tethered to a wire in part of a study on how some moths can jam the echolocation cries of bats to reduce chances of being eaten. (Photograph supplied by Nickolay Hristov, Brown University.)

the intensity of echoes returning from targets at these distances suggests that the maximum range of echolocation corresponds with the threshold of hearing in bats.

However, in natural interception, bats neither react noticeably to targets at these distances nor do they appear to react to progressively larger targets at increasingly greater distances. Instead, the onset of the approach stage, which represents the first visible reaction of the bat to the target, occurs when the bat is between 1 and 2 m away in nearly all cases. Therefore it is probable that the approach stage begins at a critical distance between the bat and the target and does not necessarily begin when the target is first detected. The transition to the approach stage is marked by the bat turning its head, especially its ears, directly towards the target and by an increase in the repetition rate of the echolocation sounds to a value of about 40 Hz. In bats such as *Myotis*, which use only frequency-modulated pulses, the pulses become shorter but the slope of the FM sweep becomes steeper so that the bandwidth of the signal is maintained (Fig. 6.9c). Species that use long CF pulses for Doppler shift echolocation do not drop the CF component during the search stage, but it becomes shorter and the small FM component increases in bandwidth.

There is thus a shift towards brief, frequency-modulated pulses at the approach stage in all species of echolocating bats. It is probable that bats are taking advantage of the greater information content of broadband FM signals as they approach the target, especially since the decision about whether or not to catch an item of potential prey is evidently made during the approach

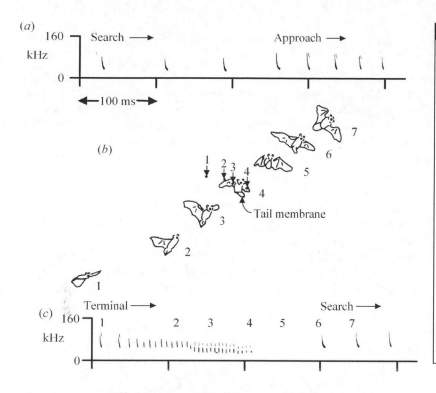

(a)

160 kHz — Search ⟶   Approach ⟶

0

◀—100 ms—▶

(b)

7

6

1   2 3
↓   ↓ ↓   4
        5

4

Tail membrane

3

2

1

(c) Terminal ⟶   Search ⟶

160 kHz
        1   2   3   4   5   6   7

0

Fig. 6.11 Interception of insect prey by the bat *Pipistrellus pipistrellus* (Vespertilionidae) in the field. Echolocation calls emitted by the bat as it searches for a target then approaches a moth it has detected are shown in the sonograms in (*a*). The approach and capture sequence (*b*) is drawn from flash photographs and the corresponding sequence of echolocation calls is shown in the sonograms in (*c*) which follow immediately from (*a*). Numbers 1–7 correspond in (*b*) and (*c*), and numbers 1–4 with arrows indicate the position of the moth at the corresponding instants. Note how the bat brings its tail membrane forward to capture the insect at 4 and bends its head down to consume the insect at 5. The search stage of flight is resumed at 6. (Redrawn from Kalko and Schnitzler, 1998.)

stage. When a mixture of living insects and similar-sized plastic discs or spheres are thrown into the air for bats to catch, the bats break off pursuit of the plastic objects at the end of the approach stage. Again, bats trained to discriminate between two targets in flight make their choice, as judged by the orientation of their ears, towards the end of the approach stage. The increase in repetition rate of the echolocation sounds is also understandable as a response to the need for increasingly frequent estimates of range and direction as the bat closes upon the target. When the approach stage goes to completion, it normally takes the bat to within 50 cm of the target.

The transition to the terminal stage is marked by an abrupt increase in pulse repetition rate, which rises to about 100 Hz or even 200 Hz in some cases (Fig. 6.11). This increased rate clearly provides a rapid updating of information about the target's position as the bat makes its final manoeuvres to capture the target. In most species, the pulses emitted during the terminal stage are FM sweeps, often with several harmonics, that are 0.5 ms or less in duration (Fig. 6.10). Only in horseshoe bats and others that exploit the Doppler shift is a CF component retained during the terminal stage; and even then the CF component is reduced to a length of about 10 ms or less. Bats do not usually capture flying insects in their mouths but rather use their outspread wing or tail membranes as a scoop for collecting the prey. They then fall silent briefly as they bend down to retrieve the insect from the wing or tail membrane (Fig. 6.11).

The auditory system and echolocation

When echolocation cries return as echoes to a bat, they are received by an auditory system that conforms to the general mammalian pattern. The sounds are collected by the external ear and enter the ear canal, where they impinge on the tympanum (Fig. 6.12a). The vibrations of the tympanum are transmitted by the bones of the middle ear to the oval window of the inner ear; in mammals, there are three bones in the middle ear rather than one as in birds. From here, the vibrations travel through the cochlea along the basilar membrane, which forms a helical ribbon, wide at the apex of the cochlea and narrow at its base. Because of its spiral construction the cochlea in mammals is much longer than it is in birds, which enables mammals to have a greater number of hair cells and to detect a wider range of sound frequencies. As in birds, the receptor potentials that are produced

Fig. 6.12 Diagrams showing the auditory system of a bat. (a) The middle and inner ear of a bat; diagram based on a horizontal section. The external ear opening is off picture to the right. Structural elements of the middle and inner ear are shown in solid black; the surrounding bone of the skull is shown in grey. Note that the bones of the middle ear are acted upon by various muscles, known collectively as the middle ear muscles. (b) The ascending auditory pathway in bats. Six brain regions contributing to this pathway are shaded and links between them are shown by arrows. The three named regions (cochlear nucleus, inferior colliculus and auditory cortex) are discussed in this chapter. (Part (a) redrawn from Sales and Pye, 1974; (b) redrawn from Neuweiler, 2000.)

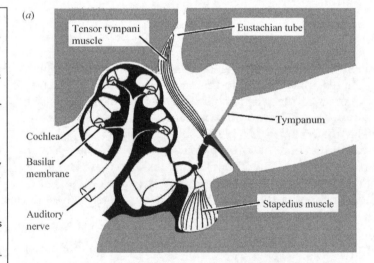

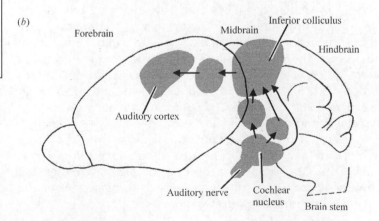

in the hair cells by the vibration of the basilar membrane are transmitted across synapses to first-order neurons with axons that carry spikes to the brain along the auditory nerve.

The information that is provided by the traffic of spikes in the auditory nerve is processed through a sequence of brain nuclei in the mammalian brain. The auditory pathway consists of a rather complex network of brain regions, but may be illustrated by three main staging posts (Fig. 6.12b). First, the cochlear nucleus in the hindbrain receives input from the axons of sensory neurons in the auditory nerve. Second, the inferior colliculus is one of a number of important nuclei in the midbrain containing higher-order auditory neurons. Third, the auditory cortex in the forebrain is the final staging post of the auditory pathway.

Echolocation is possible because a bat's auditory system has several specialisations that enable it to receive and analyse faint echoes. The first of these specialisations is that a bat's hearing is particularly sensitive to sounds that have similar frequencies to its own echolocation pulses. This is shown by testing bats with sounds of different frequencies and measuring the auditory threshold, which is the lowest sound intensity that elicits a detectable response. The results are expressed as threshold curves, in which the sound pressure level at threshold is plotted against frequency (Fig. 6.13).

In bat species that use broadband (FM) signals in the search stage, it is found that the frequencies with the lowest threshold

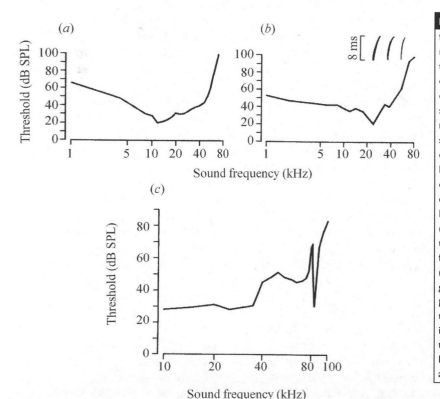

Fig. 6.13 Hearing threshold curves for three species of bat, derived from recordings of summed potentials in the inferior colliculus; thresholds are expressed as sound pressure level in decibels (dB SPL), as a function of sound frequency. (a) Curve for a non-echolocating fruit bat from southern India. (b) Curve for an echolocating bat from the same locality, with a sonagram of the echolocation signal shown above for comparison (rotated through 90°). Note the similarity of the curves in (a) and (b), apart from the tuning of the curve in (b) to the strongest frequencies in the echolocation signal (24 to 26 kHz). (c) Curve for the greater horseshoe bat (*Rhinolophus ferrumequinum*) showing the sharp tuning to 83 kHz and the notch of insensitivity to frequencies just below this. (Parts (a) and (b) redrawn from Neuweiler *et al.*, 1984; (c) redrawn after Neuweiler, 1983.)

coincide with the dominant frequencies in the echolocation signal (Fig. 6.13*b*). Apart from this feature, the threshold curve is not very different from that of a non-echolocating fruit bat (Fig. 6.13*a*). Nor is the absolute threshold of hearing exceptional by mammalian standards; if humans could hear echolocating cries made by bats, we would find them to be exceptionally loud. In bat species that use narrowband (CF) signals in the search stage, the threshold curve is much more sharply tuned to a narrow frequency band. In *Rhinolophus*, for example, hearing is very sharply tuned (Fig. 6.13*c*) and the echoes are kept close to this best frequency by the Doppler-shift compensation.

A second specialisation is that echolocating bats have highly directional hearing, in contrast to most mammals, which have good all-round hearing. This is shown well by a behavioural test carried out on restrained horseshoe bats. A loudspeaker in front of the bat's head produces 'echoes' with a shifted frequency following each of the bat's CF pulses, and the bat responds by compensating for this apparent Doppler shift. The directionality of hearing is tested by masking these artificial echoes with sounds from another speaker at different angles around the bat's head. When the bat perceives the echo, it does not show the compensation response. Hearing is most sensitive directly in front of the head, and sensitivity falls off by about 45 dB from the midline to the side. A similar fall in sensitivity occurs at angles below the horizontal, but the drop is less severe above the horizontal. Thus returning echoes are useful only if they fall within a narrow cone in front of the head, not more than 30° off the direction of flight (Grinnell and Schnitzler, 1977).

A third specialisation for echolocation in bats is that the peripheral auditory system shows a reduced sensitivity to the pulse of sound the bat makes as it cries. The echolocation sounds emitted by bats are very intense, with an absolute sound pressure level of around 110–120 dB when measured 5–10 cm in front of the head. If the bat's auditory system were directly exposed to such intense sounds, it would not be able to recover fully by the time the echo arrived. In bats using brief FM signals, such as *Myotis*, this problem is overcome by contraction of the middle ear muscles, which partially uncouples the inner ear from the vibrations of the tympanum (Fig. 6.12*a*). During echolocation, these muscles begin to contract before each sound pulse and develop maximal tension at the onset of sound emission; they then relax very rapidly, within about 8 ms, so that the response to the echo is not attenuated except at very close range. In *Myotis*, the attenuation produced by contraction of the middle ear muscles is some 20 to 25 dB.

In addition, neural attenuation takes place in the midbrain prior to the inferior colliculus. Recording with simple wire electrodes shows that the summed potential elicited by sounds emitted by the bat is smaller in the midbrain than in the cochlear nucleus, but this is not the case for sounds from other sources. Hence, there must be a central, neural mechanism that attenuates self-stimulation by the emitted sounds. This lasts only for a short time and echoes returning after 4 ms are not affected. The magnitude of the neural attenuation averages about 15 dB in *Myotis*, and so the total attenuation available from both mechanical and neural mechanisms amounts to some 35 to 40 dB (changing the signal to 0.02% of its original intensity).

Bats that use long CF signals, such as *Rhinolophus*, cannot exploit these mechanisms since the long outgoing pulse overlaps the returning echo. However, when the bat is on the wing, the emitted sound normally has a lower frequency than the echo, which is kept at a constant frequency by Doppler-shift compensation. Consequently, *Rhinolophus* is able to solve the problem of self-stimulation simply by being rather deaf at the normal emission frequency, which is a few kilohertz below the echo frequency (Fig. 6.13c).

Auditory specialisations for echo ranging

The mechanisms of attenuation outlined above are particularly important in facilitating accurate measurement of the echo delay, and hence of distance, since they prevent the auditory system from being overloaded by the outgoing sound pulses. In order to measure echo delay, the bat's auditory system must be capable of resolving very small time intervals. Most mammals are unsuited to this task because their auditory systems take too long to recover between successive sound stimuli – usually some tens of milliseconds need to elapse after a pulse of sound before full sensitivity is recovered. But in echolocation, most echoes will return within 30 ms of the outgoing pulse, given that bats track prey at distances of no more than 3 or 4 metres and that sound travels at 334 m/s.

It is therefore not surprising that the auditory neurons of bats are found to recover rapidly when stimulated with paired pulses of sound, roughly resembling the natural pairing of pulse and echo. For instance, summed potentials recorded from the mid-brain of *Myotis* show some response to a second pulse that follows the first after only 0.5 ms and full recovery is achieved in 2 ms. Similar tests of hearing in fruit bats, most of which do not use echolocation, have shown that most species are slow to recover from the first pulse; but rapid recovery is found in one genus, *Rousettus*, which has evolved an echolocation capability independently of the insectivorous bats. Since this is the only respect in which the hearing of *Rousettus* differs conspicuously from that of its non-echolocating relatives, this rapid recovery may represent the most fundamental adaptation of the auditory system for echolocation (Grinnel and Hagiwara, 1972).

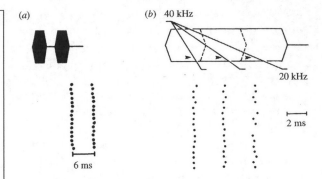

Fig. 6.14 Properties of time-marking interneurons in the inferior colliculus of bats. (*a*) Screen display showing a sound stimulus, consisting of a pair of identical, frequency-modulated pulses 6 ms apart (above), and the interneuron's response (below). Each dot represents the occurrence of a single spike in the responding neuron, and the pair of dot columns is generated by 16 consecutive presentations of the stimulus. (*b*) A similar display showing the response of a collicular neuron to a single frequency-modulated pulse, consisting of a downward sweep from 40 kHz to 20 kHz, stretched over three different durations (shown diagrammatically, above). The interneuron's response (dot columns, below) shifts in register with the occurrence of 25 kHz in the FM sweep (indicated by the arrowheads, above). This suggests that the neuron is responding to this particular frequency in the stimulus. (Part (*a*) redrawn from Pollak, 1980; (*b*) redrawn from Bodenhamer et al., 1979.)

There is more to it than this, however, if a bat is to measure the arrival time of an echo rather than merely be aware of its existence. For echo delay to be coded by the auditory system as a time interval, some classes of interneurons must be able to act as accurate time markers. Interneurons that appear to be specialised as time markers for echolocation have been found in the inferior colliculus of bats belonging to several different genera. In fact, the majority of neurons in the inferior colliculus are found to have suitable properties, the most important of which is that their response is highly phasic. When one of these neurons is stimulated with a pair of FM pulses, it responds to each sound pulse by generating a single spike, or at the most two (Fig. 6.14*a*).

Repeated presentation of the stimulus shows that the delay between stimulus and response, the response latency, remains very consistent from one presentation to the next. The neuron's recovery is also sufficiently rapid that its response to the second pulse is essentially independent of that to the first pulse. As a result, the time interval between the two pulses is coded with remarkable precision in the pattern of spikes in the interneuron (Fig. 6.14*b*). Furthermore, the response latency remains almost constant regardless of the intensity of the sound stimulus, which is a crucial specialisation for time marking in echolocation. The rate of recovery remains fast enough for the neurons to respond consistently to the second pulse, at short pulse intervals, even when the first pulse is as much as 30 dB louder than the second (Pollak *et al.*, 1977; Pollack, 1980).

Each of these interneurons shows an exceptionally sharp tuning to a particular frequency within the FM sweep used for echolocation. This can be seen in their threshold curves, which have nearly vertical slopes on either side of the best frequency rather than the more usual V-shaped curve. As soon as the intensity of the appropriate frequency rises a few decibels above threshold, the neuron produces its single spike. Consequently, the neuron is fired promptly by the same frequency component in each sound pulse and so locks on to each event with consistent precision. This is shown nicely by stretching the sound pulse used as a stimulus over a longer time: for each stimulus duration, the neuron responds at the same relative position in the FM sweep (Fig. 6.14*b*).

Echo ranging in the auditory cortex

Information about time of arrival provided by these neurons is passed to the next stage of the auditory pathway, particularly to the auditory region of the cerebral cortex. Neurons that are sensitive to time delays between the outgoing pulse and the returning echo have been found in the auditory cortex of all echolocating bat species tested. Such neurons are well suited to measuring the range of a target, based on the information provided by the time-marking neurons of the inferior colliculus. It is often the case that these cortical neurons are selective for different frequencies in the pulse and in the echo. This may reflect the fact that, during active flight, the echo will inevitably be Doppler shifted to a higher frequency than the outgoing pulse. The equivalent point in time for pulse and echo will therefore be indicated by separate time-marking neurons in the inferior colliculus.

The delay-sensitive neurons of the auditory cortex respond vigorously to paired FM pulses simulating natural pulse–echo pairs but hardly respond at all to FM pulses presented singly or to CF pulses. The response consists of a short train of spikes and occurs, in the majority of neurons, only if the time interval between pulse and echo is appropriate. If one of these neurons is presented with a sequence of simulated pulse–echo pairs, in which the echo delay is progressively reduced to simulate the bat's approach to a target, it responds strongly only to a narrow range of delays (Fig. 6.15a). By systematically varying the echo delay and recording the neuron's responses, it can be seen that each of these neurons is sharply tuned to a particular echo delay, termed the best delay (Fig. 6.15b).

In the moustached bat, *Pteronotus parnellii*, which uses combined CF–FM pulses, the auditory cortex is organised in a way that has made it possible to follow the neural processing of target range. The delay-tuned neurons are located within a specific area of the auditory cortex that is quite distinct from other areas, which are arranged tonotopically. In the echo-processing part of the auditory cortex, neurons having similar best delays are grouped together in inwardly directed columns, which are arranged systematically with best delay increasing along the cortical surface from anterior to posterior (Fig. 6.15c). The delay-tuned neurons are thus spatially arranged according to the distance they encode and so provide a neuronal map of target range.

The echolocation pulses of *Pteronotus* contain three higher harmonics in addition to the fundamental frequency (see Fig. 6.10 for harmonics in the cry of *Rhinolophus*). Each delay-tuned neuron responds to the fundamental frequency of the outgoing pulse and to only one of the three harmonics in the echo (H2, H3 and H4 in Fig. 6.15c). By varying the intensity of the simulated echo when testing these neurons, it has proved possible to measure the

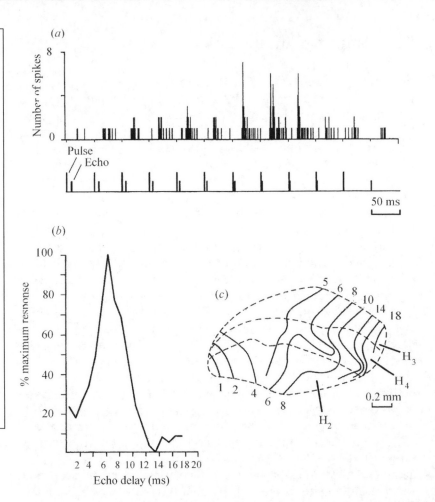

Fig. 6.15 Echo-ranging neurons in the auditory cortex of bats.
(a) Histogram (above) summarising the response of an echo-ranging neuron from *Myotis* to a sequence of pulse–echo pairs (below) simulating the natural approach to a target.
(b) The response of another neuron from *Myotis*, expressed as a percentage of the maximum number of spikes, as a function of the delay between a simulated pulse and echo of constant amplitude. (c) FM area of the auditory cortex in *Pteronotus*, showing the systematic distribution of echo-ranging neurons according to their best delay. The solid lines labelled with a number are contours of best delay in milliseconds. There are three horizontal clusters of neurons, indicated by broken lines, each tuned to only one of the three harmonics (H2, H3, H4) in the echo. (Part (a) redrawn from Wong *et al.*, 1992; (b) redrawn from Sullivan, 1982; (c) redrawn from O'Neill and Suga, 1982.)

threshold of the response at each delay tested and so to construct a threshold curve. A conspicuous result is that each of these curves has an upper threshold as well as a lower one, which means that the neuron fails to respond if the echo is too loud as well as if it is too quiet. Also, neurons tuned to shorter delays tend to have a higher threshold coupled with a narrower delay range. Each of the delay-tuned neurons thus picks out a particular combination of time delay and intensity appropriate to echoes returning from a certain distance (O'Neill and Suga, 1982).

The best delays in the FM area of the auditory cortex of *Pteronotus* cover a range from 0.4 ms at the anterior edge to 18 ms at the posterior edge (Fig. 6.15d), which corresponds to target ranges from 7 cm to 310 cm. This agrees closely with the range over which bats are observed to detect and react to targets. An especially large number of neurons, reflected in cortical surface area, is devoted to delays from 3 ms to 8 ms (50–140 cm), corresponding roughly with the approach stage of target interception.

Especially with the larger values of echo delay, it is obvious that the neuronal response to the emitted pulse must be considerably delayed if it is to reach the cortex at the same time as the response to the echo. In fact a large range of response latencies are found among the time-marking neurons of the inferior colliculus. It is therefore possible that the neuronal pathways through the colliculus are acting as delay lines and the cortical neurons are effectively coincidence detectors for particular combinations of pulse and echo latencies. The neuronal map of target range could thus be assembled in a similar manner to the barn owl's map of interaural time differences, although it is experimentally hard to establish the existence of delay lines, and in mammals time differences between left and right ears are established by a different mechanism from the one that seems to operate in birds (Grothe, 2003).

Since the FM area of the auditory cortex is sharply separated from the tonotopic area in *Pteronotus*, it is possible to inactivate them separately with drugs. As with the corresponding experiments in the barn owl's brain, this provides a test of the behavioural function of these regions. When the FM area is inactivated with the drug, fine discrimination of target range is impaired, although coarse discrimination is still possible, and frequency discrimination remains unaffected. This confirms that the FM area is involved in the perception of distance, as expected from the responses of its interneurons and their topographical arrangement, but it appears to have little to do with frequency discrimination (Riquimaroux *et al.* 1991). The opposite result is obtained when the tonotopic area is inactivated with the drug, and this area appears to be specialised for analysis of the Doppler-shifted CF signal.

However, the auditory cortex of some bats that use only FM pulses, such as *Myotis*, is not arranged in such an orderly way. The auditory cortex of *Myotis* is dominated by a single tonotopic area, in which the interneurons are arranged according to their best frequency. At high stimulus rates, most of these neurons also show delay-tuned responses to stimulus pairs. They even show an upward shift in their best delay as the pulse repetition rate increases. The functional organisation of the auditory cortex is thus able to change dynamically to meet the requirements of echolocation, and there is no clear map of best delay across the auditory cortex. This way of handling information about target range evidently represents quite a different computational strategy from that found in CF–FM bats like *Pteronotus* (Wong, 2004).

Auditory specialisations for Doppler shift analysis

All echolocating bats need to be able to analyse sound frequencies accurately. The ability to distinguish different frequencies is certainly excellent in bats that use only FM signals, such as *Myotis*, but is not so very different from those of other mammals. In contrast, bats that use long CF signals, such as *Rhinolophus*, have quite exceptional frequency resolution that greatly exceeds the abilities of non-echolocating mammals. The high resolution is for sounds with frequencies near to the CF part of the cry, particularly slightly higher frequencies. This enables CF bats to obtain an accurate measure of the Doppler shift.

Specialisations for fine frequency analysis begin at the inner ear, with the basilar membrane (Fig. 6.12a). As in birds, mammalian hair cells are tuned to particular frequencies according to their position on the basilar membrane, typically with equal lengths of the membrane being devoted to each doubling of frequency. But in the greater horseshoe bat (*Rhinolophus*), the representation of frequencies from 80 kHz to 86 kHz is greatly expanded, and the greatest expansion is found at the reference frequency of 83 kHz. This expanded representation on the basilar membrane is reflected in the number of first-order neurons that innervate the hair cells. Compared to frequencies below 70 kHz, which are not involved in echolocation, frequencies of the expanded region are overrepresented approximately ten times, with the result that 21% of all first-order auditory neurons represent the frequency range from 80 kHz to 86 kHz (Fig. 6.16a). This expanded representation is termed an 'acoustic fovea'.

The fovea is related to structural specialisations in the basal part of the basilar membrane, where the highest frequencies are represented. These structural peculiarities abruptly disappear at a distance of 4.5 mm from the oval window, and the expanded frequency region is located immediately beyond this critical point. As a result, the basilar membrane acts as a mechanical filter, which tunes a disproportionate length of the membrane to a narrow frequency band (Vater *et al.*, 1985). A consequence of this arrangement is that the first-order neurons innervating the hair cells within the foveal region are each extremely sharply

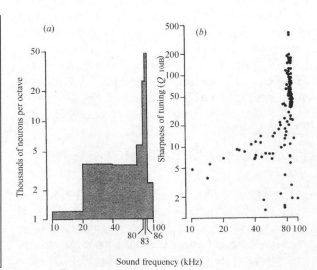

Fig. 6.16 The acoustic fovea in the greater horseshoe bat (*Rhinolophus ferrumequinum*). (a) Histogram of frequency representation among first-order auditory neurons, showing the great overrepresentation of frequencies around 83 kHz. (b) The sharpness of tuning, expressed as the Q 10 dB value, for neurons with different best frequencies in the cochlear nucleus, showing the exceptional sharpness of tuning in neurons with best frequencies around 83 kHz. See also Fig. 6.11(c). (Part (a) redrawn from Bruns and Schmieszek, 1980; (b) redrawn from Suga et al., 1976.)

tuned. A measure of the sharpness of tuning is provided by the Q 10 dB value, which is the neuron's best frequency divided by the bandwidth of its threshold curve 10 dB above the minimum threshold. Very high Q 10 dB values between 50 and 200 are found in the foveal region and, in the region of greatest expansion around 83 kHz, even values of over 400 are found (Fig. 6.16b). For frequencies below 70 kHz, the Q 10 dB values fall below 20, which is within the range found in other mammals.

This overrepresentation and sharp tuning are conserved at all higher levels of the auditory pathway. During echolocation, Doppler-shift compensation serves to clamp the echo of the CF component within this expanded frequency range as a bat flies among vegetation, in which tree trunks and branches reflect relatively loud echoes. In effect, this behavioural response creates a constant carrier frequency in the echo, on which the small frequency modulations produced by the wingbeats of flying insects are superimposed, so enabling them to be analysed by the sharply tuned neurons. A similar combination of peripheral tuning and Doppler-shift compensation has evolved independently of the horseshoe bats in the moustached bat (*Pteronotus*) and this combination therefore probably represents a general strategy for Doppler shift analysis.

The kinds of echo modulation produced by flying insects has been examined by directing at a fluttering insect a simulated horseshoe bat cry, made with a loudspeaker broadcasting a pure tone of 80 kHz. Recording the resulting sound with a microphone shows that the fluttering wings of an insect produce a strong echo or acoustic glint only when they are approximately perpendicular to the impinging sound waves, which happens for a short moment in each wingbeat cycle, but there are no acoustic glints from non-flying insects. A glint consists of a momentary increase in echo amplitude and a concomitant broadening of echo frequency, which represents Doppler shifts caused by the movement of the wings with respect to the sound source.

In echoes returning from a flying insect, glints modulate the echo at a rate corresponding to the wingbeat frequency of the insect, and this is termed the modulation frequency. The extent of frequency modulation involved is termed the modulation depth and is generally between 1–2 kHz above or below the carrier frequency. By perceiving these glints, a horseshoe bat should be able to distinguish with certainty between echoes from a fluttering insect and echoes from inanimate objects. Neurons specialised to encode the acoustic glints produced by flying insects are found among those processing the CF echo (foveal) frequencies in all the major staging posts of the auditory pathway.

At the level of the inferior colliculus, the overrepresentation of the CF echo frequencies is actually enhanced, being approximately 24 times that for frequencies below 70 kHz. Consequently, the normal tonotopic arrangement is distorted by the substantial proportion of interneurons devoted to the CF echo frequency range. Within this group, the neurons are very sharply tuned and the great majority are extremely sensitive to small frequency modulations. When stimulated with sinusoidal frequency modulations that sweep as little as ± 10 Hz around the 83 kHz carrier frequency, these neurons respond with discharges that are phase locked to the frequency modulation. The response remains phase locked at all modulation frequencies up to about 500 Hz, but some neurons show a preference for rates between 20 Hz and 100 Hz. The ability of these collicular neurons to follow the complex modulations produced by real wingbeats is confirmed by using the recorded echoes from a flying moth as stimuli in the experiments. With this natural stimulus, the neurons reliably encode the wingbeat frequency of the moth, as well as more subtle features of the echo (Pollak and Schuller, 1981; Schuller, 1984).

The encoded information is passed on to the tonotopic area of the auditory cortex, within which neurons devoted to the CF echo frequencies form a disproportionately large region, often termed the CF area. Neurons in the CF area respond to sinusoidal frequency modulations in much the same way as the collicular neurons but are more selective in the range of modulation to which their response is phase locked. Most of them prefer modulation depths around ± 1 kHz and modulation frequencies of 100 Hz or less, with a strong preference for frequencies between 40 Hz and 70 Hz. Among the nocturnal moths that are potential prey for horseshoe bats, many species have wingbeat frequencies around 40 Hz to 60 Hz, and such moths produce acoustical glints with a modulation depth of about 1 kHz. The phase locking of the cortical neurons is thus tuned to a behaviourally relevant range, in contrast to the wide range of responses found in the collicular neurons.

Behavioural observations confirm that horseshoe bats do in fact detect their natural prey by means of these modulations of the CF echo. Newly caged horseshoe bats only pursue insects that are beating their wings and the bats ignore stationary insects or those walking on the sides of the cage. The caged bats will take dead, tethered insects when these are associated with an artificial wingbeat simulator placed nearby. The crucial role of the CF signal is also shown by observations on horseshoe bats foraging in flycatcher style in natural forests: while hanging on twigs, they scan the surrounding area for flying insects and take off on a catching flight only after the prey has been detected. During this stationary scanning for prey, the bats emit long CF pulses without any FM component, indicating that prey detection is based on the CF signal alone (Link *et al.*, 1986; Neuweiler *et al.*, 1987).

Conclusions

Hearing in owls is adapted to process the noises made by their prey since owls depend on listening only, but hearing in bats is adapted to process the echoes of their own cries since bats hunt and navigate by echolocation. Bats and owls have auditory systems that are dedicated to the task of detecting and localising prey by means of sound. In both groups, specialisations are found at many levels of the auditory pathway that enable spatial information about the prey to be extracted from measurement of the intensity, frequency and time of arrival of sound at the ears. Individual neurons act as feature detectors, filtering out, for example, particular differences in the time of arrival of a sound at the two ears in an owl, or a particular difference in the time of a cry and then an echo in a bat.

Brain areas that encode sound frequency are arranged tonotopically, an arrangement that is produced by simple topographical projection from the array of receptors in the cochlea. But the representation of acoustic space or target range is synthesised by neuronal interactions and does not arise by topographical projection of the receptor array. Discrete parallel pathways extract different types of information from the simple frequency and intensity coding carried out by the receptors. Individual neurons at the higher levels then respond only to a specific combination of parameters synthesised by neuronal interactions. Thus, they act as feature detectors for specific values of behaviourally relevant information, such as angular location or target range.

Questions

What kinds of experimental test could you use to confirm whether delay lines operate in a nervous system?
What are the advantages and disadvantages of using echolocation?

Summary

- Many species of owls and bats rely on hearing to locate prey, and their auditory systems are specialised to enable them to do that.

Owls
- Owls locate sounds by comparing signals between the two ears: intensity for elevation and timing for azimuth.
- Space-specific neurons in the external auditory nucleus of the midbrain respond to sounds from particular locations. They are arranged in an orderly way, forming a map of space.

- Sensory neurons from the ear encode information about sound frequency, time and intensity. The brain uses this information to compute the receptive field for each space-specific neuron.
- Information about relative intensity and timing of sounds at the two ears is processed in a separate pathway on either side of the brain.
- Sound intensity is compared in the angular and then the posterior leminiscal nuclei.
- Sound time is compared in the laminar nucleus, in which neurons act as coincidence detectors receiving inputs from axons of the left and right magnocellular nuclei, which act as delay lines.
- The auditory map of space is calibrated by reference to a visual map in the optic tectum.

Bats

- Most insectivorous bats use echolocation, monitoring the echoes of their own cries to navigate and to find insect prey.
- In an FM cry, sound frequency alters to give a broadband signal, suitable for target ranging and description. In a CF cry, sound frequency is constant, useful for detecting relative velocity and for hunting in woodland.
- As a bat detects and then intercepts prey, the rate of echolocation pulses increases dramatically.
- A bat's auditory system is specialised to detect faint echoes that follow loud cries: it is most sensitive to frequencies near that of the cry; it is highly directional; and shows reduced sensitivity to a loud cry compared with a flowing soft echo, both in the ear and in the brain. Also, bat auditory neurons are quick to recover their responsiveness following each sound.
- Some neurons in the inferior colliculus act as accurate time markers, signalling the exact time of a cry and then of its echo.
- In the cerebral cortex, time-marking neurons provide inputs to neurons that are sensitive to particular cry–echo delays. In some bats, these neurons are arranged in an organised map in which a neuron's location is related to the distance between a bat and a sound-reflecting target.
- Bats that use CF cries have an acoustic fovea in their ears, and a large number of brain neurons dedicated to analyse echoes of the CF cry.
- In the auditory cortex of a CF bat are neurons that detect flying insects by responding to small modulations in the frequency of echoes from CF cries.

Further reading

Knudsen, E.I. (2002). Instructed learning in the auditory localization pathway of the barn owl. *Nature* **417**, 322–328. Reviews the evidence on how the auditory space map is adjusted by input from the visual system over the lifespan of the barn owl.

Suga, N. (1990). Biosonar and neural computation in bats. *Scientific American*, **262**, 34–41. Deals almost exclusively with *Pteronotus*, describing its echolocation sounds and showing how both the FM and CF components are represented in the auditory cortex.

Thomas, J. A., Moss, C. F. and Vater, M. (2004). *Echolocation in Bats and Dolphins*. Chicago University Press. Although bats and dolphins live in very different environments, both use echolocation. This volume, with chapters by many leading researchers, describes what is known about echolocation in each group.

Ulanovsky, N. and Moss, C. F. (2008). What the bat's voice tells the bat's brain. *Proc. Natl. Acad. Sci. USA* **105**, 8491–8498. A comprehensive article that reviews bat echolocation, and stresses the importance of studying animal behaviour in a natural setting.

Programmes for movement: how nervous systems generate and control rhythmic movements

Sequences of muscle activity in locomotion are basic building blocks for much of an animal's behavioural repertoire, so understanding the mechanisms which generate and control them is fundamental to a knowledge of the neuronal control of behaviour. Many movements used for locomotion are rhythmically repeating, and there are three basic questions about the control of movements such as walking, flying or swimming. First, what mechanisms ensure that muscles contract in the appropriate sequence? In walking, for example, the basic pattern is repeated flexion and then extension of each leg, with flexion of the left leg coinciding with extension of the right. Second, how does a nervous system select, start and end a particular type of movement? For example, what initiates the pattern of walking; and how is walking rather than running or swimming selected? Third, how is the basic pattern for movement modulated appropriately? Stride pattern changes, for example, when a person walks up a flight of steps or turns a corner.

Experimental approaches to these questions have often involved work on lower vertebrates and invertebrates, animals in which the parts of the nervous system that generate programmes for movement contain a limited number of neurons. This offers the opportunity to identify and characterise the individual components involved in generating a particular movement. A question that has occupied many investigators, and the one on which this chapter focuses, is to determine the source of the repetition that underlies rhythmic movements. One possible source is that proprioceptive reflexes, like those described in Chapter 2, could form a chain between different parts of a sequence. For example, a limb's position at the end of one part of a sequence could trigger both the ending of that part and the start of the next part. Alternatively, the central nervous system itself could produce appropriate patterned activity in motor nerves, and its ability to do this has been demonstrated many times in experiments where normal connections made by the central nervous system with muscles and sense organs are cut. The name **central pattern generator** is given to a network of neurons that can generate sequences of movements without the need for proprioceptive feedback from sense organs. The relative importance of central pattern generators and proprioceptive reflexes has been much debated, but both are clearly important for an animal to generate effective movements. The timing cues for activating different muscles are provided both by pattern-generating mechanisms in the central nervous system, and by proprioceptive feedback, a dual system that ensures ruggedness in the overall control of a rhythmical movement.

The concept of central pattern generators matches that of action patterns, which early ethologists developed. Specific behaviour patterns might be generated by particular networks of neurons responsible for a particular pattern of motor neuron activation. Central pattern generators use a number of different mechanisms to generate motor programmes, but there are two general ways in which rhythmical, repeating programmes are generated. The first is by small networks, such as a pair of neurons that inhibit each other so that excitation alternates between the two neurons; and the second is by **pacemaker neurons** that, like mammalian heart cells, have the intrinsic property of generating regular, clock-like waves of excitatory depolarisation followed by repolarisation of their membrane potentials without any synaptic input from other neurons. The pattern that emerges is usually not fixed, but can be altered, for example to allow an animal to steer around obstacles, to cope with changes in terrain, or to speed up and slow down.

In this chapter, we first describe the way in which regular rhythm is generated by describing research on swimming movements in newly hatched frog tadpoles. These animals have several advantages as experimental animals – for example, their nervous systems contain fewer neurons than the adults, and their spinal cord is fairly readily accessible. The existence of a central pattern generator for swimming by vertebrates was first established in experiments on dogfish in which repeated left–right alternating excitation of motor neurons occurs when muscular movements are prevented with the drug curare, which blocks synaptic transmission to motor neurons and muscle cells (Roberts, 1969). In young tadpoles, the basic mechanism responsible for repeated alternation between motor neurons on the left and right sides is well understood, and provides a relatively simple example of a central pattern generator in which networks of interneurons inhibit each other, leading to alternating activity. Second, we address the question of how power by muscles can be regulated by describing how larval zebra fish, which swim in a similar way to young tadpoles, alter their speed. Third, we examine how a behaviour such as swimming can outlast the initial stimulus that triggered it by referring to swimming by a sea slug. Fourth, we describe how neuronal networks can be reconfigured to generate new activity patterns. This is well understood by study of a small group of neurons responsible for controlling movements of the teeth and foregut of spiny lobsters and crabs, which have been particularly intensively investigated, although movements of the teeth and stomach might seem to be rather remote from the study of animal behaviour. Finally, we describe a complex rhythmical behaviour and show how proprioceptors and other sensory neurons steer a rhythmic behaviour. Our example is locust flight, which has generated several basic concepts of the way movements are controlled.

Swimming by young *Xenopus* tadpoles

The most complex patterns of behaviour occur in vertebrates, and understanding the neuronal basis of these behaviours must involve unravelling the networks of the spinal cord. To do this, Alan Roberts and colleagues have investigated swimming by newly hatched tadpoles of the clawed frog, *Xenopus* (Roberts *et al.*, 2008a, b), a motor pattern that is relatively simple compared with movement patterns of other animals. They have addressed the question of how the regularly repeating rhythm of alternating activity on the left and right sides is generated. For the first few days of life, *Xenopus* tadpoles spend almost all of their time attached to leaves and stems by a cement gland on the head. If the skin is touched, a tadpole will swim by side-to-side undulating movements of its trunk and tail until it finds a new attachment site (Fig. 7.1*a*). At the start of a swim, the body flexes to the left and then to the right up to 20 times per second, and this rate of flexion declines during a swim. Waves of movement travel towards the tail end and propel the tadpole forwards at up to 5 cm/s. A brief stimulus lasting only a millisecond can be sufficient to trigger an episode of swimming that lasts several minutes. Swimming usually ends when the tadpole's head touches a solid object, such as a plant stem or leaf.

The spinal cord of a young tadpole contains about 1000 neurons, including about half a dozen types of interneuron. Cell bodies of each type of neuron aggregate to form columns of 100–200 neurons along the length of the spinal cord (Roberts, 2000). For example most of the motor neuron cell bodies lie in a column in the ventrolateral

Photograph of a *Xenopus* tadpole at stage 37–38 of development, an age when it starts to feed. Scale bar, 1 mm long. (Photograph provided by Alan Roberts, Bristol University.)

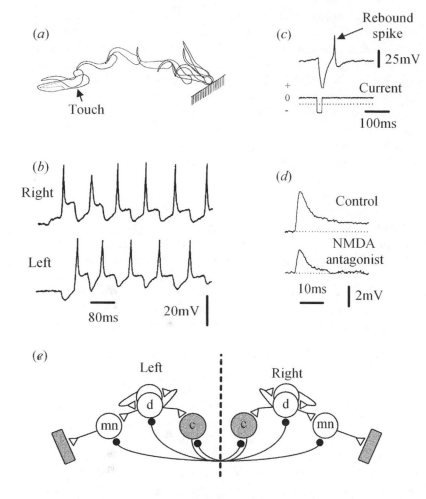

(a) Touch

(b) Right Left
80ms 20mV

(c) Rebound spike
25mV
+ 0 −
Current
100ms

(d) Control
NMDA antagonist
10ms 2mV

(e) Left Right
d mn c c d mn

Fig. 7.1 The generation of the swim pattern in *Xenopus* tadpoles. (*a*) Tracings from a video recording of a short swim by a young tadpole, initiated by touching its skin. (*b*) Intracellular recordings from interneurons of left and right sides of the spinal cord during swim-like activity. During the swim, both neurons were depolarised from the initial resting potential. The neurons spiked out of phase with each other, and received inhibitory postsynaptic potentials (IPSPs) in mid cycle. (*c*) Rebound spike in a spinal cord neuron. A small amount of steady, depolarising current was injected into the neuron, and a rebound spike was triggered when a brief pulse of hyperpolarising current ended. (*d*) Dual component excitatory postsynaptic potential (EPSP) in a motor neuron, caused by stimulating the axon of an interneuron. When the spinal cord was bathed in a solution containing an antagonist of the NMDA glutamate receptor, the slow and prolonged component of the EPSP disappeared. (*e*) Diagram of the network that is thought to generate the rhythm for swimming. The commissural interneurons (*c*) convey inhibitory signals across the nerve cord so that excitation of motor neurons (mn) alternates in the two sides, and the descending interneurons (*d*) are involved in circuits of mutual excitation. (Redrawn from Roberts, 1990.)

region of the spinal cord, and their axons gather at intervals to run along nerves, which are segmentally arranged, to muscles. In the experimental technique that Alan Roberts pioneered, a tadpole is immobilised with the snake venom α-bungarotoxin, which blocks neuromuscular transmission. Careful dissection enables access to neurons in the spinal cord. In older work, recordings were made from individual neurons by penetrating their cell membranes with sharp-tipped glass capillary microelectrodes, but a technique that yields better quality recordings is to apply a wider tipped electrode to a neuron's surface, then gently apply suction to break the cell membrane under the electrode tip and establish a tight seal between the electrode and the neurons. Whichever type of electrode is used, including a dye in the solution used to fill the electrode enables neurons that have been recorded to be stained for anatomical identification later on. With this type of experiment, recordings can be made from individual neurons while the immobilised tadpole's nervous system generates swim-like activity in its motor nerves.

When the skin is touched, left and right motor nerve activity alternates and spreads from head to tail. Because muscles cannot contract in this experiment, it demonstrates that the central nervous system contains a central pattern generator for swimming. A length of just 0.4 mm of spinal cord plus hindbrain is sufficient to generate and sustain a swimming pattern (Li *et al.*, 2006). Touching the skin excites sensory neurons called Rohon Beard cells and a single spike in one of these cells can be sufficient to elicit a bout of swimming. During a swim, each motor neuron produces just one spike per swim cycle, but when tadpoles mature sufficiently to feed, the pattern of activity during swimming becomes considerably more complex, and motor neurons generate spikes in bursts rather than singly (Sillar *et al.*, 1991). If a newly hatched tadpole is grabbed rather than just touched, it struggles by making strong side-to side movements at a slower rate than in swimming.

Both the way in which neurons are interconnected in a network and the special membrane properties of individual neurons shape the output of the central pattern generator. An essential feature of the swim pattern is that motor neurons driving the muscles on the left and right sides are excited alternately. This alternation is achieved by interneurons that have axons which cross the nerve cord from one side to the other – **commissural interneurons**. The commissural interneurons are excited in time with motor neurons on the same side as their cell bodies and dendrites and they inhibit neurons of the opposite side. Commissural interneurons on the left side, therefore, inhibit commissural interneurons on the right side and vice versa, a network interaction that is one of the most simple methods for generating a rhythmic behaviour. The mutual inhibition between the commissural interneurons ensures that muscles on the left and right sides take it in turns to be active. The inhibition by commissural interneurons is not just directed at their partners on the other side of the spinal cord, but at other types of interneuron and at motor neurons as well. The pattern of connection in which two neurons inhibit each other is called **reciprocal inhibition**. This kind of organisation was suggested in 1910 by Graham Brown to explain alternation between leg flexion and extension when mammals walk, and is commonly referred to as a half-centre model, in which the two halves of the network inhibit each other. It is a common mechanism for ensuring that **antagonistic** muscles that work against each other are activated one after the other rather than at the same time as each other.

During a swim, an interneuron or motor neuron remains depolarised from resting potential. It produces just one spike per swim cycle, and receives an inhibitory postsynaptic potential (IPSP) when neurons on the opposite side fire (Fig. 7.1*b*). The mid-cycle IPSP has a dual role. Besides ensuring that the neurons on one side of the cord are inhibited at the time when their contralateral partners are excited, the end of the IPSP triggers the next spike. A spike that is triggered at the end of an IPSP rather than by an excitatory postsynaptic potential (EPSP) is called a **rebound spike**, in which the neuron rebounds from inhibition and its membrane potential overshoots its resting level, triggering an active membrane response – a spike. A pulse of hyperpolarising current injected from a microelectrode can also be a trigger for a rebound spike. However, rebound spikes are not initiated unless the neuron is excited by steady, excitatory input that depolarises it from its normal resting potential, and the neurons involved in swimming in newly hatched tadpoles do not normally generate more than one spike at a time.

For a tadpole swim neuron to generate a rebound spike, it must receive a steady excitation to depolarise it from its resting potential. This excitation can be provided by steady depolarising current from a microelectrode upon which hyperpolarising pulses are superimposed (Fig. 7.1c) or by synaptic excitation, which is its natural source. In a swimming tadpole, the background excitation comes from excitatory interneurons. The steady depolarising potential in a swim neuron sets up a balance between ionic currents carried through two types of voltage-sensitive channels. The first type are the familiar sodium channels responsible for the rapid depolarising phase of a spike. The second type are potassium channels, through which positively charged potassium ions leave the neuron, similar to those that cause the downward phase of a spike in an axon. The current carried outwards from the neuron by potassium ions increases as the neuron depolarises, opposing the excitatory action of the sodium channels. The effect of a brief hyperpolarisation is to close both types of channel. When the hyperpolarisation ends, the sodium channels open more rapidly than the potassium channels, triggering the generation of a single, rebound spike. The steady depolarisation is necessary for the generation of a rebound spike because it enables the sodium channels to open following an IPSP and it is provided during swimming by a population of interneurons that excite motor neurons on their own side of the nerve cord. In the spinal cord, these interneurons carry excitation in the direction of the head. A similar group of interneurons in the hindbrain, however, carry excitation in both directions along the body axis, and excite each other, forming a network that helps to sustain swimming (Li *et al.*, 2006).

Another factor that sustains excitation throughout a swim is that the EPSPs which the excitatory interneurons cause have relatively long durations, up to 200 ms, which is much longer than one swim cycle. This is because the EPSPs have two different phases, each mediated by different types of receptor in the postsynaptic membrane for the neurotransmitter that the excitatory interneurons release, which is the amino acid glutamate. The two phases of an EPSP can be dissected apart by pharmacological agents (Fig. 7.1d). The first phase can be mimicked by applying the drugs kainate or quisqualate to the surface of a postsynaptic neuron, while the second, slower phase can be mimicked by applying another drug, N-methyl-D-aspartate. Both of these drugs are **agonists** of glutamate – they bind to receptors and activate their ionic channels in a similar way to glutamate. The neurons, therefore, have two different types of receptor for glutamate at their input synapses: one type causes the initial, fast excitation, while the second type, called the 'NMDA' receptor, mediates the longer-lasting excitation. The fast

EPSPs help excite the motor neuron at the appropriate time in a swim cycle, reinforcing the rebound from an IPSP; and the slower EPSPs provide the excitation needed to sustain a swim.

A characteristic of the NMDA receptor is that its channel can only open if the postsynaptic neuron is already strongly depolarised, for example through the action of other synapses that excite the post-synaptic neuron. This means that NMDA receptors respond to combi-nations of signals rather than signals from a single source, a feature that suits them for a variety of tasks including some types of learning. In the tadpole spinal cord, strong synaptic excitation by the faster acting of the glutamate activated synapses is necessary to ensure that the longer-lasting, NMDA-mediated EPSPs can then switch on.

The mechanism for generating the swim pattern, therefore, depends on both network and cellular properties. Networks of neurons are connected in a way that ensures alternation of activity between left and right sides, and mutual excitation between some of the interneur-ons helps sustain a swim episode (Fig. 7.1e). Cellular properties underlie the sustained excitation of motor neurons by NMDA receptors, and the generation of rebound spikes. Simulation by computer has shown that networks that have these characteristics can generate an output pat-tern like the tadpole swim pattern (Roberts and Tunstall, 1990; Sautois et al., 2007). Some of the features of the swim generator of young tad-poles have also been discovered in neurons of the mammalian spinal cord, including rebound spikes and long-lasting EPSPs mediated by NMDA receptors, and so could be general features of motor programme generating mechanisms. However, reciprocal inhibition between the commissural interneurons is not the only rhythm-generating mecha-nism in the tadpole spinal cord, which is shown by experiments in which the spinal cord is split longitudinally, when it is found that each half can generate rhythmically repeating activity on its own.

Once swimming has started, what stops it again? An effective stimulus to stop swimming is contact between the head, particularly the cement gland, with another object such as a twig or leaf. This mechanical stimulus is detected by sensory neurons that project into the brain. The excitation is turned into inhibition because the sen-sory neurons excite a group of interneurons that send their axons from the hindbrain down the spinal cord, and these interneurons cause IPSPS in most, if not all, of the neurons involved in generating the swim pattern. A brief current pulse to cause a short burst of spikes in a single one of these brain interneurons is sufficient to shut down swimming completely (Li et al., 2006). During swimming, before the tadpole bumps into a solid object, the brain interneurons are inhibited. When the tadpole is at rest, suspended by a strand of mucus from its head, the brain interneurons are continually excited at a relatively low level. This low level of activity is probably sufficient to suppress a tadpole's swim generator so that it remains relatively immobile and unresponsive to gentle stimuli, rendering it less obvious to potential predators such as dragonfly larvae than if it were continually moving (Lambert et al., 2004).

Speed and power control: larval zebra fish

For effective behaviour, in addition to the need to avoid obstacles or steer towards a goal, animals often need to change the way in which muscles are used, for example to alter the force that a muscle produces or its speed of contraction. This may involve a change in the pattern of movement, as happens when a horse changes from a walk to a trot and then to a gallop. Fish and tadpoles can alter their swimming speed in a smooth way without changing gait, and the way a change in swimming speed is accomplished has been revealed by examining 4-day-old larval zebra fish, in research by Joseph Fetcho, David McLean and colleagues (McLean *et al.*, 2007; 2008). The rhythm for swimming in these animals is probably generated in a very similar way to young tadpoles, and research on young zebra fish has focused on the way in which swim speed is controlled.

A zebra fish takes 30 days after hatching to develop most adult characters, and 3 months to reach sexual maturity. At 4 days old, zebra fish larvae are still transparent so that muscles and nerves can be seen and stained neurons in the spinal cord can be seen under a microscope. Larval zebra fish usually swim with left–right waves at 20–40 per second travelling posteriorly along the body and with movements of the pectoral fins assisting the tail and trunk. During bursts of speed, undulation frequency can reach 80–100 per second and the pectoral fins are folded into the body. Adult fish swim with somewhat slower movements. When a larva is startled, it starts with a burst of speed with a rapid rhythm and then gradually slows down in an escape movement that lasts only a few hundred milliseconds. The power of the movements decreases along with the swimming speed. Figure 7.2 shows recordings from motor nerves in two different segments during a swim caused by a brief shock. The experiment was similar to those performed on *Xenopus* tadpoles – the larva was laid on its side and bathed in a solution containing snake venom to block transmission between motor neurons and muscle cells. To record from a motor nerve, a short length of it was sucked into a

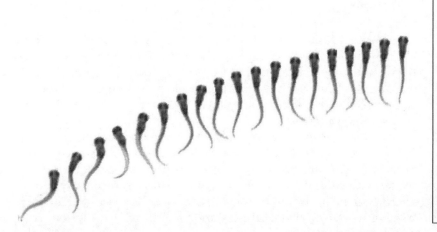

Frames superimposed from a high-speed video to show how left-to-right tail movements decline in power and rate during an escape swim by a zebra fish larva (the swim followed on from the initial fast start shown in an earlier photograph in Chapter 4). As swim speed changes, different sets of motor neurons and interneurons are activated. The frames taken from the high-speed video are 8 ms apart, so the whole event shown lasted 144 ms. (Photograph supplied by David McLean, Northwestern University, Illinois.)

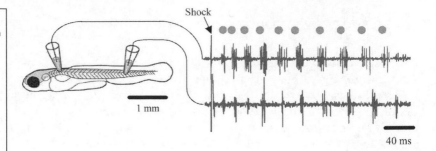

Fig. 7.2 Recording motor nerve activity from a short swim episode in a larval zebra fish. The fish was immobilised with a drug, and two suction electrodes recorded from the nerves of segments 4 (left electrode and top trace) and 24 (right electrode and bottom trace). A brief electric shock triggered a quarter-second-long swim, during which bursts of motor neuron axon spikes were picked up by each electrode. The dots indicate the start of each burst in segment 4. During the swim, the frequency of bursts declines, and the delay between a burst in segment 4 and the corresponding burst in segment 24 increases. (Redrawn from McLean *et al.*, 2008.)

small pipette containing a silver wire to act as an electrode. The grey dots in Fig. 7.2 mark the start of each burst of motor nerve activity for the most anterior muscle. During the short swim, the duration of swim cycles slows down, from 14 ms at the beginning to 30 ms at the end, and also the delay between the anterior and posterior muscles lengthens.

In the larval fish, the first motor neurons to develop in the spinal cord are called primary motor neurons, and there are two or three of these serving each segment on each side of the body (the fish has about 30 segments altogether along its trunk and tail). Subsequently, secondary motor neurons develop, and occupy a more ventral position in the spinal cord. Each half segment is served by about 20 secondary motor neurons, which have smaller cell bodies and narrower axons than the primary motor neurons. In a 4-day-old fish, a burst of swimming first involves all the motor neurons, and the larger, dorsal motor neurons are the first to drop out. As the swim episode continues and slows, more motor neurons drop out, and the order in which they drop out corresponds with a dorsal-to-ventral gradient in the spinal cord. Each motor neuron excites its own pool of muscle fibres, so that, as might be expected, the number of muscle fibres that are active decreases as swim speed decreases. In addition, the power of a motor neuron's muscle fibres is related to the motor neuron's size.

This mode of controlling the strength of a muscle's contraction is called the **size principle**, which was first established in mammals (Henneman *et al.*, 1965) but applies to many other animals, both vertebrates and invertebrates. According to the size principle, the smallest motor neurons are recruited by relatively weak synaptic excitation and they control relatively weak and slow contractions by muscle fibres. Stronger synaptic excitation recruits increasingly larger motor neurons, each of which controls increasingly strong and fast muscle fibres. Not only does this kind of recruitment in order of size enable regulation of muscle power, it also enables a considerable degree of fine control.

The size principle is shown in operation in the recordings from a larval zebra fish in Fig. 7.3*a*. Microelectrodes were used to make intracellular recordings from two different motor neurons: one of the larger, primary motor neurons, and a smaller, secondary motor neuron. In each case an episode of escape swimming activity was triggered by a brief electrical stimulus. During the episode the

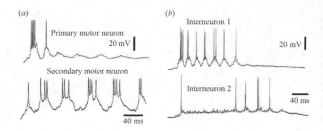

Fig. 7.3 Speed and power change during swimming in motor neuron and interneurons of larval zebra fish. (*a*) Intracellular recordings from a primary and a secondary motor neuron during a swim episode. The frequency of excitation cycles declines during the swim, as in Fig. 7.2, but the primary motor neuron is only sufficiently excited to spike during the early part of the swim. (*b*) Intracellular recordings from two different types of interneuron during a swim episode. Interneuron 1 produces bursts of spikes in the first part of the swim. Interneuron 2 depolarises at the start of the swim, but does not make bursts of spikes until the last part of the swim. (Redrawn from McLean *et al.*, 2008.)

frequency of motor neuron bursts declined, as shown in the records of motor nerve activity in Fig. 7.2. The intracellular records in Fig. 7.3*a* show that the primary motor neuron made an intense burst of spikes on the first cycle, a single spike on the second, and experienced subthreshold waves of excitation on subsequent cycles. In contrast, the secondary motor neuron spiked with roughly equal vigour on each cycle after the first two, which were too brief for it to generate many spikes. This pattern reflects a decline in the strength of synaptic excitation to the motor neurons during the swim episode. The smallest, more ventral motor neurons are most easily excited by synaptic inputs so they are the ones in which excitation persists longest in a swim episode.

A different pattern of recruitment is found among the pre-motor interneurons, with different interneurons being active at different swim speeds, as shown in the recordings in Fig. 7.3*b*. At the start of the swim episode, one interneuron was excited but later in the episode, as the rhythm slowed, the other interneuron took over. Unlike the motor neurons, the order of recruitment of interneurons is not related to neuron size, although there is a correlation between location in the spinal cord and a particular speed of swimming movement. In one series of experiments, interneurons were selectively destroyed by introducing a fluorescent dye into them and then irradiating a small area of the spinal cord with a laser beam (McLean *et al.*, 2007). As a result of this treatment, a fish lost its ability to make either fast or slow swim movements, depending on which interneurons were killed. At faster speeds, the active interneurons are thought to excite both primary and secondary motor neurons, whereas at slower speeds the interneurons that are active excite only secondary motor neurons

Triggering and maintaining escape swimming in a sea slug

Often, a brief stimulus triggers a sequence of movements that last very much longer. The way that a response is sustained following a short-lasting stimulus has been elucidated in research on swimming in another kind of animal, the large sea slug *Tritonia*. Compared with the high rate of movement when a tadpole or larval zebra fish swims, the movements by which *Tritonia* swims are leisurely. It swims by arching its body, with dorsal and ventral flexion movements

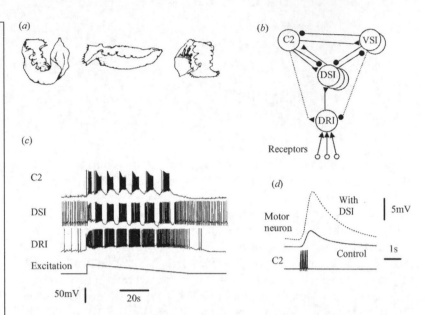

alternating, and does this to escape from starfish, the smell of which reliably triggers swimming. It is not a well-directed form of locomotion, but serves to move the sea slug upwards into a water current that would be likely to move the slug away from its adversary, a starfish. Some of the first intracellular recordings from nerve cells in an animal behaving in a more or less normal pattern were made from *Tritonia* (Willows *et al.*, 1973). By exposing the brain through a small incision and then fastening it to a support platform, intracellular recordings can be made from single neurons. A major attraction of *Tritonia* as an experimental animal is that, like some other gastropod molluscs, its brain contains some very large neurons with cell bodies more than 0.1 mm across and clearly visible near the brain surface. The animal is suspended in a tank of seawater and can perform normal muscular movements. This experimental technique allowed identification of motor neurons and interneurons that are involved in local withdrawal reflexes as well as in the dorsal and ventral movements that occur during escape swimming (Fig. 7.4).

A swimming episode lasts up to a minute, and consists of 4–7 cycles of spikes in dorsal and ventral flexion neurons. Three groups of interneurons on each side of the brain are involved in rhythm generation: three dorsal swim interneurons; two ventral swim interneurons; and one other interneuron called C2 (Fig.7.4*b*). None of these interneurons has an intrinsic capacity to generate rhythmic oscillations in membrane potential, and the rhythm is generated by a network of linked interneurons. In an isolated brain, electrical stimulation of a sensory nerve elicits a very similar pattern of intracellular activity to that which occurs in an intact animal that is swimming in response to starfish odour (Fig. 7.4*c*).

The swim neurons are strongly excited at the start of a swim. As the swim progresses, the excitation gradually diminishes and the cycle period becomes longer. A single interneuron on each side of the brain collects input and conveys the excitation to the swim interneurons (Frost and Katz, 1996). The interneuron is called the 'dorsal ramp interneuron' (DRI): the term 'ramp' refers to the way its excitation decays in a slow, descending ramp-like manner during a swim. As shown in Fig. 7.4c, excitation of the DRI precedes that of other interneurons before a swim starts. As a swim progresses, the DRI continues to spike, with a brief interruption during each ventral flexion caused by an inhibitory synapse from the ventral swim interneurons. Stimulating a DRI with enough current to make it spike at similar frequencies to those that occur during a swim will trigger swim activity in the network; and injecting hyperpolarising current into a DRI so it does not spike prevents the normal swimming response. The DRI, therefore, acts as a channel through which all sensory excitation must pass in order to activate the central pattern generator for swimming, and might be considered to be a 'command neuron' (Chapter 4).

Each DRI makes excitatory connections with the three dorsal swim interneurons (DSI), which together play a dual role in generating the programme for swimming (Katz et al., 1994). The first role of the DSIs is to contribute to the pattern of each cycle of the rhythm through their synaptic connections with the ventral swim interneurons and with C2. Their second role is to sustain the rhythm, which they are able to do because the neurotransmitter that they release, serotonin, has multiple effects within the network. Not only does it act as the neurotransmitter at the output synapses made by a dorsal swim interneuron, it also increases the excitability of C2 and enhances the release of neurotransmitter from C2. The serotonin acts as a synaptic transmitter by binding to and opening ion channels in the postsynaptic sites of target neurons. The other effects of serotonin are triggered when it binds to and activates G-protein receptors in the cell membrane, which initiates a chain of enzymic responses (Clemens and Katz, 2001). An example of the second type of effect is illustrated in Fig. 7.4d, which shows how the EPSPs that C2 mediates in a motor neuron increase in amplitude when a dorsal swim interneuron is excited. The facilitatory action which the dorsal swim interneurons exert on the output synapses from C2 is called 'intrinsic neuromodulation', because the neuromodulator is released by neurons that are intrinsic to the pattern generating network and is only released when the pattern generator is active. The neurons of the swim generator are also involved in less dramatic, local withdrawal responses. When swimming starts, the dorsal swim interneurons generate an intense burst of spikes, and it is likely that the serotonin that this releases ensures that the network is configured to generate co-ordinated, rhythmical swimming movements.

Network reconfiguration in lobsters and crabs

Compared with sea slugs, tadpoles or fish, animals with jointed limbs use a great variety of motor programmes. This increased complexity of movements might require increased diversity in the types of neuron and networks. However, the same neurons can participate in different pattern-generating networks, so an increase in complexity of movements may not require a large increase in the number of neurons involved. Research on the stomatogastric ganglion of decapod crustaceans illustrates very well how neuronal networks can be reconfigured extensively, allowing groups of neurons to participate in quite different activity patterns.

In lobsters and crabs, movements of the foregut are achieved by contraction of discrete striated muscles, more similar to the muscles that move limbs during movements of the body than to the smooth muscles that move food through the vertebrate alimentary tract. Quite complex, regularly repeating patterns of activity occur in these crustacean muscles. A major reason for investigating their control is that the neurons responsible for generating them are contained in four discrete, small ganglia (Fig. 7.5a). The most intensively investigated is the stomatogastric ganglion which, in lobsters, contains just 30 neurons. A small nerve connects it with three other ganglia of the stomatogastric nervous system. Almost all of the neurons in the stomatogastric ganglion are motor neurons and, unlike the situation for the three examples described earlier in this chapter, the motor neurons make connections with each other, making them essential components of the central pattern generator

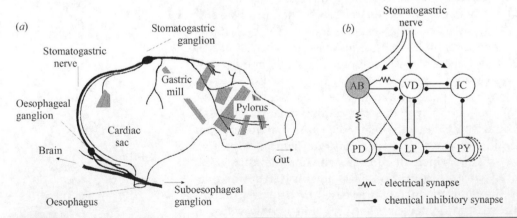

Fig. 7.5 The foregut and stomatogastric nervous system of a spiny lobster (*Panilurus*). (*a*) Foregut with its ganglia and nerves (black). The stomatogastric ganglion lies in a blood vessel (not shown) on top of the stomach. (*b*) Diagram showing some of the synaptic connections between neurons of the pyloric section of the stomatogastric ganglion. All except the anterior burster (AB) are motor neurons; the two pyloric dilator (PD) neurons control dilation of the pylorus and the other motor neurons control different phases of constriction. The stomatogastric nerve contains axons of neurons which have a variety of different effects on neurons of the stomatogastric ganglion. (Part (*b*) redrawn from Selverston and Miller, 1980.)

The stomatogastric ganglion controls two different parts of the foregut, called the gastric mill and the pyloric region. The gastric mill contains three teeth, which cut and grind food after it has been churned in a region called the cardiac sac. Muscles of the gastric mill are activated in a rhythm with a cycle period of 5–10 s. The pyloric region, into which the gastric mill empties, has a shorter cycle time, of 0.5–2 s. It squeezes and mixes food particles.

The 14 neurons responsible for the pyloric rhythm are connected in a network that includes over 20 electrical synapses and over 60 chemical, inhibitory synapses (Fig. 7.5b). One experimental approach that was particularly useful in revealing mechanisms for generating the rhythm was by isolating single cells from others in the ganglion by killing the cells that are normally presynaptic to them using a technique similar to that used in zebra fish spinal cords (Selverston and Miller, 1980). A single pair of neurons that are connected by reciprocal, inhibitory synapses is able to generate rhythmic activity (Miller and Selverston, 1982).

Neurons of the stomatogastric nerve, which contains 60–120 axons, play a vital role in pattern generation by unmasking particular properties that are intrinsic to individual cells. If this nerve is cut, the stomatogastric ganglion does not generate rhythmically repeating activity, but this can be restored by stimulating axons within the nerve or by adding certain transmitters, particularly amines or peptides, to the seawater bathing the ganglion. Because the pyloric neurons only burst in the presence of particular transmitters, they are called conditional oscillators. All of the neurons in the pyloric region are conditional oscillators, but one, the anterior burster (AB in Fig. 7.5b), has the fastest rhythm and acts as the master clock, setting the pace for the whole pyloric region.

One particular bilateral pair of neurons that project into the stomatogastric ganglion, called the pyloric suppressors (PS), have widespread effects on motor patterns (Meyrand et al., 1991; 1994). When the PS neurons are excited, the separate pyloric and gastric mill rhythms stop; and the pyloric and gastric mill neurons become active in a new, co-ordinated pattern of rhythmic activity (Fig. 7.6a). When the stomatogastric ganglion is expressing the pyloric and gastric mill patterns, the PS neurons are silent but they are excited by applying food to chemoreceptors on the valve that separates the oesophagus from the stomach. When a PS neuron is excited its membrane potential oscillates and it generates bursts of spikes. These bursts of spikes excite motor neurons which cause the valve to open, and intracellular recordings from pyloric and gastric mill neurons show profound changes in the activity patterns they express. Some neurons are tonically inhibited from producing spikes, but others, such as the pyloric and gastric mill neurons, express a new pattern in which neurons from both regions are co-ordinated.

The new pattern is probably that for swallowing, and the PS neurons also cause the valve between the oesophagus and stomach to open. During swallowing, movements of the whole foregut must be

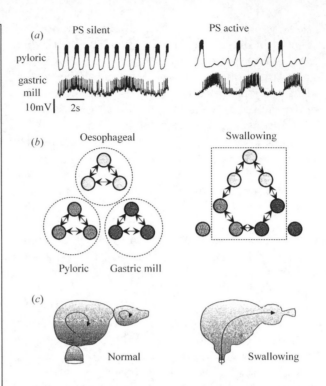

Fig. 7.6 Action of the pyloric suppressor neuron, PS, in reconfiguring the stomatogastric ganglion. (*a*) Intracellular recordings from a neuron of the pyloric region and a neuron of the gastric mill region of the stomatogastric ganglion. On the left, PS was silent and the two neurons expressed different rhythms of activity. When PS was stimulated, the two neurons became active in the same new rhythm. (*b*) Diagram to illustrate the effects of PS on the circuitry of the stomatogastric ganglion. When it is silent, networks for the oesophageal, pyloric and gastric mill rhythms operate independently of each other. Excitation of PS causes some neurons of these three networks to be incorporated into a new functional network, and silences others (dotted). (*c*) Behavioural correlates of PS activity. On the left, different foregut regions are acting independently and valves between them are closed. On the right, the lobster is swallowing. (Part (*a*) redrawn from Meyrand *et al*, 1991; (*b*) and (*c*) redrawn from Meyrand *et al.*, 1994.)

co-ordinated to move food onwards (Fig. 7.6*b*) and the swallowing pattern has a frequency between that of the pyloric and gastric mill regions. After the PS neurons have stopped firing spikes, the new pattern persists for several tens of seconds and the original two pyloric and gastric mill rhythms are slowly re-established. This observation is important because it shows that networks within the ganglion itself are reconfigured by activity in the PS neurons (Fig. 7.6*c*). If the stomatogastric ganglion neurons were all driven by bursts of spikes in the PS neurons during swallowing, we would expect the pyloric and gastric mill rhythms to be re-established as soon as the PS neurons stopped firing. The PS neurons reconfigure the networks of the stomatogastric ganglion, so that neurons which previously participated in different and unco-ordinated activities now express a new, single pattern.

Locusts and their flight

One of the reasons why insects are abundant and successful is their ability to fly. Locusts are capable of long migratory flights, covering several hundred kilometres a day at heights greater than 1000 metres. Insect flight is one of the most sophisticated forms of locomotion in the animal kingdom, allowing rapid movement in three dimensions, and involving precise co-ordination of many muscles with different actions. Insects react quickly to visual, olfactory, wind and other stimuli, and are excellent subjects for investigating how brain neurons and proprioceptors influence motor activity. Because some insects are relatively easy to maintain in laboratory conditions and are rugged experimental subjects, their flight is an important topic for investigating the neuronal control of behaviour. Like the other subjects of this chapter, insect flight is a rhythmic movement in which activity of different muscles alternates. In many insects, including flies, bees and beetles, the major power-providing muscles are stretch activated, and wing elevators and wing depressors activate each other mechanically, requiring occasional spikes in motor neurons to remain in the activated state. But in other insects, including locusts, moths and dragonflies, timing signals for each muscle contraction are provided by motor neuron spikes. Compared with human-made flying devices, locusts and other insects use a very large number of different types of sensors to regulate their flight. Several interneurons that play roles in generating and altering the flight motor pattern are known, including interneurons in the thoracic ganglia that are involved in the central pattern generator, and interneurons that carry commands from the locust's cockpit, in the head, to the flight control centres in the thorax.

The flight engine

As in most insects, flying is achieved by cyclical movements of four wings, borne on the posterior two segments of the thorax. As the wings move up and down, they twist so that they constantly generate lift to keep the insect airborne (Fig. 7.7a). Each wing is moved

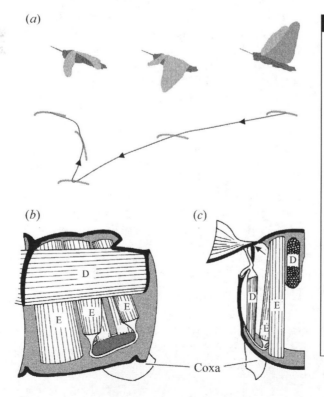

(a)

(b)

(c)

Coxa

Fig. 7.7 Movements and muscles involved in locust flight. (a) Movements during one wingbeat cycle. The drawings show three stages during a downstroke; the hindwings move slightly before the forewings. Below are shown the path of movement of the tip of the left forewing and the angle of the wing during one wingbeat cycle. The wing is twisted during the upstroke to help maintain lift throughout the cycle. (b and c) The main flight muscles of the third thoracic segment (bold outline in the left drawing, a). The right half of the segment is shown in medial view (b) and anterior view (c). D, depressor muscles; E, elevator muscles. The wing hinge is indicated by an arrow in (c). (Part (a) redrawn from Pringle, 1975; (b, c) redrawn from Snodgrass, 1935.)

by ten muscles, which can be divided into three main groups (Fig. 7.7*b*). The first group contains just one large muscle for each wing, oriented along the animal's long axis. When these dorsal longitudinal muscles contract, they distort the stiff cuticular box structure of the thorax in a manner that causes the wing tips to move downwards, and they are called 'indirect depressor' muscles. The other two groups of muscles lie upright in the thorax and pull more directly on the wing base. One group pull outside the fulcrum of the wing hinge, and so are direct depressor muscles, and the other group pull on the inside of the fulcrum and so are direct elevator muscles (Fig. 7.7*c*). The three largest direct depressor muscles of each wing control the way the wing twists around its long axis, and are important in altering the pattern of wing beat during steering manoeuvres.

The pattern of innervation of the wing muscles is straightforward because each flight muscle is usually controlled by just one or two motor neurons and a spike in a flight motor neuron mediates a rapid, strong twitch of its muscle. The motor neurons make synapses along the length of each muscle fibre, and these operate in the same way as synapses in the central nervous system. When the presynaptic terminals are depolarised by the arrival of a spike, each releases a tiny squirt of neurotransmitter, in this case glutamate. The ion channels in the muscle cell membranes that bind glutamate cause large EPSPs. Through a series of events, the electrical signal is transduced into the development of tension by individual muscle fibres.

The simple pattern of innervation of locust flight muscles contrasts with that found in many vertebrate skeletal muscles, in which large muscles are controlled by several tens or hundreds of motor neurons. The pattern of flight muscle innervation also contrasts with the innervation pattern of leg muscles in locusts because some of the leg motor neurons mediate small individual twitches that sum together gradually to develop graded, strong, slow contractions when the motor neuron generates a train of spikes. These slow motor neurons are particularly important in slow movements and in posture of the insect. Many leg muscles are also controlled by fast motor neurons that, like the flight motor neurons, generate rapid, powerful twitches. Other leg motor neurons are inhibitory and oppose the action of the slow excitors by causing hyperpolarising postsynaptic potentials in their muscle fibres.

The flight programme

Recordings that show the electrical signals that cause contractions of a muscle are called **electromyograms** (EMGs). The technique employed for recording EMGs during locust flight is similar to that

A locust making flying movements while fastened to a brass rod. The path taken by the forewing during a wing stroke is revealed by attaching a very thin light guide to the front edge of the wing. (Photograph supplied by Berthold Hedwig, Cambridge University.)

used for recording electrocardiograms in humans, except that the electrodes are fine wires inserted into the muscle through small holes in the cuticle and secured in place with glue or wax (Fig. 7.8a). It is possible to fit a locust with a tiny radio transmitter to record EMGs from pairs of muscles while a locust is flying unrestrained (Kutsch *et al.*, 1993; Fischer and Kutsch, 2000). More usually a locust is tethered, often to a solid bar but sometimes to a harness attached to counterweights so that some of the forces generated by the wing movements can be measured. It is quite easy to induce a suspended, tethered locust to flap its wings in a flight-like manner by blowing a current of air over the head that excites wind-sensitive sensory hairs. During straight, level flight the wingbeat frequency is between 15 and 20 per second. It tends to fall gradually during a long flight, perhaps because the locust needs less power after it has burned some of its fuel. Touching a foot immediately stops flight.

Each flight motor neuron usually produces one or two spikes per wingbeat and excitation of elevator motor neurons alternates with excitation of depressors (Fig. 7.8b). Over a range of different wingbeat frequencies, the delay between excitation of the elevator and depressor motor neurons is fairly constant (Fig. 7.8c) as is the duration of the depressor activity, and the most variable event during a wingbeat cycle is the duration of the burst of spikes in elevator motor neurons (Fig. 7.8c). Hindwing depressors spike 5 ms before forewing depressors, although elevators of all four wings are active synchronously. The result is that the hindwings are depressed before the forewings in each wingbeat cycle, a sequence

Fig. 7.8 The motor programme for locust flight. (*a*) The locust is tethered to a rod, and induced to fly by wind directed at the head. Fine wire electrodes are inserted into flight muscles, and are attached to amplifiers (triangle symbols). The positions of the leg attachments are shown as circles.
(*b*) Electromyograms (EMGs) from a depressor and an elevator muscle of a forewing. Usually, two motor neuron spikes are registered in each muscle per wingbeat cycle. In mid-cycle, small spikes are picked up from other muscles as cross-talk.
(*c*) During a long flight, the gradual increase in the duration of wingbeat cycles was due to an increase in the delay between spikes in depressor and elevator motor neurons. The delay between spikes in elevator and depressor motor neurons remained constant. (Part (*a*) redrawn from Horsman *et al.*, 1983; (*b*) redrawn from Pearson and Wolf, 1987; (*c*) redrawn from Hedwig and Pearson, 1984.)

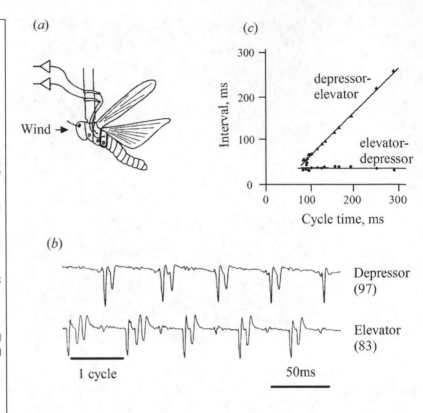

that ensures that the hindwings are not moving into the turbulent air caused by forewing movements.

Generation of the flight rhythm

In 1960 Donald Wilson showed that when he removed the wings and destroyed the sense organs of the wing bases (including the wing hinge stretch receptor described in Chapter 2), a locust could still generate in its motor neurons a pattern similar to flight. Either an air current directed at the head or a series of randomly timed electrical shocks to the connective nerves could elicit this flight-like activity, which consisted of regularly repeating, alternating spikes in wing elevator and wing depressor motor neurons. Flights were quite short in duration, and the wingbeat frequency was reduced to about half that of an intact locust. Wilson concluded that the basic programme for generating flight movements is situated in the central nervous system – a central pattern generator. Generation of the rhythm, or the co-ordinated activity of different motor neurons in the correct sequence, does not require proprioceptors to report details of the movements caused by muscle contraction. Even if all the flight muscles and the head are removed, the thoracic ganglia are capable of generating a

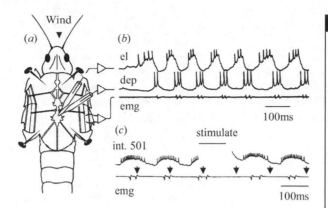

Fig. 7.9 Intracellular recordings during flight-like activity. (*a*) Method for preparing the locust, which is opened mid-dorsally and the body walls pinned, exposing the ganglia. Two glass capillary microelectrodes penetrate neurons in the ganglia while wire electrodes record electromyograms (EMGs) from muscle 112. (*b*) Intracellular recordings from an elevator (el) and a depressor (dep) motor neuron, together with an EMG recording from a depressor muscle (the one labelled in Fig. 7.7*b*). (*c*) A phase resetting experiment with interneuron 501. The upper trace is an intracellular recording from the neuron and the lower trace is the EMG. During stimulation with depolarising current, the intracellular recording could not be registered. The arrowheads indicate the times when bursts of spikes in the depressor muscle would have been expected to occur without any stimulus to the interneuron. (Parts (*a*) and (*b*) redrawn from Robertson and Pearson, 1982; (*c*), redrawn from Robertson and Pearson, 1983.)

repeating pattern of alternating excitation of elevator and depressor motor neurons, although the rhythm was half the speed of that in an intact, flying locust.

Intracellular recording and staining have been used to identify and characterise some of the neurons involved in generating the flight rhythm. In a technique developed by Keir Pearson and Meldrum Robertson, the thorax is opened dorsally to allow access to the thoracic ganglia, which are stabilised against movements by supporting them with a metal platform (Fig. 7.9*a*). Further stability is achieved by removing the legs and cutting most of the nerves that supply the flight muscles. The nerve to one muscle, usually one of the dorsal longitudinal muscles, is left intact so that EMGs recorded from this muscle provide a monitor of the flight rhythm. This nerve also carries the axons of most of the proprioceptors of the wing base. A locust prepared in this way will produce sequences of flight-like activity when wind flows over the head, or if the neurohormone octopamine is applied to the thoracic ganglia. Flight sequences are rather short in duration and there are differences in the detail of the pattern from that generated by intact animals, for instance it tends to have a slower rhythm and individual motor neurons spike more often in each wing beat cycle. However, the basic pattern of repeated, alternating activation of depressor and elevator motor neurons is present. As in an intact animal, there is a constant delay between excitation of elevator and depressor motor neurons; and a delay of 5 ms between spikes in motor neurons of the hindwing and forewing depressor muscles.

At the start of a flight sequence, elevator motor neurons depolarise and generate spikes before the depressors. In an intact animal, this would open the wings. During flight, repeated smooth oscillations in membrane potential, up to 25 mV in amplitude, are recorded from flight motor neurons (Fig. 7.9*b*). Depolarisation of the elevator and depressor motor neurons alternates and sometimes a motor neuron might hyperpolarise from its resting potential between cycles of depolarisation.

Interneurons of the flight generator

None of the motor neurons contributes to the generation of the flight rhythm, and one good functional reason is that motor neurons, and therefore muscles, can be selected for use in different combinations for different motor patterns. This is important during steering, and during some movements of the legs in which muscles that also move the wings are involved. The regular waves of depolarisation that occur in motor neurons during flight must, therefore, originate in interneurons and be communicated to the motor neurons by synaptic transmission. Robertson, Pearson and others have characterised many different interneurons which show cycles of rhythmical activity, similar to those in motor neurons, during flight-like activity. All of these interneurons generate spikes, and the interneurons generally produce a greater number of spikes per wingbeat cycle than the motor neurons. At the end of an experiment, stain is injected into a neuron so that its structure can be examined. In the thorax, about 85 different types of interneuron that are active during flight have been distinguished. All of the interneurons exist as bilateral pairs. Some belong to groups in which individual members have not been distinguished from each other so that the total number of interneurons involved in the flight rhythm probably exceeds 100. The interneurons that form the central pattern generator for flight are distributed among several ganglia, and most of them branch extensively in both the second and third thoracic ganglia.

There are two criteria for establishing whether an interneuron is part of the central pattern generator: first, during flight it must be rhythmically active at the flight frequency; and, second, injection of a brief pulse of current into it should reset the flight rhythm. If an interneuron plays a role in generating the flight rhythm, a brief pulse of depolarising current injected into it will reset the rhythm by delaying or advancing the time of subsequent bursts of spikes (Fig. 7.9c), not only in the interneuron but also in flight motor neurons. Whether a pulse of depolarising current delivered to an interneuron advances or delays the time of the next wingbeat cycle depends on the phase of the cycle in which the stimulus current is delivered. Experiments of this type are called **phase resetting** experiments. They reveal whether or not a particular neuron is involved in the clock mechanism that determines the timing of cycles in a rhythmically repeating activity. However, they do not reveal the exact mechanism for generating the rhythm.

One interneuron, number 301, is shown in Fig. 7.10a. This neuron has its cell body in the second thoracic ganglion, and many branches both in this ganglion and the third thoracic ganglion. Another interneuron, number 501, is arranged the other way round, with its cell body in the third thoracic ganglion. Intracellular recording from 501 shows that this interneuron is excited at the same time as

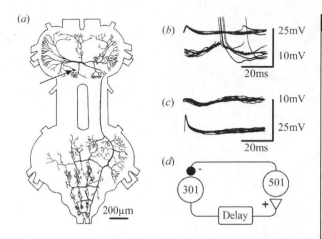

Fig. 7.10 Interneurons of the central pattern generator for flight. (*a*) Anatomy of interneuron 301 in the second and third thoracic ganglia, stained by intracellular injection of dye. The arrow indicates the cell body. (*b*) Excitatory connection from interneuron 301 to interneuron 501, demonstrated by simultaneous intracellular recordings from the two neurons. Multiple sweeps of the oscilloscope are overlain, each triggered by a spike in 301. (*c*) Inhibitory connection from 501 to 301. (*d*) Schematic circuit representing the excitatory (+) and inhibitory (-) connections between 301 and 501. (Redrawn from Robertson and Pearson, 1985.)

depressor motor neurons in each wingbeat cycle. Interneuron 301 is excited slightly before the activation of depressor motor neurons.

When intracellular recordings from interneurons 301 and 501 were examined in detail, interactions between the two interneurons were found. Consistency in these interactions is shown by overlaying several oscilloscope sweeps in which each sweep is triggered from a spike in one of the interneurons. A spike in 301 is always followed, after a delay of 6 ms, by a small depolarising potential in 501; and a spike in 501 is always followed, after 3 ms, by a brief, hyperpolarising IPSP in 301 (Fig. 7.10*b*, *c*). Allowing for time for neuronal signals to be conducted to the recording sites, the delay of only 3 ms in transmission from 501 to 301 suggests that this connection is direct, or monosynaptic. The greater delay in the connection from 301 to 501 leaves room for at least one additional neuron to be involved.

These two neurons form part of a network, shown in Fig. 7.10*d*, that could generate bursts of activity (Robertson and Pearson, 1985). Excitation of 301 leads to excitation of 501 after a short and fixed delay; and 501 in turn inhibits 301 and terminates its burst of spikes. If 301 were excited from another source, 301 would become active again once 501's excitation had died away, and the network would reverberate on its own. There is some evidence that the network can work in this way because, in some experiments, steady excitation of 301 by current injection causes steady, rhythmical activity in 501 and in flight motor neurons. However, the critical test of isolating this network from others in the thoracic ganglia has not been performed. The network is just one of many that have been found in the flight pattern generator. It is unlikely that any single interneuron is indispensable for generating the flight pattern, which makes the rhythm generating mechanism rugged and dependable, but quite hard to understand.

The mechanism by which 301 excites 501 has not been conclusively demonstrated, but a possible pathway includes interneuron number 51, which is located between 301 and 501 in the network (see Fig. 7.11). Neuron 301 inhibits 511, which in turn inhibits 501. This

Fig. 7.11 Synaptic relationships of some of the elements of the central pattern generator for flight in the locust. Excitatory connections are indicated as triangles and inhibitory connections as circles. Interneurons are identified by their numbers; and E and D indicate elevator and depressor wing motor neurons. (Redrawn from Robertson and Pearson, 1985.)

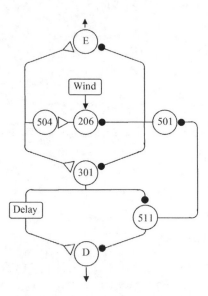

implies that the delayed excitation which 301 causes in 501 is by way of two successive inhibitions, a kind of interaction called disinhibition, as described in Chapter 2 for non-spiking interneurons controlling leg motor neurons. However, for disinhibition to work in this network, it is necessary for 511 to cause continuous, tonic inhibition of 501. This would mean that its membrane potential was continually depolarised above the threshold for its synapses to release transmitter. Only in this way can a discrete inhibitory potential in 511, caused by a spike in 301, be converted into a discrete depolarising potential in 501. Physiological characteristics of the depolarising potential in 501 suggest that it is caused by a reduction in inhibition rather than by a conventional excitatory synapse. However, we do not know whether output synapses from 511 release transmitter tonically, without requiring spikes.

The network in Fig. 7.11 represents a small part of the central pattern generator for flight, and allows glimpses into its mode of operation. Wind on the head activates pathways that excite 206. Interneuron 206 excites elevator motor neurons through 504, and causes delayed excitation of depressor motor neurons through 301. As explained above, a feature of the flight programme is that the duration of bursts of spikes in depressor motor neurons is relatively constant over a range of different wingbeat frequencies. This constancy can be explained by the actions of 301 in indirectly exciting the depressor motor neurons while at the same time removing inhibition of them from 511. This creates a discrete time slot, following a burst of spikes in 301, during which it is possible to excite the depressor motor neurons. Another feature of the flight programme, the constant delay between bursts in elevator and depressor motor neurons, can be explained by the dual action of 504 which excites the elevator motor neurons directly and excites the depressor motor neurons indirectly through interneuron 301.

Not all of the thoracic interneurons that are involved in flight participate in generating the rhythm. For example, some interneurons seem to play a role in initiating and maintaining flight, but not in the timing of wingbeats (Pearson *et al.*, 1985). A group of about four interneurons on each side of the second thoracic ganglion have axons that describe a tight loop in that ganglion and then travel anteriorly, so they cannot make direct contact with most of the interneurons of the flight generator. They are excited by air currents blown at the head; injecting depolarising current into one of them can trigger expression of a flight pattern; and injecting hyperpolarising current makes it less likely that air currents will trigger a flight. These interneurons may, therefore, be part of pathways that can start the flight motor pattern generator. During flight, they can spike tonically, so they neither contribute to nor receive timing information for wingbeat cycles. But some stimuli which initiate flight, such as wind to the tail end of a locust, do not excite these looped interneurons so they are not obligatory for starting the flight motor programme, which can be triggered by alternative pathways.

Proprioceptors and the flight motor pattern

Locusts have a large array of sensors that report back to the central nervous system on the mechanical effects caused by the commands it issued slightly earlier. These sensors are needed because the effects of muscle contraction are not totally predictable. For example, a flying locust might experience a gust of wind that causes its course to deviate, and the wings on one side of the body to move up and down more than those on the other side for a few wingbeat cycles.

Two types of proprioceptor have been studied most. The first is the wing hinge stretch receptor, which excites depressor motor neurons as described in Chapter 2. The large calibre of its axon has enabled experimenters to record from it during tethered flight and show that wing elevation excites it to fire a burst of spikes during each wingbeat cycle, with the number of spikes reporting the extent of elevation (Fig. 7.12a; also see Fig. 2.10a). The second sense organ is called the tegula ('tile'), a small dome-shaped piece of cuticle that bears about 40 stout hairs on the outside and a proprioceptor called a chordotonal organ inside, containing about 30 receptor cells. Sensory neurons of the tegula are excited during wing depression. It seems likely that the exact way the wing moves during depression could be registered by differences in the pattern of excitation of the different sensory receptors of a tegula, but it is difficult to record from individual sensory cells of this organ. Other aspects of wing movement are also monitored by sense organs; even the tiny wing veins bear tension transducers on their surfaces.

Both the stretch receptor and sensory cells of the tegula make direct, monosynaptic connections onto flight motor neurons and interneurons (Burrows, 1975; Pearson and Wolf, 1988). The stretch

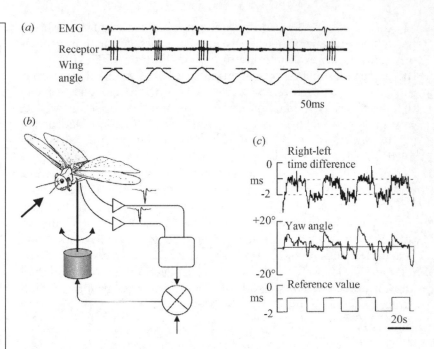

Fig. 7.12 Proprioceptors and locust flight. (a) Spikes in a forewing stretch receptor recorded simultaneously with up and down movement of the wing and an electromyogram from a hindwing depressor muscle. (b) Arrangement of an experiment where a locust controls its angle of flight into the wind by varying the interval between spikes in two flight motor neurons, which is measured by a computer and controls the motor to alter the direction of flight relative to the wind. (c) Recordings from an experiment similar to that in (b), in which spikes from right and left motor neurons 129 were recorded. The reference value for the right–left delay switched every 25 s between 1 ms and 2 ms. Quite rapidly, the locust adjusted the delay between right and left spikes to follow the reference value, and, after an initial swing, it maintained a flight direction more or less straight into the wind. (Part (a) redrawn from Möhl, 1985; (b) and (c) redrawn from Möhl, 1988.)

receptor excites almost all of the depressor motor neurons on the same side of the body as its wing. This means that when the wing is elevated, excitation of the stretch receptor will enhance excitation of the depressor motor neurons whose action opposes the elevation. If the wing is elevated more strongly than usual, the action of this proprioceptive reflex is to ensure that the depressors are excited more quickly than usual. The effect is to restore wing movement to the usual or preferred pattern. Similarly, the tegula excites wing elevator motor neurons when the wing is depressed.

In early experiments, attention was focused on the effects that stretch receptors have on the flight rhythm. When stretch receptors were removed, the flight rhythm dropped, and was accelerated again by stimulating the axon of a stretch receptor. Later experiments showed that these, and other proprioceptors, play a vital role in ensuring that the flight motor output is appropriate for achieving stable locomotion. Intracellular recordings from flight motor neurons show characteristic features that are due to input from particular proprioceptors (Pearson and Wolf, 1988; Pearson and Ramirez, 1990). Elevator motor neurons, for example, show two phases of excitation during a wingbeat, with synaptic inputs from the tegula sensory neurons preceding excitatory input from interneurons. The tegula acts to stabilise the flight rhythm. If its nerve is stimulated with relatively weak electrical shocks, which would excite a relatively small number of its sensory axons, the flight rhythm frequency increases. However, if the shocks are stronger, so that most of its axons would be stimulated, the flight rhythm slows down.

The wing proprioceptors are, therefore, integral parts of the neuronal network that generates the flight pattern; they work together with the central pattern generator to ensure that the animal's motor output is continually adjusted to maintain its desired course. The flight control system includes considerable plasticity in its mode of operation, rather than being exactly engineered to work in a precisely regulated way. This flexibility is illustrated by some experiments conducted by Bernhard Möhl (1993). If a locust is tethered to a holder that allows it to swivel to left or right, the locust will continually adjust its angle of yaw so that it flies straight into an air stream. In his experiments, Möhl did not allow the locust to provide its own power for changing its angle of yaw, but a small motor attached to the holder twisted it (Fig. 7.12b). A computer controlled the motor's rotation. It also measured the time interval between spikes in the EMGs from two different flight muscles and compared this with a reference value selected by the experimenter. The locust, therefore, could control its own yaw angle by slight variations in the interval between spikes in the two motor neurons. In the experiment shown in Fig. 7.12c, the first reference value chosen was a time difference of 2 ms between spikes in the left and right 129 motor neurons, so that the locust remained on a flight course straight into the wind as long as it maintained this interval. The reference value was then switched to 1 ms, and the locust quite rapidly reset the interval between spikes in the left and right motor neurons. Whenever the reference value was altered, the locust reset the interval between spikes in the two selected motor neurons.

The locust cannot know which pair of motor neurons the experimenter has selected for a particular experiment. This means that the locust must be able to compare the results of varying the timing among different pairs of motor neurons before discovering which ones are effective in maintaining course. Cutting the pair of muscles selected destroys the ability of the locust to maintain course in this way, which shows that proprioceptive feedback about the mechanical effects of muscle contraction is required. It is not known which proprioceptors are involved: the stretch receptor and tegula hairs both report movements of skeletal elements rather than individual muscle contractions. Möhl's experiments show that small, continual variations in the motor pattern are important because they allow the nervous system to experiment at finding the most effective pattern at maintaining course. They suggest that the central nervous system does not generate an unchanging motor pattern, which is tuned through the action of proprioceptors. Instead, we can view one important role for proprioceptors as participating in selecting the most effective programme for maintaining a desired course. The ability to make continued, small adjustments to the motor output means that the pattern can be continually updated, and rapidly adjusted to particular conditions. For a locust, these conditions would include wind direction, and turbulence caused both by the wind and neighbouring locusts in a swarm. A similar strategy is sometimes adopted by engineers in designing control systems for robots, because it provides a way to compensate for inevitable inaccuracies in manufacturing. It is difficult, for example, to ensure that the friction in wheel axles on the left and right sides is exactly balanced.

Steering and initiating flight

Flying animals require sensory mechanisms both for maintaining a certain flight course, and for altering it so that they avoid collisions or predators. In the locust, a number of neurons have been identified that carry sensory information from the brain to the thoracic ganglia and can influence flight. Several of these neurons are **multimodal**, which means that each responds to a number of different sensory modalities. Many of them are excited by wind flowing over the head, which is detected by groups of short hairs mostly situated dorsally, between the compound eyes.

One wind-sensitive interneuron is called the 'tritocerebral commissure giant' (TCG), after the small nerve branch or commissure that runs between the left and right connective nerves between the brain and suboesophageal ganglion (Fig. 7.13a). This commissure contains just two axons, an unusual feature that has enabled Jonathan Bacon and Bernhard Möhl (1983a, b) to record spikes from the TCG in loosely tethered, flying locusts. Dendrites of a TCG in the brain receive signals from various sense organs, especially wind-sensitive hairs but also from the eyes. Its axon runs through the suboesophageal and thoracic ganglia, branching to contact motor neurons and interneurons in each ganglion (Fig. 7.13b). Each wind sensitive hair is bent, and responds best to wind directed towards the

Fig. 7.13 A locust brain neuron, the tritocerebral giant (TCG), and flight control. (a) Structure of the TCG neuron, stained by intracellular injection, in the brain and suboesophageal ganglion. The brain is viewed as if from the front of the head. (b) TCG axon and its branches in the mesothoracic ganglion, which contains most of the forewing flight muscles. (c) Diagram of a locust head, viewed from the side. The two arrows point to two groups of wind-sensitive hairs, and also indicated are the right compound eye, and the right and median ocelli (shaded). (d) Recording of TCG spikes in a locust tethered and flying as in Fig. 7.12b, together with an electromyogram from a right forewing depressor muscle. Wind direction was varied systematically, as indicated. (Part (a) redrawn from Bacon and Tyrer, 1978; (b) redrawn from Bacon and Möhl, 1983a; (c) redrawn from Bacon and Möhl, 1983b.)

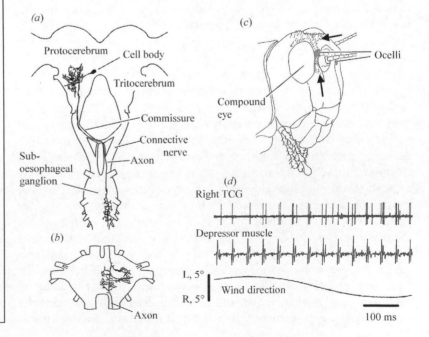

direction of its bend. Hairs with similar best directions tend to be grouped together: those on top of the head respond best to wind from the front, whereas those just in front of a compound eye respond best to wind from below (Fig. 7.13c), an arrangement reminiscent of that of the filiform sensilla of cockroach cerci (Chapter 3). Each hair is attached to a sensory neuron; some of these excite the TCG, and others inhibit it. The right TCG responds most vigorously to wind from the front and to the right of the locust, and the left TCG to wind from the front and the left. If the locust yaws so that it is no longer facing directly into the wind, excitation increases to one TCG and decreases to the other (Möhl and Bacon, 1983a, b). Stimulating one TCG electrically can induce yaw movements during tethered flight. Stimulating a TCG when a tethered locust is flying also causes a shift on the timing of wing depressor motor neuron spikes during wing-beats, and the TCG makes synaptic connections with some flight motor neurons, so it probably plays a role in making sure a locust flies straight into the prevailing wind direction.

In a locust that is flying into a steady air stream, the TCG generates a brief burst of spikes to coincide with the end of elevation during each wingbeat cycle rather than spiking at a steady rate (Fig. 7.13d). The reason is that movement of the wings and nodding motions of the head create their own air currents, which interact with wind caused by the locust's movement forward through the air. This interneuron is, therefore, rhythmically active during flight. Stimulating it artificially, to induce extra spikes, can reset the flight rhythm by lengthening or shortening the cycle in which the stimulus is delivered. Therefore, in addition to playing a role in steering flight, the TCG meets the same criteria for being part of the central pattern generator as some of the local, thoracic interneurons. The TCG neurons can also initiate flight during a jump, when they are excited by the rush of wind over the head. Electrically stimulating the TCG axon is an effective way of initiating flight activity (Bicker and Pearson, 1983).

Many of the brain neurons that are involved in flight control have their cell bodies in a region of the brain called the proto-cerebrum. Altogether, there are about 25 relatively large proto-cerebral neurons that could connect with the flight system (Hensler, 1992). Their influence on flight has been studied by stimulating individual axons with trains of electrical pulses. If a locust is carefully prepared in a way that allows it to move its wings, abdomen and head freely, steering movements can be produced by these stimuli (Hensler, 1992). A train of spikes lasting a few hundred milliseconds causes a change in the relative latency of action potentials in motor neurons of steering muscles within one wingbeat cycle and causes the head to roll after a slightly longer delay. It would be interesting to know how these neurons work in a flying locust – how they combine signals from different sensory modalities and whether they are rhythmically active in the same way as the TCG. One sensory pathway that plays a significant role in regulating flight attitude involves the three ocelli, or simple eyes, that locusts have in addition to the image-processing compound eyes. The ocelli have limited image-processing capability, but have enormous fields of view and monitor changes in horizon position by measuring changes in the total amount of light they receive (Wilson, 1978). The ocellar pathway includes the largest neurons to carry information into the brain, providing a fast and direct pathway between sense organs and flight motor neurons (Simmons, 1980, 2002).

The DCMD neuron of locusts (Chapter 5) is a prime example of an image-processing neuron in the compound eye pathway. It is excited by objects approaching on a direct collision course, and there is good evidence that it acts to trigger last-chance collision avoidance manoeuvres such as attack by a predatory bird or collision with another locust in a swarm. The response that a tethered, flying locust makes to the image of an approaching object – the stimulus that the DCMD is tuned to detect – is to hold its outstretched wings elevated, interrupting the rhythmic flight movements (Santer *et al.*, 2005). This response would cause it to dive rapidly, while enabling it easily to resume flying after losing height. Interestingly, a number of insect species respond in this way to bat cries, including noctuid moths (Roeder, 1962) as well as locusts (Robert, 1989; Dawson *et al.*, 2004). A number of lines of evidence show that the DCMD is responsible for causing this behavioural response by locusts (Santer *et al.*, 2006). To be effective, the neuron needs to produce a vigorous burst of spikes (shown in Fig. 2.11*c*), and this response must be timed to occur just as the locust begins the upward phase of a wingbeat: it does not seem to be able to reverse a wing downstroke into a diving glide posture.

Overall view of locust flight

The overall picture we now have of the flight control system of the locust is of a series of intermeshed, overlapping neuronal loops (Fig. 7.14). It contains an array of different types of neurons, including motor neurons, interneurons, proprioceptors and other sense organs. Some of the thoracic interneurons form a central pattern generator, but the pattern it generates has a slow repeat period and a pattern of motor activity capable of moving the locust through the air requires participation by sensory neurons. So the proprioceptors and some of the brain interneurons, too, must be considered to be part of the flight generator itself. They are able to reset the flight rhythm in the same way as some of the thoracic interneurons, and are rhythmically active during flight activity. So the flight generator is highly redundant – it contains many elements with similar or overlapping functions, none of which is indispensable to its normal operation. This redundancy has advantages: first, the flight system is rugged and not easily perturbed; and, second, the pattern produced is flexible and able to adapt rapidly to changing demands from the environment.

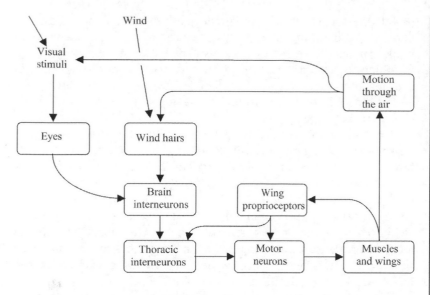

Fig. 7.14 Flow diagram to show the relationships between some of the elements involved in generating the flight motor programme in a locust. Some, but not all, of the thoracic interneurons included here are involved in generating the rhythm. Rhythmic input to motor neurons and interneurons is also derived from wing proprioceptors and brain neurons. As a result of the forward and rather irregular movement through the air, both the wind currents and the visual stimuli that the locust experiences are modified, so that the locust receives stimuli that are a combination of those caused by the external environment and those resulting from the locust's own movements.

Flight involves additional elements, not shown in Fig. 7.14. One of these is that early on during flight the neurohormone octopamine is released into the haemolymph. Octopamine affects the contractile properties of muscle cells and increases the supply of fuel to muscles. In addition, it can affect the properties of some of the interneurons of the flight motor programme generator, making them liable to fire spikes in cyclical bursts (Ramirez and Pearson, 1991), as well as counteracting habituation in the DCMD collision-warning neuron (Chapter 5).

Conclusions

Networks of neurons that can generate programmes for movement are often called pattern generators. Central nervous systems contain networks of interneurons that are capable of producing patterned activity in motor neurons but, as illustrated by the research on the flight system of the locust, the central pattern generators work together with sensory feedback from proprioceptors to generate a programme that is effective at ensuring the animal moves in the intended way. In general, rhythmic movements are generated by a number of different mechanisms operating together, which makes the pattern generator robust and able to continue to work properly under different conditions.

Usually, motor neurons are driven through synapses from inter-neurons and sensory neurons rather than being directly involved in the generation of rhythms for movements. This arrangement allows motor neurons to be excited independently of each other, so that their muscles can be used in different movement patterns. The research on speed change in zebra fish indicates that some of the pre-motor interneurons, too, might be reserved for particular speeds or pattern of rhythm. But consistent generalisations about how the networks controlling movements are organised are quite hard to make. This is illustrated by the crustacean stomatogastric ganglion in which, in contrast to the other examples we described in this chapter, motor neurons are heavily involved in pattern generation, and individual neurons participate in generating rhythms with different frequencies.

There are a number of different ways in which the excitation of a rhythm generating network can be sustained. In tadpoles, excitatory interneurons in the hindbrain and spinal cord are responsible, both by mutually exciting each other and by activating long-lasting post-synaptic excitation. Neuromodulators can play important roles by strengthening synaptic connections, as in *Tritonia* swimming, or by switching on bursting properties, such as in the pyloric rhythm of lobsters or, perhaps, locust flight. In locust flight, sensory activity, particularly from wind-sensitive neurons that are excited as the locust moves through the air, is important.

Networks that generate programmes for movement are not 'hard wired' and inflexible. One way in which this is evident is the manner in which sensory feedback works in controlling locust flight. At its simplest, sensory feedback allows a motor programme to compensate for changing demands on muscles as the nature of the terrain alters. Another more subtle role is to select between patterns that differ in their effectiveness, for example in moving an animal along a straight path. Continual small variations in motor output allow the motor programme to be updated continually, allowing compensation for changes in the mechanical properties of an animal's skeleton and muscles as it grows or is injured. Interneurons can also cause radical changes in the way that networks are configured. This is well illustrated by the way in which a single interneuron can reconfigure the lobster stomatogastric ganglion so that the neurons participate in new groupings and patterns of activity.

Questions

What problems would a pattern generator face in sustaining functionality throughout the life history of an animal?
What differences between animals that swim, fly and run would you expect to find in the complement of proprioceptors that each has?

Summary

- Nervous systems often contain central pattern generators (CPGs): networks of neurons that can generate programmes of activity in motor neurons. CPGs normally work together with proprioceptors and other sense organs to produce functional movement control programmes.

- Two general mechanisms for generating rhythms are: pacemaker properties of neurons; and networks, such as reciprocal inhibition between a pair of neurons.

Swimming by young tadpoles

- If skin is touched, the body flexes left–right about 20 times per second in an anterior–posterior travelling wave.
- When muscles are inactivated the spinal cord continues to generate the swim pattern in motor neurons. This shows the spinal cord contains a CPG, and it allows intracellular recordings to be made to investigate the mechanisms of rhythm generation.
- Motor neurons and some interneurons spike once per cycle, with an IPSP between spikes.
- Both network properties and membrane physiology of individual neurons shape the output of the central pattern generator:
  o Network – commissural interneurons carry mid-cycle inhibition across the spinal cord, and reciprocally inhibit each other.
  o Membrane – when depolarised, motor neurons and interneurons generate single rebound spikes, triggered by release from inhibition.
  o Network – during a swim, steady excitation comes from excitatory interneurons that excite each other as well as motor neurons.
  o Membrane – EPSPs in spinal neurons during swimming have two components, both caused by glutamate: a fast one, and a slower, more prolonged one that helps sustain swimming.

Swimming in larval zebra fish

- During a swim episode, speed and power usually decline.
- Motor neurons are recruited according to the 'size principle' – smaller ones deliver least power and speed, and require relatively weak excitatory synaptic input to excite them.

Swimming in a sea slug

- Brief exposure of *Tritonia* to starfish odour triggers a swim that can last tens of seconds.
- A network of interneurons generates the rhythm of alternating dorsal and ventral flexion.
- Sensory input is delivered to it through the dorsal ramp interneuron.
- Dorsal swim interneurons play a dual role: through synaptic connections they play a role in the swim pattern network; the serotonin they release also causes intrinsic neuromodulation of synaptic strengths, helping to sustain the rhythm.

Crustacean stomach movements

- The stomatogastric ganglion of lobsters and crabs contains 30 neurons and generates two rhythms: a slow gastric mill rhythm that moves the teeth and a faster pyloric rhythm.

- The pyloric network is generated by properties of networks and of membrane physiology. Pairs of neurons connected by inhibitory synapses can generate reciprocal activity; and most of the neurons are conditional oscillators, which generate pacemaker potentials when exposed to neurochemicals released by axons in the stomatogastric nerve.
- The pyloric suppressor neurons are activated by chemoreceptors in the gut and reconfigure both the pyloric and gastric mill networks, generating a new pattern, for swallowing, that involves neurons from both networks.

Locust flight

- Each wing is moved by about ten muscles including elevators, and direct and indirect depressors.
- Most of the wing muscles are innervated by single motor neurons that cause fast twitches.
- Electromyograms show that in steady flight each muscle contracts 15–20 times per second.
- The flight rhythm and programme are generated by a central pattern generator together with feedback from proprioceptors.
- In the thoracic ganglia, many tens of interneurons participate in networks that generate rhythmical activity and communicate it to motor neurons. Phase resetting experiments can determine whether an individual neuron contributes to rhythm generation.
- Important proprioceptors include the wing hinge stretch receptor and the tegula. These influence both the rhythm and phase of activation of different motor neurons.
- The flight programme is plastic, allowing the animal to make adjustments to it for maintaining a desired course.
- Many of the brain neurons that play roles in steering flight are multimodal, integrating information from different modalities including wind hairs and the ocelli. Some play a role in rhythm generation.

Further reading

Burrows, M. (1996). *The Neurobiology of an Insect Brain*. Oxford: Oxford University Press. A book that describes in detail how the thoracic ganglia control many of the types of movement a locust makes.

Grillner, S. (2006). Biological pattern generation: the cellular and computational logic of networks in motion. *Neuron* **52**, 751–766. This review concentrates on swimming in lampreys to show how different levels of analysis can be applied to an understanding of how motor programmes are selected and generated.

Marder, E., Bucher, D, Schulz, D. J. and Taylor, A. L. (2005). Invertebrate central pattern generation moves along. *Current Biology*, **15**, R685–R699. A summary of how research on invertebrate central pattern generators provides insights into the ways in which rhythmic motor patterns can be generated and controlled, emphasising the crustacean stomatogastric ganglion, and swimming and heartbeats in leeches.

8

Changes in nerve cells and behaviour: learning in bees and rats; swarming in locusts

One of the most important and intriguing aspects of animal behaviour is that it continually changes. As they move around actively exploring their surroundings, animals rarely behave in completely predictable ways. That is evident just by observing flies walking on a table top: they rarely take more than a few steps in one direction, but often turn and stop for short times as they walk. Animals are programmed to change their behaviour when their environment alters, which is an efficient strategy that enables them to react appropriately to a wide variety of possible situations. Some of the changes are elements of the processes of development and maturation while others allow an animal to learn about alterations in its environment so it can make and modify predictions based on experience, for example to predict that a particular action will be followed by a rewarding or an aversive event. Sometimes learning particular features only happens during restricted time periods of an animal's life history, or **critical periods**. These are part of a programme of the normal development of behaviour – for example an owl develops its auditory map most easily during the first few weeks after hatching (Chapter 6), and a young song bird needs to hear the songs of adults during the first few weeks of its life so that when it matures it sings a song that is effective in attracting a mate (Chapter 9).

The ability to change is a fundamental property of many nerve cells and their interconnections. Over the short term, changes occur in the excitability of neurons and the effectiveness of synapses. In the longer term, neuronal processes and synapses grow, multiply or are pruned, and this involves new protein synthesis. Much recent research has focused on events at the molecular and cellular level that could underlie learning or events during the development of a nervous system (see Carew, 2001, for a good introduction to this, particularly in the sea hare, *Aplysia*). Many of the molecules involved are signalling molecules that most animal cells possess, and the cellular events underlying learning are similar between animals as diverse as molluscs and mammals (Pittenger and Kandel, 2003). Some of the cell signalling molecules can be activated in a number of ways through the electrical signals used by neurons, for example when there is a temporal coincidence between different sources of excitation to a neuron. A major challenge in neuroethology is to relate the changes in cellular properties of neurons to alterations in animal behaviour. The big challenge is not so much finding out what events happen within individual cells, but in identifying the neurons that are involved in laying down, retaining and recalling particular memories.

While examples of changes in an animal's behaviour are easy to notice, it is harder to develop experimental strategies to investigate the events at the neuronal level that underlie these changes. What is required is to be able to record from and manipulate individual neurons while an animal reliably exhibits learning or other behavioural changes. In this chapter, we describe three experimental studies that relate alterations in the properties of individual neurons to changes in animal behaviour, in honey bees, rats and locusts. The first two studies involve learning about the layout of an animal's environment, in which quite specific changes to behaviour occur; and the third example involves widespread changes in the behavioural landscape of individual animals. Honey bees, which are the subject of the first study, are programmed to forage effectively for food. Once a bee has found flowers that provide a good source of nectar and pollen, she remembers their characteristics. She can also communicate its location to her hive mates. A honey bee's brain occupies a space of about 1 mm^3, and contains nearly a million neurons. One of those neurons has been studied in some detail, and has been shown to be able to play a role in learning particular attributes of flowers. Rats, which are the subject of the second study, are extremely good at learning the spatial layout of their environment. As a rat wanders about in new landscapes, the properties of individual neurons in one particular brain region, the hippocampus, alter and these changes in responsiveness of individual neurons are very likely to be part of the rat's spatial learning. Locusts, which are the subject of the final study, can form immense swarms that cause major damage to wide areas of food crops, so study of locust biology has economic as well as scientific significance. Individual locusts, when isolated from others, tend to lead secretive, non-social lives. But particular stimuli trigger a change in behaviour to a different, gregarious phase, so that these locusts behave quite differently from their solitarious cousins.

Associative learning and the proboscis extension response in honey bees

Why do many plants invest in flowers that have bright colours, striking shapes, and scents? The reason is similar to the way packets and cans of food in a store are brightly advertised, to enable people to recognise and remember them. Flowers advertise for particular animals to visit them, in the process of which they pick up pollen from one flower and transfer it to another, starting the process of seed development. A variety of animals carry pollen between flowers, including bats and hummingbirds, but 65% of flowering plants are pollinated by insects. During a burst of evolution 65 million years ago, flowering plants diversified along with bees, wasps, butterflies, moths and some other insects. By visiting a flower, an insect can drink nectar, which is a rich fuel source, and pollen itself is an important source of protein. Roughly a fifth of insect species rely on flowers as their principal source of food. One species of honey bee, *Apis mellifera*, is the best known pollinator insect, and under domestication has spread to most of the world, including to America (where there are no native honey bees). The number of species of honey bee is generally agreed to be about seven, with many subspecies. A honey bee lives in a highly structured society in which there is clear division of labour. Usually, a hive contains a single fertile female, the queen, and her eggs are fertilised by a male, or drone, during a nuptial flight which is a rare occasion when queens emerge from the hive, and is triggered when the colony grows beyond the control of a single queen. The labouring tasks in the hive, which include caring for larvae, fanning to cool the hive, and scouting for and collecting food are performed by infertile females, or worker bees.

A worker honey bee, *Apis mellifera*, extending its proboscis to drink sucrose solution offered at the tip of a hypodermic needle. The bee's body is inside a plastic tube and held in place with black tape in preparation for an experiment to test its ability to distinguish different odours, and to associate particular odours with a sucrose reward. This way of studying bees led to the discovery of a particular brain neuron that is important in associative learning. (Photograph by Peter Simmons; bee supplied by Geraldine Wright, Newcastle University.)

When a foraging worker visits a flower that is rich in nectar, she remembers it and, after returning to the hive which may be a mile or more away, lets the other workers know by dancing on the vertical honeycombs in the hive to show how to find the source of food. The meaning of the dance was first decoded by observations and experiments by Karl von Frisch in the 1940s (von Frisch, 1967), and his conclusions were verified later by showing that workers could follow instructions provided to them by a dancing robot bee (Michelsen *et al.*, 1989). For navigation between food and hive, honey bees memorise landmarks (Menzel *et al.*, 2006; Collett, 2007). Karl von Frisch analysed the disposition of bees to learn to associate colours and shapes of flowers with a good source of food, and he exploited this to investigate several features of the sensory capabilities of bees, work for which he was awarded a Nobel Prize jointly with Lorenz and Tinbergen in 1973. Bees learn aspects of a flower's shape, colour, pattern and odour just before they visit it to collect nectar, and bees identify that type of flower by the combination of these characters. A flower that looks the same as a previously rewarding one but has a different smell will usually be ignored by a foraging bee; she needs to relearn all its new attributes together.

Just as humans have long- and short-term memories, so too do honey bees. While foraging in a patch of vegetation where one particular flower is a good source of nectar, bees need to retain the memory of that flower for only a few seconds at a time. The memory is strong when tested 9–10 minutes later, but for 2–3 minutes after the initial training the memory is not so good, indicating that during the flight back to the hive, the memory switches from a short- to a longer-term form (Menzel and Erber, 1978). Bees remain a favoured subject for behavioural experiments because they are quite easy to obtain and to train in laboratory conditions, and it is also possible to study the roles of some individual neurons in a bee's brain during learning.

One of the simplest behaviours that a honey bee performs is to extend its proboscis, or tongue, in response to a drop of sucrose solution applied to chemosensory hairs on the proboscis. A bee will occasionally extend its proboscis without any obvious stimulation, or if a tiny puff of a particular odour such as the smell of carnation or orange is blown at an antenna. If the bee has recently tasted sucrose, the likelihood that it will extend its proboscis in response to a subsequent odour puff increases – stimulation with sucrose is said to **sensitise** the proboscis extension response. A much greater enhancement of the response to a puff of odour occurs, however, after the odour puff has been paired with delivery of a drop of sucrose. For maximum effect, the odour puff must be delivered between 1 and 3 seconds before the drop of sucrose (Bitterman *et al.*, 1983). The next time that the odour is directed at an antenna, there is a very high chance that the bee will extend its proboscis (Fig. 8.1*a*). In the bee's brain, an

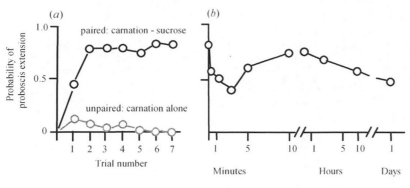

Fig. 8.1 Conditioning the proboscis extension reflex in honey bees (*Apis*) to odours. (*a*) Graph to show the acquisition of the conditioned response to carnation odour. Groups of bees were trained by pairing a puff of carnation with a sucrose reward 2 s later. After a single pairing procedure, about 80% of bees tested responded to the carnation odour with proboscis extension. In comparison, few untrained bees responded to the carnation when not paired with a sucrose reward. (*b*) Time course of memory retention. Each bee in a group was trained by a single trial to associate an odour with a sucrose reward, and each bee was then tested with odour alone at one interval following the training trial. (Redrawn from Menzel, 1999.)

association has been made between the odour and the likely presence of sucrose. The order in which the two different stimuli are delivered is vital for the formation of memory: if the sucrose is applied to the proboscis just before the odour is blown over an antenna, a second puff of the odour delivered a minute later is unlikely to cause proboscis extension. The two antennae are not equally effective in making associations: a bee learns better if odours are directed at its right antenna than at its left (Letzkus et al., 2006). Any asymmetry in the bee's brain that underlies this left–right difference would be interesting because, as described in Chapter 9, brain asymmetry in birds and mammals is associated with well-developed learning capacities.

The proboscis extension response is a good example of **conditioning**, a type of behavioural change often studied in vertebrates. The odour stimulus becomes conditioned so that after it has been paired with a sucrose reward, it reliably evokes a stimulus that was previously only rarely linked with this stimulus. In learning terminology, the odour is the **conditioned stimulus** and the sucrose the **unconditioned stimulus**. A single pairing of a puff of carnation with a taste of sucrose is sufficient to enhance the new coupling between carnation and proboscis extension for many hours. However, the coupling will weaken and become extinguished if several odour puffs are delivered with no sucrose as a reward, and can be replaced by a new association between another odour, such as orange, and proboscis extension. The memory of the association between an odour and sucrose goes through different phases in time, so that the initial short-term memory is transferred into longer lasting medium- and long-term memories (Fig. 8.1*b*; Hammer and Menzel, 1995; Menzel and Giurfa, 2001).

Neuronal pathways and conditioning

Taste-sensitive sensory neurons that detect sucrose send their axons from the proboscis into the suboesophageal ganglion,

Fig. 8.2 Morphology of the brain and suboesophageal ganglion of the bee to show structures, pathways and neurons involved in conditioning the proboscis extension reflex to odours. (*a*) Information about odours travels to the protocerebrum by different routes, one direct and the other through the mushroom bodies. The structure of one Kenyon cell is shown in the mushroom body on the right. (*b*) Diagram to show the extent of the innervation pattern of neuron VUMmx1. Its cell body and dendrites are in the suboesophageal ganglion. (Parts (*a*) and (*b*) redrawn from Hammer, 1993.)

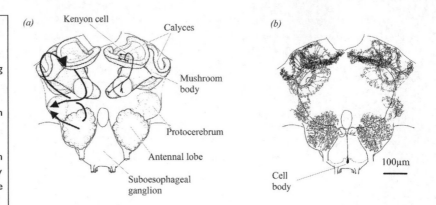

which is the first ganglion in the nerve cord after the brain and contains the motor neurons of the proboscis muscles. Odours are detected by sensilla on the antennae, which send axons to project into antennal lobes of the brain. After processing in the brain, information about odours is carried to the suboesophageal ganglion by neurons that originate in the protocerebrum (forebrain). During conditioning, information about the sucrose reward must be carried to the brain from the suboesophageal ganglion and cause modification of the pathways that link olfactory sensory receptors with neurons in the protocerebrum of the brain. The pathways involved are shown in Fig. 8.2*a*.

About 60 000 olfactory receptors run from each antenna to a discrete area on either side of the brain called the antennal lobe. Each sensory neuron terminates in a spherical, compact structure called a **glomerulus**, and olfactory information is processed both within and between the 160 glomeruli in an antennal lobe before being distributed by projection neurons to other brain regions. Some neurons project directly from glomeruli to the protocerebrum, and others project to another brain region, called the **mushroom body**. The mushroom bodies were named after their resemblance to some kinds of horn-shaped mushrooms, and have been implicated in the more complex types of insect behaviour, including learning and social behaviour. They are particularly well developed in bees, and in some other types of insects, including cockroaches in which they may play roles in spatial memory (Mizunami *et al.*, 1998). Each mushroom body contains a parallel array of neurons called Kenyon cells (one is drawn in Fig. 8.2*a*). About a third of all the neurons in the brain of a bee are Kenyon cells, with 170 000 in each mushroom body. It is difficult to make recordings from them because they are small and tightly packed together. Their dendrites are arranged in structures called calyces, and in each calyx different incoming sensory modalities are directed to distinct regions: olfactory information to the outermost rim region; visual information to the collar region; and mechanosensory and chemosensory information to the basal ring. Kenyon cell axons branch into two, and the two branches enter different lobes of their mushroom body. One of the major tasks of Kenyon cells is to integrate information from different sensory modalities. Various output neurons gather information from many Kenyon cells and deliver information from the mushroom bodies to other parts of the brain.

Two types of experiment suggest that both the antennal lobe and the mushroom body play roles in associative conditioning. In one, a small needle-like probe was used to cool local regions of the brain to 1–5 °C for 10 seconds, transiently blocking or reducing activity (Erber et al., 1980). Cooling either an antennal lobe or a mushroom body soon after training reduced conditioning when tested 20 minutes later. The effect was not simply to reduce sensory activity because conditioning was blocked if several seconds or even minutes elapsed between delivery of the odour stimulus and the start of cooling. The experiments suggested that the process of establishing memory is not localised to one restricted brain region. They also suggested a sequence of activity in different brain regions because cooling did not disrupt the memory if applied to the antennal lobes later than 2 minutes after training, to the alpha lobes later than 3–4 minutes, or to the mushroom body calyces later than 5 minutes. In the second type of experiment, a small amount of the neuromodulator octopamine was injected into a brain region just after delivery of an odour pulse. The octopamine substituted for the effects of stimulation of taste receptors with sucrose, and when the odour was presented later on its own it evoked proboscis extension. Octopamine had this effect when injected into the antennal lobe or calyx of a mushroom body, but not elsewhere.

Kenyon cells in the mushroom bodies are a likely site for some of the changes that occur during learning, and support for this idea comes from experiments in which recordings were made from one of the mushroom body output neurons, called PE1, that receives input from many Kenyon cells (Mauelshagen, 1993). PE1 fires a burst of spikes when an odour, such as carnation, drifts over the antennae. But following a pairing of carnation with a sucrose reward, the next time the carnation odour is delivered the spiking response by PE1 is smaller than to the first exposure to carnation. This reduction is specific to the pairing of carnation followed by sucrose: it is not found if sucrose is given just before a puff of carnation, nor when a series of carnation puffs with no sucrose rewards are delivered to the bee. Recording from PE1 over several hours has confirmed that its firing rate decreases as the likelihood of a learned behavioural response increases, suggesting that the mushroom body exerts some kind of general inhibition over motor responses, and that learning involves a relaxation of the inhibition to specific responses (Okada et al., 2007).

The role of an identified neuron in conditioning

A bee's brain must be able to form associations between a sucrose reward and many different other stimuli, including a variety of odours. We could, therefore, expect that any neuron that plays a key role in conditioning would have branches distributed in many different regions of the brain. One such neuron is VUMmx1 (Fig. 8.2b), an unpaired neuron that has branches that are arranged symmetrically on both left and right sides of the suboesophageal ganglion and brain. Its branches overlap the pathways that link olfactory stimuli to the proboscis extension response at three different locations: in the antennal lobes; in part of the protocerebrum; and in the mushroom body calyces. VUMmx1 appears to contain and release octopamine. Martin Hammer (1993) demonstrated that this neuron can induce conditioning by experiments in which he showed that exciting VUMmx1 by injection of depolarising current can substitute for a sucrose reward in conditioning the proboscis extension response.

Delivery of a drop of sucrose to taste sensilla on the proboscis excites VUMmx1, and the neuron continues to spike for at least half a minute after the sucrose has been removed. VUMmx1 is also excited, but only for a brief time, by various odours directed at the antennae. However, Hammer found that when sucrose was delivered just after a particular odour, the next time the odour was delivered alone VUMmx1 responded with a brief, intense initial burst of spikes followed by excitation that continued for 20 seconds, a similar response to that given by VUMmx1 to the first sucrose stimulus. Like proboscis extension, the enhancement in response to the conditioned odour depended strictly on the odour and the sucrose being delivered within a short time and in the correct order.

Exciting VUMmx1 by injecting current into it through a microelectrode can substitute for stimulation of the bee's taste receptors with sucrose in conditioning (Fig. 8.3). In forward pairing between an odour and an electrical stimulus to VUMmx1, the odour stimulus started 2 seconds before the electrical excitation, and during backward pairing, VUMmx1 excitation preceded the odour by 5 seconds. Activation of the proboscis extension muscle was monitored in an electromyogram recording. After forward pairing, a test pulse of odour 10 minutes later produced a large response from the muscle (Fig. 8.3b). In contrast, backward pairing of VUMmx1 excitation with odour did not lead to any enhancement of the response to the test odour.

This experiment clearly shows that VUMmx1 can condition the proboscis extension response to a particular odour. When a bee smells an odour and then tastes sucrose, the sucrose will strongly excite VUMmx1, triggering conditioning of the proboscis extension response to that odour. However, Hammer's experiments do

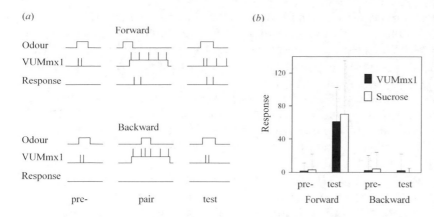

(a)

(b)

Fig. 8.3 Electrical excitation can substitute for stimulation with sucrose in conditioning the proboscis extension response to odours. (a) Outline of the protocol used in training experiments. First, any response by proboscis extension muscles to a puff of odour was measured in a pre-test. Next, a puff of the odour was paired with electrical excitation of VUMmx1 by a pulse of depolarising current. In forward pairing, the odour preceded the start of the current pulse, and in backward pairing, the current pulse started before the odour. Finally, responses to a test pulse of odour were recorded. (b) Responses to a test puff of odour after training with either electrical excitation of VUMmx1 or with a sucrose reward. The response was measured as the number of spikes in a proboscis extensor motor neuron during 10 s after a test or pre-test puff of odour, and the histograms plot the median response from groups of 11–15 bees. The error bars indicate the maximum responses from each group. (Parts (a) and (b) redrawn from Hammer, 1993.)

not show that VUMmx1 is solely responsible for conditioning, which could be tested experimentally by removing or poisoning VUMmx1, or by hyperpolarising it during conditioning to prevent it from spiking. Besides VUMmx1, there are other unpaired, octopaminergic neurons in the suboesophageal ganglion, and these neurons may work alongside each other during conditioning. VUMmx1 overlaps extensively with olfactory neurons, and may be specialised for making associations between odours and a good source of food, while the jobs of forming associations between flower colour or pattern and food are likely to be left to other neurons. In a different experimental approach, bees were prevented from making receptors for octopamine by injecting them with double-stranded messenger RNA for the octopamine receptor (Farooqui *et al.* 2003). In these bees, VUMmx1 or other neurons were not removed or inactivated, but those that released octopamine could no longer communicate with their target neurons. The bees did not learn about pairing of odours with a sucrose reward, although they could still respond to odour and to taste stimuli. This experiment indicates that neurons that release octopamine, such as VUMmx1, are needed for conditioning.

The same type of flower is not always the best source of fuel for bees, so the association a bee makes between flower characters and food is a changeable one. A bee initially learns to associate one particular odour with food, and also learns that another is not associated with food. The pattern of response in VUMmx1 reflects this. First, the bee is trained to associate carnation odour with sucrose. In early training trials, the neuron produces a weak or no response to carnation, and a strong burst of spikes in response to the sucrose. After training, the carnation odour on its own produces a brisk response of spikes, but the following sucrose reward causes little if any response from VUMmx1. The neuron acts as if it signals a mixture of expectation and surprise – after training, the expectation is that the learned odour, carnation, will be followed by a sucrose reward. So the sucrose that follows is no surprise. Both during and after training with carnation odour the neuron's responses to a different odour such as orange are relatively weak. But if, after training to associate carnation with sucrose, a puff of orange is given just before a sucrose drop, VUMmx1 expresses its surprise at the unexpected sucrose drink with a burst of spikes not present when sucrose, as expected, follows a carnation puff.

Because it has an extensive network of branches, VUMmx1 can deliver information about a sucrose reward to the antennal lobes and the mushroom body, the sites that the experiments with local cooling and octopamine injection showed are important in conditioning. A key event in conditioning is that excitation of VUMmx1 is closely associated in time with stimulation of particular odour-activated pathways, but it is not known what the cellular basis for this is. In fact, details of the pathway that links sensory responses to sucrose with VUMmx1 are rather sketchy. The mechanism by which VUMmx1 establishes conditioning is likely to involve intracellular messenger molecules such as cyclic AMP and cyclic GMP. In *Drosophila*, mutations that affect production of these molecules also cause impairment of memory (Davis, 2005). Conditioning of proboscis extension in bees is quite complex because a large range of possible stimuli can become associated with the response. It is probable that the complexity of the mushroom bodies is in some way related with the need to be able to deal with a large number of possible stimulus configurations.

Rats, burrows and mazes

An albino mutant of the brown rat, *Rattus norwegicus*, is known, for good reason, as the laboratory rat. The reason why this rat finds a place in our book is that activity of individual neurons can be studied as a rat changes its behaviour, learning about its surroundings. The albino mutant is less aggressive and breeds faster than its wild cousin, and has been studied in medical, psychological and other research since the mid nineteenth century. There are various varieties, including the Lister hooded rat, which is often used in studying spatial behaviours because it has particularly good vision. Brown rats originated in China and are now widespread, having accompanied humans to establish themselves on all continents except Antarctica. *Rattus norwegicus*, despite its species name, arrived in Norway after most of the rest of Europe, including Britain. Another species, the black rat, *Rattus rattus*, is even more commonly found near to humans, and is more likely to climb above ground level than the more secretive brown rat.

In wild conditions, rats are social animals, and build extensive burrow systems (Fig. 8.4). The burrows have several entrances, which rats presumably learn to find rapidly while foraging outside, and many of the entrances can be concealed in various ways. The tunnel system includes many narrow tunnels, and two kinds of chambers: food stores and various types of nests, which can

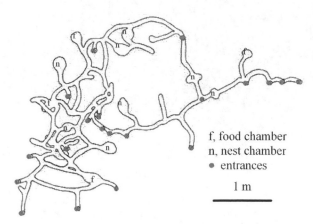

Fig. 8.4 Natural burrow system of *Rattus norwegicus*, with 20 separate entrances, and nest and food chambers. (Drawn from Hanson, 2004.)

f, food chamber
n, nest chamber
• entrances

1 m

accommodate up to seven adults at a time, usually females. In the laboratory, an ability to find its way around has encouraged experimenters to develop many different kinds of mazes to investigate rat behaviour. Undoubtedly, a rat's ability to learn the layout of its surroundings has contributed to its success as a species. The extent of wandering by rats is illustrated by one study (Russell *et al.*, 2005). A single male rat, released onto a rat-free island off New Zealand and wearing a radio collar, wandered freely over the 600 by 250 metre island, and evaded capture for 4.5 months after swimming 400 m to another island, possibly looking for potential mates. The rat's ability to find its way around mazes in the laboratory led the psychologist Edward Tolman (1948) to propose that as rats explore and learn about their environment they build up something like a 'field map' in their brains. This idea is more commonly expressed as a '**cognitive map**', and implies that an animal can work out novel routes to get from one place to another without having to learn the individual routes by direct experience of moving along them. It is a useful and powerful strategy: exactly how a rat uses its navigational abilities under natural, rather than laboratory, conditions could well be a fruitful subject for research.

Local navigation and the hippocampus

A region of the cerebral cortex that is implicated in spatial memories is the **hippocampus**. There is evidence that a similar part of the brain of fish, amphibians and reptiles is important in spatial behaviours, although these animals lack a distinct structure like the mammalian hippocampus (Rodríguez, 2002; Jacobs, 2003). Birds also have a hippocampus, and it is particularly well developed in species such as jays and chickadees that cache seeds or nuts and need to remember where their food stores are (Shettleworth, 2003). In humans, London taxi cab drivers, who need to pass an exam on their detailed knowledge of the layout of streets, have differences in the relative sizes of parts of their

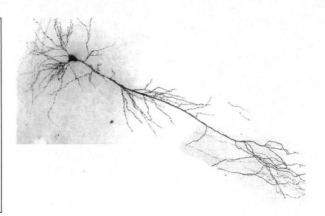

Photograph of a pyramidal neuron in area CA1 of the hippocampus of a rat. Its structure was revealed by injecting into it an enzyme called neurobiotin. It is likely that place cells, which rats use for navigation, are this type of neuron. In the plane of this photograph, the axon is not visible, but the cell body and many dendrites are. The cell body was 15 μm across. (Photograph supplied by Mark Cunningham, Newcastle University.)

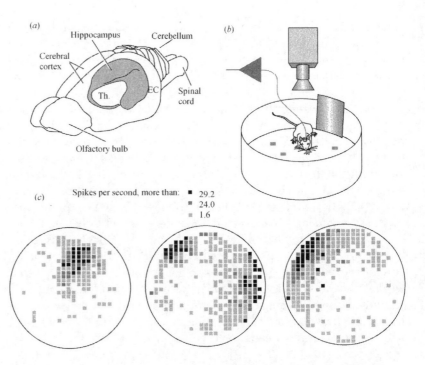

Fig. 8.5 Rat hippocampal place cells. (*a*) Diagram of a rat's brain showing the left side of the hippocampus (shaded), and locations of the thalamus (Th) and entorhinal cortex (EC). (*b*) The set-up for recording from individual place cells. The rat wanders in a circular arena that has a few food pellets on its floor. A sheet of grey card is drawn to indicate a spatial landmark. An electrode with its tip in the hippocampus is connected through a lightweight cable to an amplifier, and the rat's movements are recorded by a video camera. (*c*) Place fields of three different hippocampal neurons. The arena is divided into several small squares, and the depth of shading in each indicates the frequency of spikes by one of these neurons when the rat is at different locations. (Part (*a*) redrawn from Amaral and Witter, 1995; (*c*) redrawn from Muller *et al.*, 1987.)

hippocampus compared with other people. In rats, the hippocampus is relatively large, and is shaped a little like a pair of joined bananas, one on each side of the brain (Fig. 8.5). If it is damaged, rats perform poorly in some learning tasks, including learning about their surrounding landscape.

Neurons in the hippocampus have properties that make them suitable for roles in spatial learning: they respond in appropriate ways, and their responses are modifiable. In rats, one task that is often used to assess the effects on behaviour of treatments applied to the hippocampus, including lesions by surgery or chemicals or applications of drugs, is called the 'Water Maze'. The water maze was invented by Richard Morris (1981) and includes a circular pool about a metre across. Despite its name, it is not really a maze, but rats learn to use visual landmarks around the pool as they swim in it. Rats are good at swimming, but prefer to rest on a dry platform, particularly if the water is chilly. A small platform in the pool is invisible to the rat because it is just below the surface and the water is made opaque by mixing milk powder into it. When a rat is placed in the pool, it swims around until it finds the platform, climbs onto it and sits on it. The rat uses landmarks around the pool to learn where the platform is. Normal rats learn this quickly and on subsequent trials swim directly from any place that they are put into the pool, as do rats with damage to various parts of the cerebral cortex. But if the hippocampus or structures near to it are removed or damaged, rats fail to learn where the platform is, and on repeated trials swim around for some time until they find it. Rats with a damaged hippocampus learn to find a platform at the edge of the pool near to a single visual landmark such as a pole, which indicates that the hippo-campus is particularly involved in learning spatial layout of several objects rather than a particular response to a single cue, such as 'turn to the left when you reach the pole'.

The hippocampus is intimately connected to other brain regions, which work alongside it to give an animal the ability to find its way around, and it receives information from several different sensory modalities. There are many ideas about the way it works. It might be a complex switching device, directing particular signals to appropriate locations elsewhere to be stored and later retrieved as memories; it might itself store memories; or may act in both of these capacities.

Place cells

About 10 years before the water maze was introduced, John O'Keefe and Jonathan Dostrovsky (1971) reported that individual hippocampal neurons, which they called '**place cells**', were excited when a rat was in a particular part of its environment. Most of these place cells are probably pyramidal cells. O'Keefe and Dostrovsky devised a way to record from individual neurons in the hippocampus while rats were running on a cross-shaped elevated maze, learning by using six visual cues which arm of the cross had a food reward at its end. After the rat had been trained, but not before, individual neurons spiked most vigorously when a rat was on a particular part of the cross. Each neuron has its own 'place field', analogous to receptive fields of sensory neurons but built up gradually as the rat wanders around experiencing its environment. Later, the discovery was extended to rats walking in circular or rectangular arenas, and various techniques are used to display activity in individual neurons or in ensembles of many neurons along with location in the arena. Often, food pellets scattered in the arena encourage a rat to wander over the whole arena; it does not learn to do a particular task. Each of these neurons has its own place field, and the neuron spikes most vigorously when a rat wanders into the neuron's own place field. The spikes are often produced in sharp bursts, with 2–10 spikes separated by as short as 4 ms within a burst.

Over time, individual neurons in the hippocampus develop their own characteristic place fields. When a rat is in a given location, most hippocampal neurons are silent, but those whose place field covers the location spike at frequencies up to about 50 Hz. A single location might excite hundreds or thousands of individual place cells. Different groups of place cells, therefore, spike in sequence as a rat moves around. As shown in Fig. 8.5c, the fields of individual place neurons vary in their location within the environment, and also in their shape. Some cover crescent-shaped regions, others blob-shaped regions, and a few place cells have fields split into different regions. The hippocampus is not arranged as a topographic map as is the owl's auditory cortex (Chapter 6), so there is no regular, predictable change in the field of one place field compared with its neighbour, other than that place cells at the dorsal end have small fields whereas those at the ventral end have very much larger fields. This organisation might be expected for a structure that needs to record features of an unpredictable environment.

It is the overall configuration of landmarks that matters to a place cell: the field is not altered if a single landmark among half a dozen is moved, and background visual landmarks seem to be more significant than nearby ones. Rats can be trained to find their way around different arenas, surrounded by different config-urations of landmarks, and then place cells undergo extensive remapping. The location of a cell's new place field is not predict-able from its original location. When a rat has got used to two different arenas, the same place neuron can have a field that is in a quite different location in the two different arenas. What this suggests is that a rat's location is represented by the activity of many different place cells, which need to be combined together somehow.

Place cells mark location irrespective of the direction in which a rat is heading or how fast it is moving. They gain their information from a variety of different sources, including vision, the balance organs of the inner ears, proprioceptors, and the whiskers – vital sensory structures for rats, particularly at night or while in a burrow. If a rat learns its way around an arena in the light, the fields of individual place neurons are retained when the rat wanders around the arena in darkness, showing that both the rat and its place cells keep track of where it is by using whatever information is most readily available. In darkness, over a time of several tens of minutes, the correspondence between the visually learned location and the place field does begin to deteriorate. But when landmarks are not changed around, an individual place cell retains its place field for several months, with the longest recorded remaining stable for at least 153 days before the experimenters moved the electrode to sample another neuron (Thompson and Best, 1990). Some place cells anticipate turns that a trained rat will make to find a food source in a maze. Place cells are not just active when a rat wanders about, but sometimes when the rat is asleep the cells spike in a similar way to when it is exploring its environment. This observation accords with the importance of sleep in consolidating memories in mammals.

The hippocampus might be a bit like a warehouse, able to store new memories ready to be delivered to other locations in the brain. If the place cells are like the shelves, they are not specifically labelled for memories of particular locations in the rat's surroundings. Such an open design is essential to enable the rat to learn the layouts of lots of different kinds of places that it might encounter: a rigid design would severely restrict what it could learn. But where do place cells gather their information from? Two things a rat needs to do to keep track of where it is are to know in which direction it is moving at each instant, and to measure how far it has moved. Two different types of neurons in cortical regions nearby to and connected to the hippocampus have been discovered that may help to serve these functions, and these cells are called head direction cells and grid cells.

Head direction cells

Head direction cells, which were first discovered by James Ranck (Taube *et al.*, 1990*a*, *b*, 2007), spike vigorously when the rat's head is pointing in a particular direction within a particular environment. A head direction neuron responds to local cues in the environment rather than a large-scale feature, such as the Earth's magnetic field or direction of the sun in the sky. As for place cells, responses by head direction neurons can be studied in rats that are free to wander around an arena, picking up spikes by using lightweight recording devices attached to the head. They have been found in half a dozen different brain regions, all closely connected with the hippocampus. Responses by a head direction neuron can be summarised in a tuning curve (Fig. 8.6). For most head directions, a neuron produces no response, but it will typically spike when the rat's head is pointing over a range of angles spanning an arc of between 60° and 150°. Within that range, the response climbs steeply until the head is pointing in the neuron's preferred direction. At the preferred direction a maximum spike frequency is produced, which is up to 200 spikes/s in some neurons, but as low as 5 spikes/s in other neurons.

Head direction neurons do not adapt strongly to a sustained stimulus, but their responses are usually sustained when a rat

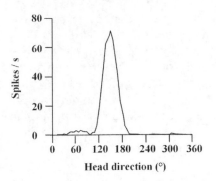

Fig. 8.6 Tuning curve for a rat head direction cell. Spikes were recorded from the cell in a freely moving rat, as in Fig. 8.5b. For most heading directions within the test arena, the cell produced few spikes. For a heading direction of 150° it responded briskly, and responded with progressively fewer spikes for heading directions about 35° either side of its best direction. (Redrawn from Jeffery and Burgess, 2006.)

remains for some time with its head pointing in a particular direction. Head direction cells respond to a variety of sensory cues, including visual cues, movements registered through the inner ears, and sources of odour. Information about movements that the rat itself makes are also important for head direction cell function. When a rat wanders into a new environment, a head direction cell maintains its preferred direction but if the rat is instead moved on a small cart over the same pathway the cell starts from scratch in establishing a new preferred head direction. When moved on the cart, the rat's inner ears provide the same information as when the rat moves itself, but it does not issue its own motor commands or receive proprioceptive information about limb movements.

In a similar way to place cells, as a rat experiences a particular landscape each head direction neuron develops its own preferred direction. Once a rat has learned its way around a landscape, when lights are turned off a head direction cell continues to indicate head direction, although its preferred direction within an arena drifts gradually from the original. When lights are turned on again the neuron's original preferred direction is restored rapidly. Different cells encode different directions, and all directions are represented by similar proportions of cells. In general, the speed at which the rat is moving makes little difference to these cells, although one group of head direction cells (in the thalamus, an early stage in processing visual and other sensory inputs) slightly anticipate a rat's movements, so responses to clockwise and anticlockwise movements made by the rat differ a little.

Grid cells

Grid cells have been mainly described in layer 2 of the medial entorhinal cortex, one of the major regions that provides input to the hippocampus. Like place cells and head direction cells, each is excited in particular ways as a rat moves around, but these **grid cells** are excited at regularly spaced, multiple locations,

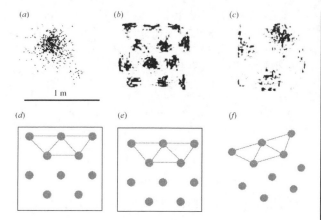

Fig. 8.7 Fields of one place cell and five grid cells in a rat. In (*a–c*), a rat wandered in a square arena, and each small black dot records a spike from the recorded cell at a particular location in the arena. (*a*) Place cell: its field is slightly displaced towards the left top of the arena. (*b* and *c*) Two different grid cells. Each spiked when the rat was in several different locations that were regularly spaced in the arena; the spacing for the cell in (*b*) was smaller than for the cell in (*c*). (*d–f*) Schematic representation to show differences between fields of three different grid cells with similar spacing between their triangle corners. The grey dots indicate locations of the rat in a small square arena in which each cell spikes, and lines show the geometric layout. Compared with (*d*), in (*e*) the fields are displaced and in (*f*) the grid is aligned at a different angle. (Parts (*a–c*) redrawn from Moser *et al.*, 2008.)

which reveal themselves if a rat is allowed a relatively large arena to wander around (Hafting *et al.*, 2005; Jeffery and Burgess, 2006; Moser and Moser, 2008). Each grid cell spikes vigorously as the rat moves over the corners of a series of invisible triangles. When the locations of these corners are plotted for a particular neuron, they form a clear grid-like representation of space (Fig. 8.7*b*), hence the name grid cell for this type of neuron. This contrasts with the fields of place cells, which are generally excited when the rat is within one particular location (Fig. 8.7*a*).

The grid for each cell is a bit like graph paper: it could provide a useful navigational tool for a rat to measure how far it has moved in a particular direction. There is some correspondence between the properties of grid cells and their location within the entorhinal cortex, with the more dorsal cells having a relatively fine grid spacing of about 30 cm within a landscape, and the more ventral cells having a coarser grid spacing near to 70 cm. Fields for two grid cells with different spacings are shown in Fig. 8.7*b*, *c*. Neighbouring cells with similar grid spacing and orientation have the locations of the angles of their triangles offset relative to each other in a random way (Fig. 8.7*d*, *e*). Grid orientation varies between different cells (Fig. 8.7*f*) and rotates gradually from one cell to the next. Grid cells respond to visual landmarks, but also work well when the rat moves around in darkness, showing they have access to a variety of types of sensory cues. A grid pattern is sustained as a rat runs, constantly changing its direction and speed, so there needs to be some way in the brain that constantly updates the animal's current location.

The properties of a grid are not completely fixed; for example, if one wall of an arena that a rat is used to is stretched, spacing of a grid is also stretched in that direction. There is some evidence that grid cell properties are influenced by feedback from the hippocampus. Exactly how the grid pattern of individual cells is established is not known; grid cells could be like the auditory place cells in a fledgling owl's brain, receiving instructions from a connecting region of the brain (Chapter 6).

It is possible to imagine how a particular location in space could be represented by combining signals from overlapping grids with different spacings, orientations and offsets. If a cell in the hippocampus receives connection from a number of different grid cells, then adjusting the relative balance of the strengths of the synaptic connections could create a place field for that hippocampal cell. In order for a mechanism such as this to work, the properties of synapses within the hippocampus need to be modifiable. Indeed, hippocampal synapses show quite a lot of plasticity, changing their strengths in various ways.

Hippocampus and long-term potentiation

Hippocampal place cells are largely pyramidal neurons, similar to the cortical pyramidal neuron shown earlier in Fig. 1.3a. For them to develop and alter their fields as a result of experience there must be changes in the neuronal networks that deliver information to the place cells. The most likely changes are that some synapses become stronger while others become weaker with usage, so that the relative balance of excitation onto a particular place cell from the different neurons that converge on it alter as a rat explores its environment. Ramon y Cajal suggested that changes in the strengths of connections between neurons could be a basis for learning and later, in the 1940s, the Canadian neuroscientist Donald Hebb proposed that a mechanism for learning could be that synapses would strengthen if both partner neurons in a particular synaptic connection were excited simultaneously. Such use-dependent changes in synaptic strength have been investigated in detail in the rat hippocampus, and a great deal is known of the physiological and molecular events that underlie them. It is likely that they are vital in the mechanism for establishing place fields, although completely definitive evidence is extremely hard to establish.

One reason why the hippocampus is particularly amenable to physiological study is that thin slices of it can be removed from the brain and, when bathed in an appropriate solution, they can continue to function for long periods to allow detailed investigation of their properties. The hippocampus also has a very regular internal structure, with its principal neurons lined up, as indicated by thin grey bands in Fig. 8.8a. Many different pathways interconnect neurons within the hippocampus, and connect the hippocampus with other brain regions. One pathway that is often studied conducts signals across three sequential, excitatory synapses. The pathway starts in axons from the entorhinal cortex (where grid cells are found), which excite granule cells. The granule cells send axons to dendrites of pyramidal cells in region CA3. The axons of the CA3 pyramidal cells divide: one branch leaves the hippocampus, and the other connects with dendrites of pyramidal cells in

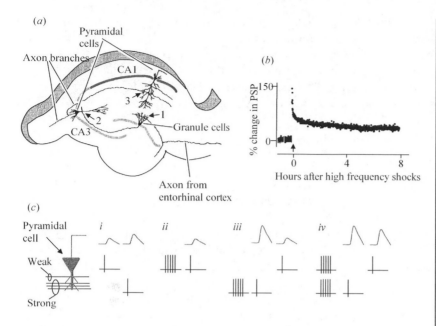

Fig. 8.8 Hippocampal pathways and long-term potentiation (LTP). (a) Summary diagram showing three of the principal neuron types in the hippocampus, granule cells and pyramidal cells in areas CA1 and CA2. Numbers 1–3 indicate three synaptic connections in one pathway, from axons from the entorhinal cortex to granule cells then to CA1 pyramidal neurons then to CA3 pyramidal neurons. The dendrites of the pyramidal neurons are directed towards the centre of the hippocampus; their axons are directed outwards from the cell bodies. (b) Long-term potentiation of a postsynaptic potential in CA3 pyramidal cells, from an experiment explained in the text. (c) Summary of important properties of LTP. The top traces show EPSPs recorded from a pyramidal cell in response to stimuli delivered to a small bundle of presynaptic axons providing a weak synaptic input to the pyramidal cell (stimulus pulses in the middle trace), or to a wider bundle of axons providing a strong synaptic input (stimulus pulses in the bottom trace). (Parts (a, b) redrawn from Nicholl et al., 1988.)

region CA1. Each CA3 axon contacts very many CA1 pyramidal cells, and each CA1 cell receives synapses from axons of many individual CA3 pyramidal cells. The CA3 pyramidal cells also make extensive connections among each other – by carefully examining adjacent sections under the electron microscope, it has been shown that each CA3 pyramidal cell receives about 16 000 individual synaptic contacts, of which three quarters are from its sister CA3 pyramidal cells.

Use-dependent change in synaptic strength in the hippocampus was first described in 1973 (in rabbits) by Tim Bliss and Terje Lømo. The phenomenon they described is called **long-term potentiation** (LTP), in which synaptic strength increases. Several other use-dependent changes in hippocampal pathways have been described since 1973, some causing increases in synaptic strength and others decreases. Some characteristics of LTP are shown in Fig. 8.8b, c. In one experimental arrangement, EPSPs are recorded from CA1 cells while a bundle of CA3 cell commissural axons is stimulated with electrical pulses that are each large enough to trigger a spike synchronously in each of the axons of the bundle. At a low repeat frequency of stimuli, one every 20 s, each stimulus pulse causes an EPSP in the CA1 cells that receive synapses from the CA3 cells. But if a burst of stimuli 10 ms apart and lasting 1 s is delivered, the synaptic connections are potentiated, or strengthened, so that the next single pulse stimulus elicits an EPSP that is significantly larger than the original EPSPs. This is what occurs at the arrow at time 0 in Fig. 8.8b. The increase in synaptic strength gradually declines, but one brief burst of high frequency stimulation is sufficient to boost synaptic strength over many hours.

In a slightly more complex experiment (Fig. 8.8*c*), two different bundles of presynaptic axons are stimulated, one wider than the other so containing many more axons. A single brief shock to the smaller bundle provides a relatively weak synaptic input to the pyramidal cell because it activates a relatively small number of individual synaptic contacts, and a single shock to the larger bundle activates a larger number of synaptic contacts, so provides a stronger synaptic input to the pyramidal cell (Fig. 8.8*c,i*). If the small bundle receives a high frequency shock, its synapses are not potentiated – a test shock 20 s after a short burst evokes an EPSP that is no larger than an EPSP before the burst (Fig. 8.8*c,ii*). However, as expected, a burst of high frequency shocks to the larger bundle does enhance the size of subsequent shocks delivered to the large bundle (*iii*). The difference between *ii* and *iii* in Fig. 8.8*c* is explained by a threshold in synaptic input strength for potentiation to occur – only EPSPs that reach a certain size are potentiated.

The explanation for a threshold for potentiation lies in the nature of the neurotransmitter receptor and its associated ion channel in the postsynaptic neuron, which is the same NMDA receptor that is a key player in swimming in tadpoles (Chapter 7). The NMDA receptor's ion channels are only activated if the postsynaptic neuron is already excited when the neurotransmitter (glutamate) arrives at the cell surface. So the NMDA is a coincidence-detecting device: to respond, it needs both an electrical signal to excite its neuron, and it needs a chemical signal, the glutamate released by the presynaptic neuron. As in the tadpole spinal cord, other receptors for glutamate at the same synapses respond more quickly to the transmitter, causing brief EPSPs in the postsynaptic neurons. If the excitation through these first-activated receptors is strong enough, then the NMDA receptors will also be effective at exciting the postsynaptic neuron, causing a larger and more long-lasting EPSP. In the experiment, the smaller axon bundle does not excite the postsynaptic pyramidal cell sufficiently for its NMDA receptors to respond, but the large axon bundle does do that. The NMDA receptor and its channel form a molecular device that underpins the cellular mechanism for learning proposed by Hebb. Successful activation of NMDA receptors then triggers a sequence of events within the cells that result in a lasting increase in synaptic strength.

Potentiation is specific to activated synapses. Following a high-frequency stimulus burst to the larger bundle, although the next shock to that bundle causes a much larger EPSP than before, the size of an EPSP evoked by the smaller bundle is not enhanced (Fig. 8.8*c,iii*). But if both the small and wide bundles are given a high-frequency burst simultaneously, EPSPs in both pathways are potentiated (Fig. 8.8*c,iv*).

Is LTP a mechanism for establishing place fields? There are very strong lines of evidence that it is, but none have conclusively shown that it is both necessary and sufficient. One of the problems in demonstrating precisely how place cells acquire their fields is that the exact pathways involved need to be teased out from the many possibilities within the hippocampus – some experiments show, for example, that CA1 cells may be able to become place cells with no involvement of CA3 pyramidal cells and their commissural axons (Brun *et al.*, 2002). Another gap in knowledge is that it is not known what kinds of signals the entorhinal and hippocampus neurons generate in a rat that is exploring a new set of landmarks. Characteristics of LTP have largely been explored in experiments where the experimenter selects and controls the pattern of spikes in presynaptic axons, but we really need to know what happens when the pattern is determined by a rat's natural behaviour.

One good line of evidence that LTP is involved in rats' navigational behaviour is that drugs that block NMDA receptors interfere both with LTP and with a rat's ability to perform in the water maze (Morris *et al.*, 1986). Similarly, old rats forget faster and LTP decays faster in them than in young rats. More recently, genetic engineering has shown that affecting particular molecular targets in specific cell types affects both LTP and place cell properties. But none of these techniques is specific to unique targets – drugs can diffuse outside their target areas, and changes in molecular function have diverse metabolic effects. Neves *et al.* (2008), reviewing the lines of evidence that LTP is involved in establishing place cells, suggest that a different experimental approach may be needed. Instead of interfering with LTP, artificial stimulation of appropriate brain cells, such as groups of neurons in the entorhinal cortex, might create novel place cells and novel behaviours in rats.

Phase change in locusts

A swarm of locusts is a spectacular sight, and a large swarm can cover hundreds of square kilometres with enormous potential to devastate crops. About 20 of the 12 000 species of grasshoppers are known as 'locusts', a term that means they sporadically form dense swarms containing very many individuals. Two of the most commonly investigated locusts are *Schistocerca gregaria* and *Locusta migratoria*, species that are often the subjects of investigation into various aspects of neuroethology (Chapters 2, 5, 7). When population density increases, locusts switch their behaviour from a solitary, cryptic lifestyle, called the solitarious phase, to a lifestyle where they aggregate into swarms, called the gregarious phase. These changes are triggered by particular sensory experiences, and the stimuli and sense organs responsible for the switch have been identified in a series of research projects initially led by Steve Simpson. A change from the solitarious to the

Behaviour and morphology of locusts alters according to how crowded they are. The photographs show late stage larvae of *Schistocerca gregaria*: upper, gregarious phase; and lower, solitarious phase. Below is a photograph of the middle thoracic ganglion in which serotonin-containing neurons, which are involved in the switch from solitarious to gregarious, are stained with a fluorescent labelled antibody. They are few in number, but their branches reach almost all areas of the ganglion. (Photographs of locusts supplied by Steve Rogers, Cambridge University; photograph of ganglion supplied by Paul Stevenson, Leipzig University.)

gregarious phase is called gregarization. Changes in behaviour are often followed later by changes in form, colour and physiology. An animal that can take on different sets of characteristics is termed polyphenic, and a locust proves to be an excellent animal in which to investigate how different sets of genes become activated under different circumstances, giving rise to two different forms of the same organism with contrasting characters. In some cases, changes in the properties of individual neurons and synapses that must underlie changes in behaviour have been identified (Matheson *et al.*, 2004; Rogers *et al.*, 2007).

In nature, the switch from a solitarious to a gregarious phase is triggered by changes in vegetation density, itself often triggered by changes in weather. When vegetation is relatively sparse, locusts remain solitarious, avoiding contact with each other, hiding and moving relatively infrequently. Their colour blends in with the background. When vegetation is abundant, young locusts, or hoppers, will move towards other hoppers; the insects move around constantly and, in some species, take on vivid patterns that warn birds away from attacking them – the locusts often eat plants containing alkaloids that are toxic to birds and mammals. The population density of solitarious locusts is very much less than that of gregarious locusts: 3 compared with 100 000 per 100 m^2. Gregarious locusts are constantly on the move, competing with each other for food and, possibly, trying to avoid being food for their fellow locusts. Offspring of a gregarious female tend also to be gregarious, their phase being determined by a chemical the female deposits in the protective foam that surrounds egg cases.

The first changes to occur during gregarization are in behaviour, and these start to appear within a few hours of crowding several solitarious locusts together. Other changes, including those in morphology and colour, take longer. The switch from gregarious to solitarious can take longer than a single generation. Because the first changes to occur during gregarization are in behaviour, Simpson and colleagues devised a way to assay an individual locust's behaviour and give it a gregarization index, from 0 as fully solitarious to 1 as fully gregarious (Roessingh et al., 1993). This assay is important because it provides a way to compare the relative effectiveness of different stimuli in switching a locust's phase. Experiments are usually conducted on hopper, or larval, locusts in their fifth instar, the one before the adult. The major difference between an adult and a larva is that the adult has functional wings and can fly, as well as mate and reproduce, but in most other respects, as a larva moves from one stage to the next it grows larger without significant changes in morphology. In a behavioural assay, a hopper is introduced into the centre of a 30-cm-square box. The two sides of the box are painted white, and the ends are sheets of clear plastic through which the subject of the investigation can see other locusts or moving objects. About a dozen different behavioural traits are measured and then combined to give the index; they include direction and speed of walking, and the frequency of resting, hopping and grooming.

The sensory trigger for gregarization

The essential sensory stimulus for gregarization turns out to be remarkably simple: mechanical stimulation of the outer surface of the large hindlegs. When vegetation is abundant locusts jostle each other, like people elbowing each other in a crowd. A locust will often accidentally rub its own head and other parts of its body on leaves and twigs while clambering over them, but touching and pushing the hindlegs is more likely to be caused by brushing against neighbours while feeding side by side in company with other locusts. In one series of experiments

Fig. 8.9 Phase change in locusts.
(*a*) Side view of a fifth instar solitarious *Schistocerca*, indicating the effectiveness of stroking different locations in triggering behavioural phase change.
(*b*) Behaviour of solitarious hopper before and after gregarization. The hopper was released into the centre of a 30 cm by 30 cm box with a transparent window through which it could view a group of other hoppers (3 are shown; 20 were used in an experiment). The locust was released into the centre of the box and its path during 500 s from release time was traced. Before gregarization, it walked slowly away from the other hoppers; after, it was strongly attracted to join them and walked more quickly. (*c*) Cell bodies of serotonin-containing neurons in a locust metathoracic ganglion, revealed using an antibody raised against serotonin. (Part (*a*) redrawn from Simpson *et al.*, 2001,
(*b*) redrawn from Anstey *et al.*, 2008,
(*c*) redrawn from Tyrer *et al.*, 1984.)

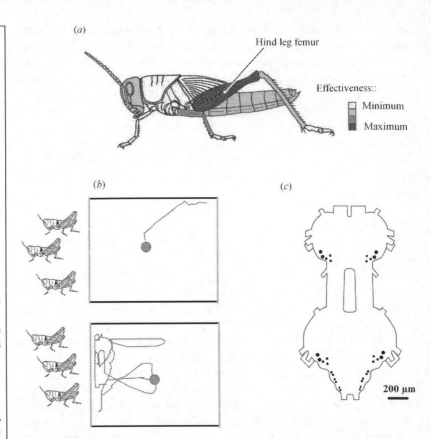

(Roessingh *et al.*, 1998) locust hoppers were exposed to various different types of stimuli that a hopper in a crowd of locusts might experience through its chemoreceptors, eyes and mechanoreceptors. Neither the site nor the smell of other locusts on their own would cause solitarious locusts to change phase, although these two types of stimuli did have some effect when combined with each other. But when a solitarious locust was constantly jostled for an hour or so by 1-cm-wide paper balls, continually rolled around on a tilting table, its behaviour when tested in the standard way changed dramatically. It became increasingly restless, walking and hopping towards a group of hoppers at the end of the test arena.

Later experiments, in which scientists tickled different parts of a locust's surface once every minute for up to four hours, established that it is specifically the femur of the hindleg that must be stimulated (Simpson *et al.*, 2001; Fig. 8.9*a*). As little as a quarter of the outer face of the femur needs to be exposed to a brushing stimulus for gregarization to occur (Rogers *et al.*, 2003). The anterior, dorsal part is the most effective. In solitarious locusts, the number of sensory hairs on the hindleg femur is about a third again as the number on the hind femurs of gregarious locusts, although the other legs and other segments of the hindlegs have slightly fewer hairs in solitarious locusts.

This points to the outer-facing part of the hindleg femur being specially adapted to respond to gregarizing stimuli. But the hindleg hairs alone provide an insufficient sensory stimulus because no gregarization occurs if a locust's leg is held immobile while it is brushed. The leg needs to be mobile, so that brushing it will cause some movement and stress to the exoskeleton and joints of the leg. This conclusion is reinforced in experiments in which electrical stimuli were applied to axons in various branches of one of the major leg nerves. Stimulating one of its major branches with shocks in a similar pattern and duration as the sensory stimuli is effective at gregarizing a locust. Small branches that supply only sensory hairs or only proprioceptors are not, on their own, effective when stimulated. But stimulating larger branches closer to the ganglion, which contain axons both of hairs and of proprioceptors, is effective at causing gregarization.

The role of serotonin

Having identified the sensory stimuli and sense organs responsible, the next stage in understanding the sequence of events in triggering gregarization is to identify neurons in the pathway underlying this switch. A significant clue to their identity comes from the discovery that the neurotransmitter, serotonin, is necessary and sufficient for gregarization (Anstey *et al.*, 2008). This is the same neurotransmitter that is important in changes in the dominance status of crayfish (Chapter 4) and other social behaviours. In establishing the role of serotonin, solitarious locusts were crowded together for periods of 0, 1 or 2 hours to cause different degrees of gregarization as assayed in the behavioural tests described before (Fig. 8.9b). Only a handful of neurons in each locust thoracic ganglion contain serotonin (Fig. 8.9c), so each of these is a strong candidate for playing a part in gregarizing behaviour.

The concentration of serotonin in various parts of the nervous system was measured. It increased during gregarization in the thoracic ganglia but not in the brain. In a fully gregarized locust, serotonin concentration in the thoracic ganglia tripled, and the increase was proportional to the change in the behavioural index rather than to the duration for which an individual was stimulated. Furthermore, when initially solitarious hoppers were stimulated with either repeated leg stroking, electrical stimuli to leg nerves, or by a combination of the sight and odour of a crowded group of hoppers, the increase in serotonin concentration was well correlated with the change in gregarization index.

So there is a good correlation between an increase in serotonin levels and a change in behaviour; but is this coincidence, or is there a causal link between serotonin and a change in behaviour? Applying specific drugs into the thoracic ganglia to block the known actions of serotonin on target neurons prevented gregarization in response to stimuli that were usually effective. The drugs were injected into thoracic ganglia together with a harmless green dye to confirm the site of injection and, in control experiments, a solution of the dye alone did not reduce the degree of gregarization. Locusts in which another drug was used to block the enzyme that synthesises serotonin also failed to gregarize normally. These experiments suggest that serotonin manufacture and release is necessary for gregarization. To test whether an increase in the concentration of serotonin within the thoracic ganglia is sufficient to induce gregarization, the thoracic ganglia were bathed in a solution of serotonin for two hours. The serotonin was injected into a bath made of petroleum jelly surrounding the ganglia, so the serotonin did not reach other tissues within the thorax. This treatment was consistently effective at triggering gregarization within two hours. Drugs that mimic the action of serotonin were also effective at causing gregarization. Finally, when the biochemical precursor for serotonin was injected into the body cavity, stimulus periods that were briefer than usual caused strong gregarization, presumably because the increase in substrate concentration enhanced the rate at which serotonin levels rose. This is an important experiment because it indicates that two steps are involved in linking sensory stimulation to gregarization: an increase in serotonin levels, and then an effect mediated by the serotonin. The rise in serotonin declines over about 24 hours, and serotonin is not needed to maintain the gregarious phase once it has been triggered – in fact, established gregarious locusts have lower serotonin levels than solitarious locusts. Serotonin may act to regulate particular sets of genes that, once expressed, generate an altered, but relatively stable, phenotype characteristic of the gregarious phase.

Conclusions

In each of the three examples in this chapter, animals change their behaviour as a result of experience – the experience of a particular flower type providing food for bees, of exploring landmarks in new surroundings for rats, or of finding itself in a crowd for locusts. Techniques have been developed that allow study in the laboratory of the roles of particular neurons in these behaviours, making it feasible to study changes in behaviour at the cellular level. In none of the examples chosen, however, are the complete neuronal pathways that are involved known. Also, although the roles of neurons such as VUMmx1 in bees or place cells in the rat hippocampus in guiding changes in behaviour seem to be demonstrated clearly, the procedures used have been designed by an experimenter selecting appropriate electrical stimuli to create meaningful effects. We do not know what electrical signals the neurons generate during natural behaviours, and how effective those are at triggering changes in behaviour compared with the experimentally selected signals. In fact, it could be quite a leap to extrapolate from a laboratory to a natural situation – for example, place, grid and direction sensitive neurons in the rat are normally recorded when a rat wanders in an arena about a metre across, whereas rats naturally forage over several hundred metres.

The changes in behaviour are highly adaptive, allowing an animal to modify its behaviour in response to environmental alterations. The mammal hippocampus and insect mushroom bodies might be viewed as sophisticated instruments allowing the animal to adapt appropriately to a wide range of possible and unpredictable changes. The memory trace in one animal is very unlikely to be the exact copy of the trace for the same memorised features in another animal – so what are the rules that a hippocampus or mushroom body use to determine where the information they process is directed to; and how is the information retrieved? Developing the tools needed to address these kinds of questions is a major challenge in understanding brain mechanisms.

Questions

Can you fill a memory up? Is this a sensible question to ask, and if so, how could you test it experimentally?

How would you investigate the evolutionary origin of phase transition in locusts?

Summary

- Animals are programmed to alter their behaviour as a result of experience, or as part of their normal life cycle.
- Changes in the properties of neurons and synapses are fundamental properties, but it can be hard to identify which neurons are involved in learning and other behavioural changes.

Honey bee proboscis extension response

- Honey bees learn to associate various characteristics of flowers, including odour, with a nectar reward.
- A single training of an odour followed by a drop of sucrose solution to the proboscis is sufficient to condition the response to the odour, forming a new stimulus-response association.
- Conditioning requires the odour to precede the sucrose and the interval between the two stimuli is brief.
- Odours are transduced by sensilla on the antennae with axons that enter glomeruli in the antennal lobes. From the antennal

lobes, some axons travel to the mushroom bodies where they connect with many very small Kenyon cells, which connect with a smaller number of larger extrinsic neurons.

- Local cooling and octopamine injection suggest the antennal lobes and mushroom bodies are involved in associative conditioning.
- The unpaired suboesophageal neuron VUMmx1 is involved in conditioning to floral odours, and:
  o branches extensively to overlap neurons involved in processing olfactory stimuli
  o contains and probably releases octopamine
  o is excited by sucrose applied to the proboscis
  o is excited by a particular odour once that odour has been paired with a sucrose reward
  o substitutes for a sucrose reward in conditioning the proboscis extension reflex to a novel odour.

Rat place cells

- Rats are good at learning spatial layout, often studied in various mazes.
- Their performance is impaired by damage to the hippocampus, but not other brain regions.
- Place cells in the hippocampus spike when the rat is in their place field, and:
  o a place field develops and changes as a result of experience
  o a place cell gains information from several sensory modalities
  o a place field may be established if a rat keeps track of its heading direction and measures how far it has travelled.
- Head direction cells respond most vigorously for a particular heading direction with respect to landmarks, and:
  o their best direction is altered by experience, and they are multimodal.
- Grid cells respond most vigorously when the rat is at the intersections of regular triangles in a grid-like array in the environment, and:
  o vary in the spacing and orientation of their grids
  o are located in the entorhinal cortex, which sends axons to the hippocampus.
- Hippocampal neurons make synapses with properties that change strength in predictable ways in response to particular stimulus patterns.
- Synapses between CA3 and CA1 pyramidal cells show long-term potentiation: synaptic strength is potentiated for several hours following a single high frequency burst, and:
  o potentiation involves NMDA receptors for the neurotransmitter glutamate.
- It is likely that the plastic properties of hippocampal synapses are very important in establishing the fields of individual place cells.

Phase change in locusts

- Locusts have two distinct behavioural, morphological and physiological phenotypes: solitarious and gregarious.
- In nature, the change from solitarious to gregarious is triggered by an increase in food availability.
- Phase change is triggered by stimulating particular sensory receptors, especially hairs and proprioceptors on the hindlegs.
- The extent of change can be measured in behavioural tests, in which the attractiveness of groups of hoppers is assessed.
- A gregarizing stimulus causes a significant increase in serotonin in the thoracic ganglia.
- An increase in serotonin in the thoracic ganglia is both necessary and sufficient to trigger gregarization.
- Only a handful of serotonin-containing neurons exist in the thoracic ganglia.

Further reading

Giurfa, M. (2007). Behavioral and neural analysis of associative learning in the honeybee: a taste from the magic well. *J. Comp. Physiol. A* **193**, 801–824. A review that covers many aspects of learning by honey bees.

Kandel, E. R. (2001). The molecular biology of memory storage, a dialogue between genes and synapses. *Science* **294**, 1030–1038. This review is a detailed account of Eric Kandel's research and shows how the molecular mechanisms that underlie learning are similar in diverse animals.

Simpson, S. J. and Sword, G. A. (2008). Locusts. *Current Biology* **18**, R364–366. An account of how and why locusts and some other insects swarm.

9

Nerve cells and animal signalling: songs of crickets, electric fish and birds

When an animal attracts a mate or declares its ownership of a territory, it has to advertise. Signals used in communication usually have clearly defined functions, and signalling behaviour is sometimes associated with clear specialisations within the nervous system. Not only does the animal that sends signals need to make the meaning of its signals clear, but also the animals for which the signals are intended need to be able to capture, interpret and respond appropriately to those signals. So, during evolution, the motor signals that individuals of a species use to create signals need to be matched by appropriate sensory filters to recognise the same signals. For these and other reasons, communication is a fruitful area for investigation by neuroethologists.

In this chapter, we shall describe three different groups of animals that specialise in making and receiving particular types of signal. Two of these, crickets and song birds, use songs – extended patterned sounds – for communication. The other group are electric fish, which broadcast waves or pulses of electrical voltage in the water rather than audible sounds. The function of these electrical signals is not only to communicate with each other, but also to investigate their nearby surroundings. Sound and electric signals share the advantage that they can be used in conditions when sight is of limited use, such as at night or in undergrowth. Sound is useful for advertising because a small animal can use it to send messages rapidly over relatively long distances, but electric fields fade out over relatively short distances. Both the temporal structure and different pitches within a song can carry signals and that is true also for the waves broadcast by some electric fish. A disadvantage of broadcasting signals, as with all advertising, is that it can attract the attention of an unwelcome audience in addition to an intended mate. It is not surprising that a number of animals, including birds and lizards, hunt singing crickets, and some small parasites can locate cricket hosts by this means.

The behaviours described in this chapter are relatively complex, but many of the topics and concepts covered in previous chapters are relevant to understanding the way nerve cells are involved in communication behaviour. First, songs often contain repeated, patterned elements. The motor control mechanisms responsible for generating them include central pattern generators, which are subject to sensory feedback in the same way as those involved in locomotion. Second, there need to be mechanisms for switching the song pattern generators on and off, and these mechanisms include command neurons. Third, sensory filtering is important, both in the physical design of sense organs and in the wiring of the neuronal networks associated with them, to enable an animal to pick out a significant song from a conspecific animal against the barrage of other sensory information it receives. Fourth, both the motor control systems and the sensory pathways involved in communication are modified in ways that integrate the business of communication with other activities of the animal. Bird song, the final topic in the book, provides a particularly sophisticated example of integration with life cycles because there are critical periods within the bird's life when it needs to have particular experiences, such as hearing another bird sing, in order for its nervous system to generate an effective song when it matures.

Finally, a feature of the neuronal control of song in crickets and birds is that the nervous system generates a pattern not only for controlling the muscles responsible for making a signal, but makes a copy of that pattern for internal reference. This type of internal record is called a '**corollary discharge**' (Poulet and Hedwig, 2007). One function for corollary discharge, illustrated by work on crickets, is that it enables an animal to modify the properties of a sensory system in a way that is appropriate to whatever it is doing. In the case of a singing cricket, auditory responses are damped so that the sound of its own song does not deafen it to externally generated sounds of potential importance. A different function for a corollary discharge in song birds is to help the bird construct its adult song. A bird models its own song on memories of songs it heard soon after hatching, and when it becomes adult it tries to match persistent memories of snatches of remembered songs with an internal representation of its own song.

Cricket song

The songs of two insect orders are often noticed by people, although many kinds of insects, including flies such as *Drosophila* as described in Chapter 1, produce much less noticeable songs. One of these orders, the homoptera, includes cicadas, which sing by repeatedly buckling a specialised area of the cuticle called the

A male cricket (*Gryllus*) singing. His forewings are raised from his body and he sings by rubbing a scraper on one wing over a toothed vein, or file, on the other. (Photograph by Peter Simmons.)

tymbal (Simmons and Young, 1978; Young and Bennett–Clark, 1995). The other order is the Orthoptera, which includes grasshoppers and crickets, that sing by rubbing body parts together, an action called **stridulation**. Grasshoppers stridulate by rubbing their hindlegs against a hardened vein on one wing, and crickets sing by rubbing one forewing against the other. Altogether there are just over 2600 species of crickets, including bush crickets (also known as long-horned grasshoppers or tettigoniids) and mole crickets.

A song is easily recorded with a microphone, and the signal is displayed on a screen. It is almost always male insects that sing, and the primary function of the 'calling' song is to attract females for mating. A calling song consists of distinct pulses of sound, usually referred to as 'syllables'. In the European field cricket *Gryllus campestris*, the syllables are usually gathered into short groups of 3 or 4, each group called a 'chirp' (Fig. 9.1a). Once a female is nearby, crickets switch to a softer 'courtship' song that probably keeps a female interested while the two insects locate each other using a number of cues including pheromones plus perhaps vision and touch. Courtship song is much quieter than calling, with less frequent syllables. Another function of song for some species of cricket is declaration of territory, and nearby males may compete with each other with loud and insistent 'rivalry' songs. The way syllables are arranged in a calling song varies between cricket species, and is often more complex than in *G. campestris*, as shown in Fig. 9.1b for an Australian species, which produces extended series of syllables, called trills, in addition to chirps. A single syllable consists of an almost pure tone of sound, which rises and then declines in volume.

To sing, a male cricket first raises his forewings clear of his body, and then alternately closes them inwards and opens them outwards

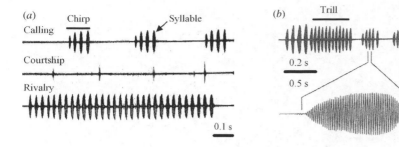

Fig. 9.1 Sonograms of cricket songs, in which each trace indicates how sound loudness varies with time. (*a*) Three song types of the European field cricket, *Gryllus campestris*. The calling song consists of a series of chirps, each containing 3–4 syllables. In courtship, the cricket produces single, short and rather quiet syllables. Rivalry is louder, consisting of a longer unbroken series of syllables. (*b*) Calling song of the Australian cricket, *Teleogryllus commodus*, which includes trills of rapidly repeated syllables as well as chirps. One syllable is expanded, below. (Part (*a*) redrawn from Kutsch, 1969.)

(Fig. 9.2). During each closing movement, a hard structure on one wing called the scraper, which is a bit like a guitarist's plectrum, slides against a toothed structure called the file on the other wing. Each time the scraper pings on a tooth the wings shake, generating a sound wave. The spacing of the teeth and the speed of movement of the scraper against the file result in a particular number of vibrations per second, which sets the fundamental frequency for the frequency of the song. Song frequency is characteristic of the species; in *G. campestris* it is between 4.6 kHz and 4.8 kHz. The muscles that move the wings during singing are also involved in flying: a muscle that closes a wing during song acts to elevate it during flight, and a muscle that opens it during song depresses it during flight (Fig. 9.2*b*). When flying, both pairs of wings are moved, the forewings are held out from the body rather than elevated directly above it, and the wings twist as they propel the insect along.

Crickets are quite small animals, but their songs can be very loud at close range, up to 105 dB, which is significantly louder than the sound level at which consistent listening to music damages human hearing. The wings operate as very efficient resonators. A resonator is a device that tends to vibrate at a particular frequency, like a bell or a tuning fork, and in the case of a field cricket's wings the resonant frequency is the same as the song's tone. A special area on each wing, called the harp (Fig. 9.2*c*), has been shown to be responsible for resonating and radiating the sound. The rate at which energy is delivered from the ping of each file tooth against the scraper corresponds with the resonant frequency of the wings, so the sound is amplified. In addition to vibration of the wings themselves, the raised wings are curved downwards at their tip, which has the effect of trapping air against the top of the thorax with just the right volume to resonate at the frequency of the cricket's song.

Electromyograms from a singing cricket show that, as in flight (Chapter 7), each of the major wing muscles is activated by a single motor neuron spike for each wing movement cycle, and the response is a fast, strong twitch by the muscle (Fig. 9.2*d*). The sound in a syllable is produced during wing closing, although opening can also produce a much softer sound.

Adult male crickets will often sing spontaneously when kept in captivity. In Eastern Asia, singing crickets in cages are popular pets, and one pastime is to bet on the outcomes of fights between males.

Fig. 9.2 The mechanism of singing by male field crickets. (*a*) Male cricket in his singing posture, with forewings raised clear of the body and drawn here in the closed position. (*b*) A diagrammatic section cut transversely through the thorax showing the way the file on one wing and scraper on the other interact, and showing two of the muscles that open and close each wing. (*c*) Right forewing showing the main structures involved in producing sound. (*d*) Electromyograms showing activation of a wing opener and a wing closer muscle and sonogram of the sound to show how each spike in the closer muscle is followed by a syllable. The syllables become progressively louder as the chirp progresses. (Part (*a*) redrawn from Dambach and Rausche, 1985; (*c*) redrawn from Michelsen and Nocke, 1974; (*d*) redrawn from Kutsch, 1969.)

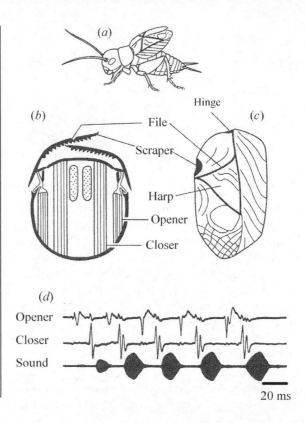

For study in the laboratory it is useful to be able to switch on a male's song at a time convenient to the experimenter. Franz Huber, in the 1950s, discovered that electrical stimuli applied locally to particular brain areas are effective at doing this, and small injections of drugs that mimic the neurotransmitter, acetylcholine, at these sites are also effective. Berthold Hedwig (2000) later extended this discovery to individual neurons in the brain. On either side of the brain, he discovered one that fulfils criteria as a 'command neuron' (Fig. 9.3*a*). He used dye-filled glass microelectrodes so that he could record from and stimulate the neuron, and then stain it to check its identity in different experiments. When depolarising currents were injected into the neuron, it spiked and caused a cricket to start singing a calling song (Fig. 9.3*b*). The cricket raised its wings and then sang with the usual pattern of groups of 3–5 syllables in a chirp. When Hedwig changed the amount of excitatory current injected, he found that the greater the spike rate in the interneuron, the greater the rate of chirps: for example changing the spike rate from 30 Hz to 120 Hz changed the chirp rate from 2.4 Hz to 4.5 Hz. This means that the interneuron does not just switch on singing, but plays a role in the central pattern generator. However, the interneuron did not affect syllable rate, which implies that there are separate pattern generators determining the rate of syllables and of chirps. Further indication that the interneuron is involved in the chirp spacing is that its spiking frequency was modulated slightly in time with chirps,

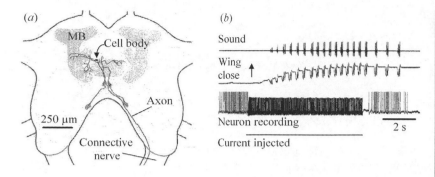

(a)

MB
Cell body
Axon
250 μm
Connective nerve

(b)

Sound

Wing close

Neuron recording

2 s

Current injected

Fig. 9.3 A command neuron for singing in the cricket. (a) Structure of the neuron in the brain, viewed on to its back surface, revealed by injecting the neuron with a fluorescent dye after recording from it. The cell body and dendrites are in the protocerebrum. Shaded are the mushroom bodies (MB) and, between them, the central body. (b) Exciting the neuron by injecting positive current into it increased its rate of spiking, and caused the cricket to raise its wings and start singing. Singing was monitored both by using an optical device to measure wing movements and by recording the sound produced. (Redrawn from Hedwig, 2000.)

possibly by sensory feedback from the cricket's cerci. As in the cockroach, cerci of the cricket bear hairs that are exquisitely sensitive to air currents, and they pick up air movements caused by singing movements of the wings.

Thus, excitation of this interneuron is sufficient to trigger calling song; but does it fulfil a second characteristic of a command neuron by being necessary for the calling song? Hedwig's experiments indicate quite strongly that it does. Once a cricket had sung several times in an experiment, it was sometimes sufficient to inject currents lasting just 1.5 s into the interneuron to excite it, after which the cricket would continue singing for several tens of seconds, although at a much lower chirp rate than normal, and the interneuron continued spiking. So the interneuron is both sufficient and necessary for a cricket to call, and can be considered to be a command neuron. Because it affects chirp rate and its spike rate changes in time with chirps, this suggests that it is embedded within the singing pattern generator in addition to commanding the calling song. Unlike the crayfish lateral giant or Mauthner neurons (Chapter 4), its commands can be overridden. A sudden puff of air to a cercus makes a cricket stop singing and, just as a cockroach does, to turn and run away. In Hedwig's experiments, air puffs to a cercus interrupted singing, but not the spiking of the interneuron. Finally, although this interneuron commands calling songs, it never triggers courtship or rivalry songs. Although the way these are triggered in crickets is not known, in grasshoppers different neurons are responsible for commanding these three separate types of song (Hedwig and Heinrich, 1997).

The basic programme for singing can be generated without the need for wings to move and provide proprioceptive feedback. This can be demonstrated by cutting all the nerves that supply wing muscles, and recording spikes from the axons of motor neurons when a cricket is induced to sing by local brain stimulation. So the thoracic ganglia contain a central pattern generator for cricket song. But proprioceptors are vital for a male to sing a song that is of sufficient quality to fulfil its function of attracting females. This was shown by destroying groups of cuticular strain detectors called campaniform sensilla near the wing base, which did not prevent a male from singing but caused subtle changes in the sound he produced.

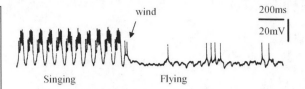

Fig. 9.4 Intracellular activity recorded from a local interneuron in the middle thoracic ganglion of a male *Teleogryllus* cricket. At the beginning of the recording, the cricket was singing and the interneuron produced clear rhythmical activity and played a role in generating the pattern of singing. A puff of wind switched the cricket's behaviour from singing to flying, and the interneuron was silenced despite the fact that the same muscles and motor neurons move the wings during the two behaviours. (Redrawn from Hennig, 1990.)

Some of the interneurons involved in generating the song programme and communicating it to motor neurons have been identified by Mathias Hennig (1990) in experiments similar to those on locust flying. In these experiments, a singing cricket could be made to switch its programme to flying pattern by blowing wind onto its head. Different interneurons are involved in the two activities. For example, in Fig. 9.4, the recording shows an interneuron producing clear rhythmical activity during singing, but not after flight was initiated by a puff of wind. This finding was surprising: it had been assumed that the same interneurons would be involved in generating the rhythms for flying and for singing because the two behaviours are similar. This would make efficient use of the limited number of neurons available in the thoracic ganglia. However, the evolution of singing behaviour must have involved the development of new neuronal circuitry.

Recognising and finding a mate

When a receptive female cricket hears a male cricket song, she must recognise it as an appropriate signal to react to, and then locate the source of the sound so she can move towards her suitor. In order to move towards the source of a song, a cricket turns in a way that matches signals received by its left and right ears. Although rarely observed in the wild, captive receptive female crickets will react to the sound of a male's song at some distance away from the sound source, and respond by running, or in some species flying, towards the source of sound. The behaviour of orienting towards a sound source is called **phonotaxis**, and mature, receptive female crickets readily perform phonotactic behaviour in the laboratory. In fact, one of the earliest uses of technology in studying animal behaviour was the demonstration in 1913 by Johann Regen that a female cricket runs towards the sound of a male cricket delivered through a telephone line. In more recent experiments, the parameters of a male song that provide the relevant sensory cues for eliciting and guiding the behaviour have been identified, and some of the neuronal mechanisms involved have been studied. In one type of experiment, female crickets run across a room in which two different loudspeakers deliver synthetic songs, so the female chooses between them. In others, the female's body is fixed, and she runs on a ball suspended in a stream of compressed air (Fig. 9.5a). The movements of the ball can be tracked in a way that reproduces the path the cricket would have taken had she been free to run and turn.

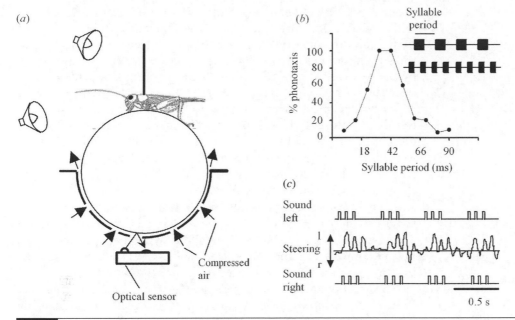

Fig. 9.5 Running movements of a female cricket on a track-ball towards calling songs. (*a*) A female cricket is fixed to a thin rod and runs on a light plastic ball that floats on compressed air. An optical sensor monitors movements of the ball. Loudspeakers either side of the cricket deliver recordings of songs or artificial songs. (*b*) A graph that shows the effectiveness of different syllable repeat rates at attracting running females. Synthetic songs, consisting of 250-ms-long chirps, were played to the cricket. The syllable period is the time between the start of one syllable and the start of the next, and was always half the syllable duration. Syllable periods of 36 ms to 42 ms (about 25 per second) consistently evoked running towards the active speaker (100% phonotaxis), and shorter or longer syllable periods were less effective. (*c*) Fast steering responses towards individual syllables from left and right speakers. (Redrawn from Hedwig, 2006.)

One cue that a female uses to recognise the song of a male of her species is the tone, or sound frequency. But a female does not walk towards a continual 4.5 kHz tone; to be effective, the sound must be pulsed in the same way as it is in the syllables of a male's calling song. The pattern of syllables, perhaps augmented by the pattern of chirps and trills, is an important cue for species that live in the same locality to recognise songs of their own male. For *G. campestris* or its close relative *G. bimaculatus*, song structure is relatively simple, with 3–5 syllables making a chirp. An artificial song in which a 4.5 kHz tone is broken into repeated pairs of syllables is effective at attracting a receptive female to a loudspeaker. By using two speakers, the relative effectiveness of different song elements can be compared. A large number of studies have used that technique, and one important feature of songs that makes them attractive to females is the rate at which syllables are delivered (Fig. 9.5*b*). Syllable length does not seem to matter. Berthold Hedwig and James Poulet (2004) found that a cricket would turn towards each individual syllable in a song. They showed this by delivering two repeated and identical chirps from different directions, timed so that syllables from the two speakers alternated with each other. In the recording from an experiment shown in Fig. 9.5*c*, the first syllable came from the left, and after about 55 ms, the cricket moved the trackball as if it was turning to the left. A syllable from the right then caused the cricket to try and turn to the right; the next syllable from the left caused it to try to turn to the left and so on. Over time, the path a cricket takes is determined by the relative balance between syllables coming from the left and from the right.

A cricket consistently turns towards the first syllable in a song. As the song continues, the size of turns towards each subsequent syllable increases gradually, reaching their maximum size after 2 s, or four chirps. It is as if the cricket is becoming primed to react to sounds and becomes more reactive as a song gets going. Poulet and Hedwig (2005) showed that this arousal is a separate process from steering by inserting abnormally long, 200 ms syllables into artificial songs. When played on its own, a long syllable is no more attractive than a normal length syllable. But if a long syllable is inserted into a normal song pattern, the long syllable evokes a very strong, sustained turning response towards the side from which it was delivered.

The neuronal pathways responsible for rapid steering are located in the thorax. Because the cricket can make rapid steering movements to each syllable at normal song rates, it is unlikely that there is sufficient time for signals to travel to the brain for processing. Also, because the cricket turns in response to the very first syllable of a song, it has not had the opportunity to measure the interval between syllables before deciding where to turn towards. What Hedwig and Poulet suggest is that, during a song, a pattern recognition mechanism that reacts to the correct syllable repetition rate is activated. That recognition mechanism then enhances connections between auditory neurons and the neurons that control turning movements of the legs, perhaps by releasing a neurohormone in an appropriate part of the neuropile. The turning reaction is activated by any pulse of sound with a pitch of around 4.5 kHz, but without enhancement by the pattern recognition mechanism, the turning each pulse of sound elicits is small. Prior to these discoveries, it had been thought that the brain is involved in both song recognition and delivering steering control commands and that a cricket turned towards the side that received the best song pattern measured over a period of several syllables.

The ears and auditory neurons

A cricket's ears are on its front legs, near the top of the tibiae just below the knees. Sensory receptor axons run from the ear in the auditory nerve to the prothoracic ganglion, which also controls the front legs. The axons make a small number of branches in a local region on their side of the ganglion, where they contact a number of different interneurons. Some of the auditory interneurons are local to the prothoracic ganglion, and others have axons in the nerve cord. Two particular interneurons that carry auditory information to the

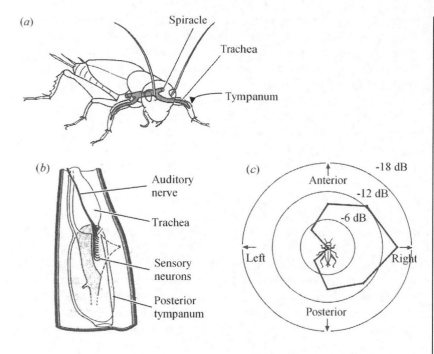

(a) Spiracle
Trachea
Tympanum

(b)
Auditory nerve
Trachea
Sensory neurons
Posterior tympanum

(c)
-18 dB
Anterior
-12 dB
-6 dB
Left
Right
Posterior

Fig. 9.6 A cricket's ear. (a) A female cricket, showing the first thoracic spiracles and the tracheae in the forelegs. The tympanum is just below the knee. Sound waves reach its outside surface directly and its inside surface via the spiracles and tracheae. (b) The main components of an ear. The top part of a foreleg tibia is cut open to show, in anterior view: the posterior tympanum; the leg trachea; and the sensory neurons. The sensory neurons each have an axon in the auditory nerve; they are covered by a tent-like membrane (dotted lines). (c) Directional tuning curve of neurons in the right auditory nerve. Sound pulses of constant loudness and at the same tone as the song of the species were delivered from different directions around the cricket. The curve plots the loudness of the sound needed to just elicit a response. Sound loudness is expressed on a decibel scale in which 0 dB was the minimum loudness for a sound from the left to elicit a response from the right ear, so the softest effective sounds came from the right (about −15 dB). (Part (a) redrawn from Schmitz et al., 1983; (b) redrawn from Young and Ball, 1974; (c) redrawn from Hill and Boyan, 1977.)

brain have been identified on each side of the cricket: AN1 responds to sounds with the same pitch as a male cricket's song; and AN2 responds to the high frequency sounds made by bats. When AN2 is stimulated electrically it causes evasive movements in a flying cricket (Nolen and Hoy, 1984).

Both the structure of the ears and the way their neuronal pathways are wired are clearly adapted to the functions of recognising the calling song of the species, and localising a source of song. Each ear consists of a thin oval-shaped tympanum, or eardrum, on the posterior surface of the leg and there is also a smaller anterior tympanum. Inside the tympanum is an air-filled sac joined to the main tracheal trunk of the leg. The tracheae are major parts of the insect's respiratory system, and open to the outside air by way of a pair of spiracles on each body segment. The leg trachea is continuous with other major tracheae, including the trachea from the other front leg. The nearest spiracles to the tympanum on the right leg are those in the prothoracic segment on the right and left of the cricket (Fig. 9.6a). This arrangement means that sound pressure waves can reach each tympanum by two different routes: directly, from the outside; and indirectly via the trachea from the spiracle on each side of the prothorax. The importance of this for localising a singing male is described below.

Associated with each tympanum is an array of 60–70 sensory neurons that have their cell bodies near to the tympanum (Fig. 9.6b). Each has a short dendrite, which is probably where vibrations are transduced into receptor potentials. The other side of the cell body is attached to the axon, which projects along the auditory nerve from the ear into the prothoracic ganglion. Each sensory neuron is most sensitive to sounds of a particular frequency. This has been determined by recording spikes from an axon while sounds of a particular frequency are played. The intensity at which that sound frequency just elicits a spike is recorded, and the procedure is repeated for different frequencies.

Most of the cricket's auditory receptor neurons are sharply tuned to respond to a narrow range of sound frequencies, and will not respond to sounds that differ by much from this range. When different receptors were compared, it was found that the ear is arranged tonotopically, with receptors responding to relatively low frequencies located towards the body and receptors responding to higher frequencies located further towards the foot. Most of the receptors responded to sound frequencies near to 4–5 kHz, covering the song frequency for the species. Other receptors respond to higher sound frequencies, and enable a flying cricket to hear and react to the echolocation calls of predating bats. Because the tuning curves of most receptors are matched to the sound frequencies close to those in the songs of male crickets, these are the sounds that the cricket's nervous system can process and use to control its behaviour. Sensory filtering, therefore, begins at the level of the sensory receptors. Sounds that are not relevant to the cricket are not processed by the central nervous system because they are ineffective at exciting the receptors.

In order to determine whether it should turn to the left, to the right or carry on straight ahead to find a singing male, a female cricket needs to make use of the signals provided by its two ears. The ears are only about 10 mm apart, so the time taken for a sound to travel between the two is only about 30 ms and it is unlikely that the time difference of signals in the left and right ears is used by the cricket. However, the sensory neurons of each ear on its own are sensitive to the direction from which a sound comes. This was determined with similar experiments to those used to measure tuning curves, but in this case a loudspeaker was moved to deliver sounds from different directions around the cricket. For each direction, sound intensity gradually increased until the threshold sound intensity that just elicited a response from that direction was measured. The resulting graph is heart-like in shape (Fig. 9.6c) and shows that the auditory neurons are extremely sensitive to sounds coming from the same side as their leg, but are insensitive to sounds coming from the opposite direction.

This is a consequence of the physical arrangement of the tympanum, tracheae and spiracles which enables sound pressure waves to reach the tympanum on both its outside and inside surfaces (Fig. 9.6a). As a source of sound vibrates, it creates waves of compressed and rarefied air and the waves travel away from the source at the speed of sound. If the sound is a pure tone, the waves are spaced evenly from each other. For a sound of 5 kHz (the song tone for the species that was studied), the distance between adjacent compressions, or its wavelength, is 70 mm. For a sound from the left of a

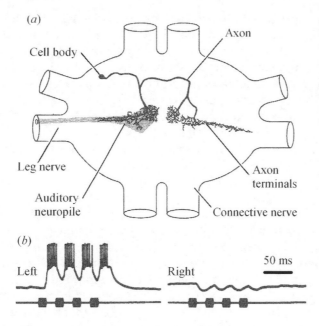

(a)

Cell body

Axon

Leg nerve

Auditory neuropile

Axon terminals

Connective nerve

(b)

Left

Right

50 ms

Fig. 9.7 The omega neuron, an auditory interneuron in the first thoracic ganglion of crickets. (*a*) A drawing of an omega neuron that was stained by injecting an intracellular dye through a microelectrode. This neuron receives synaptic input signals from auditory neurons of the left ear, which terminate in the auditory neuropile, and its axon carries the spikes generated in the omega neuron to the right auditory neuropile. A mirror image of this neuron collects input from the left ear, and makes outputs in the left auditory neuropile. (*b*) An intracellular recording of responses by an omega neuron to four syllables of a model calling song delivered to the left and then to the right ear. When delivered to the left ear, each syllable excited the neuron to produce a burst of spikes; when delivered to the right ear, each syllable caused inhibition. The source of the IPSPs was the right omega neuron. (Redrawn from Wohlers and Huber, 1982.)

cricket, pressure waves reach the left tympanum both directly on the outside and also on the inside after travelling through the spiracles and leg tracheae. The extra distance from the prothorax and along the inside of the leg is 35 mm, or half the wavelength of sound at the song frequency. This means that each compression in a sound wave will push on the outside of the tympanum at the same time as a rarefaction pulls on the inside, and vice versa. So sound waves arriving by the outside and inside routes reinforce each other, causing the tympanum to vibrate strongly. For sounds coming from the right, the waves will cancel each other out because for this direction of travel the distance to the left tympanum is the same for the two routes, so the tympanum will be pushed outwards and inwards at the same time.

Enhancement of left–right differences between auditory signals continues in neurons of the auditory pathway. Two interneurons that receptor axons connect with are a left–right pair called 'omega' neurons from their resemblance to the Greek letter (Fig. 9.7). Each omega neuron is excited by auditory neurons from the ear on the same side as its cell body, and spikes that result in the omega neuron are carried across the ganglion to output branches of its axon. Simultaneous recordings from left and right omega neurons show that they inhibit each other: a spike in the left neuron, for example, causing an IPSP in the left (Fig. 9.7*b*). This inhibition sharpens any difference in the strength of the neuronal response between the two ears.

One characteristic of the omega neurons is that they adapt quickly to a sound pulse, producing two or three closely spaced spikes just after the start of a sound pulse, which quickly reduces to a lower, sustained spike rate. This adaptation ensures that just after the start of a loud sound the omega neurons become less responsive to sounds. Gerald Pollack (1988) suggested that this is a kind of 'cocktail party' effect by helping to filter out background sounds so the neuron can remain responsive to a male's song – the sound a receptive female cricket is primarily interested in.

A second consequence of rapid adaptation is that the omega neurons could be an effective recognition mechanism for syllable repetition rate (Nabatiyan *et al.*, 2003). For syllables repeated at a faster than normal song rate, adaptation ensures that the response to the first syllable is greater than responses to later ones. For syllables repeated slowly, the response to the start of each syllable is brisk, but there are not many syllables so the overall spike count is low. For syllables repeated at the calling song rate, the separation between adjacent syllables is just sufficient for adaptation to fade, so each syllable generates the maximum number of spikes in the neuron. The omega neurons therefore have response characteristics that make them suitable for controlling the rapid leg steering movements in response to each sound pulse, as shown earlier in Fig. 9.5c. But it is not clear whether they are directly involved because a cricket in which the omega neurons have been killed by dye injection can still find its way towards a singing male.

Maintaining hearing while singing: corollary discharge

The sound made by a singing male cricket is, quite literally, deafening. But males do need to hear their rivals, and auditory cues may help them evade predators – it is hard to creep up on a singing cricket unnoticed. To protect its ability to hear sounds, a male cricket switches its auditory neurons off while it is singing. Recording from auditory interneurons such as an omega neuron or the AN1 interneuron that carries information to the brain shows that their responsiveness to continually played sounds is reduced during the wing movements of a cricket (Poulet and Hedwig, 2003). This happens even if one wing is removed, so the cricket sings silently, which suggests that the reduction in sensory responsiveness is caused by the central pattern generator for singing. Poulet and Hedwig (2006) have identified an interneuron responsible for linking the motor pattern to neurons of the auditory system (Fig. 9.8a). The activity pattern of the interneuron is a kind of reference copy of the motor pattern, called a corollary discharge. The idea that animal nervous systems make use of corollary discharges originated early in the twentieth century to explain how a person's perception of the visual world remains stable during rapid saccadic movements of the eyeballs. Discovery of a particular interneuron that carries a corollary discharge signal is an important step forward in verifying this idea, and in discovering how corollary discharges work.

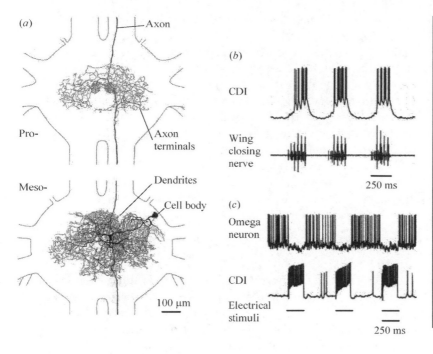

(a) Axon

Pro-

Axon terminals

Meso-

Dendrites

Cell body

100 μm

(b)

CDI

Wing closing nerve

250 ms

(c)

Omega neuron

CDI

Electrical stimuli

250 ms

Fig. 9.8 Structure and function of the singing corollary discharge neuron in the cricket *Gryllus bimaculatus*. (*a*) Structure of the neuron in the prothoracic and mesothoracic ganglia. (*b*) When the cricket sang, the corollary discharge interneuron (CDI) was excited and spiked in time with each wing closure, as indicated by a recording of motor neuron spikes. (*c*) The CDI inhibits responses to sound by the omega neuron. Continual sound at the song's tone excited the omega neuron. When the CDI was excited by injecting positive current into it, it interrupted a spike train the omega neuron was producing in response to a repeated sound stimulus (not shown). (Redrawn from Poulet and Hedwig, 2006.)

The interneuron has its cell body and dendrites in the middle thoracic ganglion, which is also responsible for controlling the forewings during singing or flying, providing a location for synapses to deliver signals from the singing central pattern generator to the interneuron. Its axons travel to all the segmental ganglia and the brain, and they branch extensively, suggesting that the neuron exerts widespread effects throughout the body. In the prothoracic ganglion its branching pattern fits the region occupied by the omega and other sound processing interneurons. When a cricket was induced to sing by applying a drug to the appropriate part of its brain, the interneuron was excited in time with each chirp, with a burst of excitation coinciding with each wing closure movement (Fig. 9.8*b*). Injecting current into it to increase the number of spikes it produced did not affect the motor pattern of singing, so the interneuron is not part of the central pattern generator for singing, but is excited by it. The interneuron was not excited by sound signals played to the cricket, and when a cricket was induced to generate a flying rather than singing motor pattern the interneuron was inhibited. Thus it appears to be dedicated to providing a signal that says to neurons within the nervous system that 'the cricket is singing'.

Paired recordings from the interneuron and potential targets showed that it makes connections directly both with auditory interneurons (Fig. 9.8*c*) and with the synaptic terminals they receive from auditory receptor axons. This arrangement is similar to the way inhibition acts both presynaptically and postsynaptically in the sensory pathways that excite the crayfish lateral giant interneuron and fish Mauthner neuron (Chapter 4). Reducing the excitation of presynaptic terminals protects the auditory system against activity-induced changes in the strengths of the synapses and means that the interneurons are fully responsive to external sounds once the cricket has finished a chirp. When a cricket was not singing, exciting the interneuron with a microelectrode reduced responses by an omega neuron to sounds. Conversely, when a cricket was singing, injecting negative current into the interneuron reduced the amount of inhibition the omega neuron received. These experiments demonstrate that it is this left–right pair of interneurons that are responsible for the corollary discharge.

Cricket song and neurogenetics

In cricket singing, the morphological structures and the neurons that are involved in producing the song are different from those that receive and interpret it. How are senders and receivers matched during evolution? In crickets, some of the same genes influence both the pattern generators for singing in males and the pattern recognition mechanisms in females. David Bentley and Ronald Hoy (1972) showed this by crossing two species of Australian cricket, *Teleogryllus commodus*, a southern species, and *T. oceanicus*, a northern species. The two species overlap in southern Queensland, but do not normally interbreed. Females normally prefer the songs of their own species to the other; the pitch of the two species differs, as well as the syllable structure of the songs, which are more than those of *Gryllus*.

When the two species were crossed, the offspring sang songs that were clearly intermediate in structure between those of the two parent species, in both pitch and syllable structure. Further careful crossing experiments showed that a number of genes influence song, which is not surprising, and that they are carried on the X chromosome (female crickets have two X chromosomes; males have one). More surprising was the discovery that females of particular crosses always preferred the songs of their brothers to songs of different hybrid crosses. This indicates that both the male's pattern generator and the female's recognition mechanism for song share genetic instructions. It has been proposed that the same clock mechanism is involved in song generation and in song recognition, so that both the motor pattern generator and the song recognition mechanisms may include some of the same neurons. Although there is no firm evidence for this, a further observation is consistent with the idea. Song pattern changes quite markedly with temperature. When offered a choice between recordings of songs made from males at two different temperatures, the female always chooses the song appropriate for its own temperature (Pires and Hoy, 1992).

Nerve cells and cricket song

At this point, it is worthwhile to reflect on the neuroethological concepts behind our understanding of the mechanisms of communication in crickets. A male cricket sings to advertise his desire to mate or to declare ownership of territory. The muscle actions in singing are controlled by a central pattern generator, which is a network of interneurons in the thoracic ganglia. For a behaviourally effective song, proprioceptive feedback onto the song controlling neurons is needed. Besides driving motor actions, the central pattern generator excites corollary discharge interneurons, which suppress responses in and protect auditory neurons while the cricket is singing. A receptive female of the same species will walk or fly towards a singing male, and this phonotaxis is released by sound signals with an appropriate tone and syllable structure. Sensory filtering to recognise an appropriate song stimulus starts at the level of auditory sensory receptors, many of which are tuned to the song's tone, and filtering continues in a network of interneurons in the prothoracic ganglion and brain. Sensory adaptation in prothoracic interneurons probably plays a significant role in recognising syllable repetition rate. A female finds a song source by comparing signals received by its left and right ears, a process that involves physical properties of the auditory system as well as synaptic interactions between interneurons.

Electric fish

When an animal in water uses neurons or muscles, some of the electrical current that is generated escapes into the surrounding water. A number of vertebrates have sense organs that are sensitive to these small currents, and use them to find prey that is hidden in sand, mud or under pebbles (Fortune, 2006). Because this electric sense occurs sporadically in unrelated groups of vertebrates, including teleost fish, sharks and rays, some amphibians, and the platypus (a mammal), it is thought to have arisen independently a number of times during evolution. A few genera of fish also include members with specialised organs that generate electric fields around the fish. A small number of these electric fish can deliver brief but extremely powerful shocks used for stunning potential prey or for defence. These include electric rays and stargazers, which are marine fish; and the South American electric eel and the African catfish, which are freshwater fish. Most species of electric fish generate much weaker electric fields around their bodies, and use these for communication, for detecting prey, and for gaining information about their environment (Hopkins, 1999). Weakly electric fish, which live in fresh water, come from two different groups: the mormyrids in Africa; and the gymnotids in South America (Moller, 1995). Many of

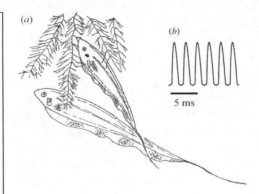

Fig. 9.9 Male and female glass knife fish (*Eigenmannia viriscens*) from South America. (*a*) The female, who is pointing upwards, was guarding a patch of waterweed in which she later laid eggs; the male was courting her. (*b*) Inset is a short recording of the wave electrical signal from *Eigenmannia*; an individual fish can generate waves like this with almost no variation in frequency for several hours. (Fish drawn from a photograph in Hagedorn and Heiligenberg, 1985.)

Current flow

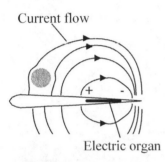

Electric organ

Fig. 9.10 Mechanism for electrolocation illustrated by a diagram looking down onto an *Eigenmannia*. The electric organ is a small block of modified muscle towards the back of the animal, and during waves the polarity of the organ switches. Here the front end is shown positive. It generates current flows around the fish, and the pattern of voltage these generate across the skin is monitored by electroreceptors. An object nearby distorts the pattern of current flow, and so changes the pattern of voltage across the skin.

these fish generate signals as brief pulses, but a number of South American species and one from Africa generate their signals as continual waves.

The weakly electric, wave-producing species that has received most attention is *Eigenmannia viridescens*. Its English name, the glass knife fish, reflects the knife-like shape and transparency of its body. *Eigenmannia* is quite a common fish for aquaria, although it tends to be secretive and is nocturnal. It breeds during the tropical rainy season, which makes it difficult to study in its natural habitat, but will breed in captivity if care is taken to replicate natural conditions. In the breeding season, females defend nesting sites in floating weed (Fig. 9.9), and males are fiercely territorial, and will butt or bite opponents.

The electric organ of *Eigenmannia* is made of modified muscle at the base of its tail (Fig. 9.10). The modified muscle cells are activated as usual by motor neurons, but they lack contractile proteins and are modified instead to generate electrical currents in the form of waves, which have a frequency of 240–630 Hz. In the Amazon basin, several species of wave-producing fish can live in the same stretch of water, and each species probably has a different band of frequencies, allowing individuals to distinguish waves from their own species. An individual fish can sustain a wave with almost no variation in frequency periods of several hours, probably the most accurate living clock mechanism. However, during social interactions individuals alter the discharge of their electric organs in particular ways. For example, during courtship a male will make continual chirps in which the wave frequency increases in short pulses. Gravid females can be induced to spawn by playing recordings of male chirps to them, delivered through electrodes in an aquarium. Individual fish differ in wave frequency. Dominant males, which tend to be largest, have relatively low wave frequencies. Interactions between males include chirping, slower rises and decreases in wave frequency, and periods of electrical silence. Dominant females tend to have waves of higher frequency than subordinates.

The electric organ discharges also provide a fish with information about its immediate surroundings. The electric organ is in line with the longitudinal axis of the fish and, when it discharges, sets up current flow in a field right around the fish. The electric current fluctuates with the signal produced by the electric organ, and the current crosses the skin of the fish at right angles to its surface. Small electrosensitive organs are distributed throughout the skin, particularly around the head, and these monitor the local voltage across the skin caused by flow of electric current from the electric organ. When an object is near to the fish, it distorts the pattern of currents impinging on the skin surface (Fig. 9.10). The distortion depends on the conductivity, size and shape of the object. For a fish that is nocturnal and inhabits streams that are often turbid, the information provided in this way enables it to find its way around by electrolocation and to detect small prey such as water fleas. In darkness, electric fish in aquaria are extremely good at keeping their bodies exactly in the centre of a plastic pipe when an experimenter moves it around.

Two different kinds of electroreceptor organ embedded in the skin monitor the electric fields; they are called ampullary and tuberous organs. Both types are extremely sensitive to voltage across their local region of skin, and up to 15 of these sense organs can occur in 1 square millimetre of skin surface. They are particularly densely distributed on the head. Each contains 25–35 small receptor cells at the base of a capsule that opens through the skin by a pore, and the electroreceptor cells synapse at the base of the pit with a sensory neuron that sends its axon into the brain. The ampullary receptors detect static or slowly changing voltages, whereas the tuberous receptors detect changing signals, particularly those with the same frequency as the fish's electric organ discharge.

The jamming avoidance response

Both for electrolocation and communication, it is important that a fish can distinguish its own electric waves from those of neighbours. If two fish with electric organ discharges that are similar to each other come close, their signals can interfere with each other. This is referred to as jamming because it is similar to the way that a radio station can be jammed by broadcasting on a wavelength that is close to that normally reserved for the station. *Eigenmannia* has a well-defined behaviour that allows an individual to maintain its own private frequency line, the jamming avoidance response. If two fish with similar frequencies come close, both react to increase the difference between their electric organ discharge rates: the one with the higher frequency increases its rate while the one with the lower frequency decreases its rate. This reaction is important not just to allow a fish to distinguish its own signals from those of neighbours, but also because two similar waves will interact, causing beats, and these regular beats have been shown to interfere with electrolocation and with prey detection. Some dominant fish can sometimes use their own signals to jam their neighbours deliberately.

Fig. 9.11 Jamming by neighbouring fish. (*a*) The larger fish, on the left, generates a wave with a slightly greater frequency than its neighbour. The electric current from the larger fish crosses its own skin at right angles to the long body axis right around the fish, as shown by the arrows. Its neighbour's signal, indicated by dotted line arrows, crosses the skin at right angles in the lateral part of the body (ii), but grazes the skin at a shallow angle at the dorsal (i) or ventral parts of the body. (*b*) As a consequence, the signal from the neighbour is negligible at location i; but relatively strong at location ii, where the two signals sum to give beats (dotted line) at location ii. The rate of spikes from a P-electroreceptor unit give a rather irregular measure of signal amplitude, as shown at the bottom. (*c*) The way that T-units at the two locations in the skin respond during a minimum between beats. Notice how the relative timing of spikes in the T-units at i and ii alter. (*d*) Is the same as (*c*), except in this case the neighbouring fish had the higher wave frequency.

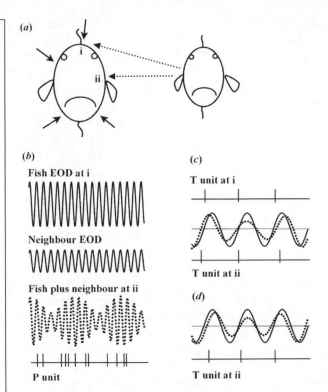

Walter Heiligenberg and colleagues (Heiligenberg *et al.*, 1978; Heiligenberg, 1991) showed that the jamming avoidance response is controlled by sensory information received by electroreceptors at different locations on the body surface. One kind of experiment was to record activity from motor neurons of the electric organ in fish in which the neurotoxin, curare, was applied to prevent the motor neuron spikes from activating the electric organ without affecting the pattern of spikes generated by the motor neurons. Two different pairs of electrodes in the aquarium delivered electric wave stimuli to the fish, one pair creating an electric field in line with the long axis of the fish, mimicking its own electric organ discharge, and the other off to one side, mimicking a neighbouring fish. The fish responded by making appropriate adjustments to the signals in its motor neurons, as if to avoid jamming between the two stimulus waveforms.

The electric current produced by the fish's own discharge crosses the skin at right angles to its surface right around the fish, so the signal strength is the same whether measured dorsally, ventrally or at either side (Fig. 9.11). In contrast, the strength of the signal received from a neighbouring fish differs according to the location on the surface of the fish. The patch of skin closest to the neighbouring fish will receive a strong signal, whereas at other locations the current from the neighbour will glance across the skin, hardly causing any voltage signal across the skin. In the example shown in Fig. 9.11, the neighbour is positioned near the left side of the fish, so the skin on that side (location ii) receives a relatively strong signal from the neighbouring fish, but dorsal skin (location i) receives hardly any signal from the neighbour. Consequently, the signal picked up by dorsal electroreceptors is from the fish's own electric organ discharge, but the signal picked up by electroreceptors on the fish's left side is a mixture between the electric organ discharges both of the fish and its neighbour.

The mixed signal at location ii shows regular beats in its strength as the signals from the two fish drift into and out of phase with each other. When the two signals are in phase, their peaks and troughs coincide with each other, so they add to give a large combined signal. When they are out of phase, the peak in one signal coincides with the trough of the other, so the two signals tend to cancel each other out and produce a small combined signal. The rate of beats in the summed signal is the same as the difference in the frequencies of electric organ discharge by the two fish, so that if two fish differed in frequencies by 10 Hz, the frequency of beats would be 10 Hz. Also the combined signal would be identical whichever of the two fish had the higher frequency. To produce a jamming avoidance reaction, the fish needs to know not just the difference in frequency of its own and its neighbour's electric signals, but also whether its own frequency is greater or less than its neighbour's. In Fig. 9.11 the fish (drawn on the left) has a slightly lower frequency than its neighbour.

Information about the rate and timing of the electric stimuli at the skin surface is provided by two different kinds of axons from tuberous electroreceptors (Heiligenberg and Partridge, 1981; Rose and Heligenberg, 1985). One type is called a P-unit because the *probability* that it will produce a spike depends on the voltage signal, so it tends to spike more often when the signals from the two fish are in phase and less often when they are out of phase (Fig. 9.11b). An individual P-unit provides a record of beats in the combined signal at its patch of skin, but the record is smudgy because although the beats are very regular, its pattern of spikes varies from one beat to the next. The other type of tuberous organ axon belongs to T-units, which provide information about the *timing* of voltage signals across their patches of skin; a T-unit spike marks very precisely the time at which a wave across its patch of skin starts – it is phase locked to the waveform in a similar way to cochlear sensory neurons and others in auditory systems (Chapter 6). This division into two separate information streams, signal strength and signal timing, is reminiscent of the way in which, in the owl, sound signals are channelled into two separate processing pathways (Konishi, 2006). As in the owl, the two pathways converge to provide information used by the animal to control its behaviour.

Figure 9.11c shows the reactions by T-units at two different locations on the body surface. A small portion of the signal in Fig. 9.11b is shown, corresponding to a minimum strength of the combined signal. For a patch of skin at the dorsal surface of the fish (i) the signal is mainly from the fish's own electric organ (solid line), and a T-unit here marks the start of each of the fish's own waves.

For a patch of skin at the side (ii) the signal (dotted line) has pronounced beats, and the T-unit here marks the start of each combined wave. While the combined signal strength is decreasing, each wave in the combined signal slightly precedes each wave in the fish's own signal; but when the combined signal strength is increasing, each wave in the fish's own signal comes first. This is shown by drawing the signals at the two patches of skin together in Fig. 9.11b: at first, the signal at ii (dotted) precedes the signal at i (solid); the signals then coincide in time as the signal at ii reaches its minimum amplitude; and then the signal at i precedes that at ii as the signal amplitude at ii starts to increase. These changes in time are reported by the spikes in T-units that innervate the two skin regions. If the fish's neighbour had the higher rather than the lower frequency, the situation would be reversed, as shown in Fig. 9.11d. So information about whether its neighbour has a higher or a lower frequency than itself is available to the fish by combining information from different receptors. P-units provide information about signal amplitude, and comparing the spikes from T-units in different parts of the skin provides information about the relative times of the signals from the fish itself and, by referring to the combined signal, from its neighbour.

Jamming avoidance and the fish brain

By following the neuronal pathways into the fish's brain (Fig. 9.12), it has been shown that this information is, indeed, combined in a way that allows it to be used to control the jamming avoidance response (Rose *et al.*, 1988; Metzner, 1999). Axons of P- and T-units terminate in a part of the brain called the electrosensory lobe in which individual neurons each have a receptive field that monitors one region of the skin. Two types of neurons project from the electrosensory lobe to the torus semicircularis, the next processing stage in the pathway, and these are called pyramidal neurons and spherical neurons. There are two types of pyramidal neuron: the E neurons are excited when strength of the combined sensory signal is increasing during a beat; and the I pyramidal neurons are excited whenever the strength of the combined signal decreases. Individual P-units make chemical excitatory synapses with dendrites of E pyramidal neurons; and they excite small granule neurons which, in turn, make inhibitory synapses with I pyramidal neurons. Responses by a pyramidal neuron are more sharply defined than those of individual P-units because each pyramidal neuron combines signals from a number of different P-units. Each spherical neuron is excited through electrical synapses with a number of T-units and with neighbouring spherical cells. These synapses function to ensure that each spherical neuron fires a spike if several of the T-units in its receptive field spike within a brief time window, so a spherical neuron spike is a very exact report of the start of a signal wave in its patch of skin.

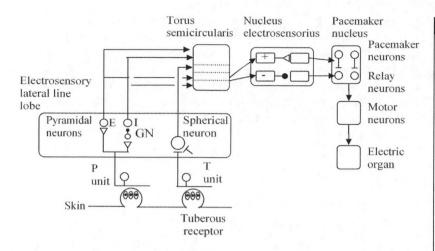

Torus semicircularis

Nucleus electrosensorius

Pacemaker nucleus

Pacemaker neurons

Relay neurons

Motor neurons

Electric organ

Electrosensory lateral line lobe

Pyramidal neurons

E I GN

Spherical neuron

P unit

T unit

Skin

Tuberous receptor

Fig. 9.12 The jamming avoidance pathway in *Eigenmannia* brain. Electroreceptor cells located within tuberous receptors respond to voltage signals across the skin and synapse with P-units and T-units that send axons into the electrosensory lateral line lobe. P-units make excitatory synapses with some pyramidal cells (E) and with granule neurons (GN) that inhibit other pyramidal cells (I); and T-units make electrical synapses with spherical neurons. The axons of pyramidal and spherical neurons project to various layers of the torus semicircularis, where information from different skin regions is brought together in a way that determines whether the fish's own signal frequency is higher or lower than a neighbour's. Then, the nucleus electrosensorius organises and delivers signals that control the frequency of the pacemaker that drives the electric organ motor neurons; one part is responsible for increasing frequency and the other for decreasing it. (Partly based on Rose, 2004.)

The torus semicircularis has eight layers, or laminae, each of which maintains the topographic representation of body surface that is found in the electrosensory lobe. Spherical neurons send their axons to lamina 6, and the pyramidal neurons send their axons to all layers except lamina 6. Some neurons in lamina 6 combine information from different parts of the body, and some respond when the fish's signal precedes that of a neighbour while others respond when the neighbour's signal comes first (Heiligenberg and Rose, 1985). That information is carried to lamina 8 of the torus where it is integrated with information about the amplitude of waves, provided by neurons in other laminae that process signals from the E and I pyramidal neurons. Thus lamina 8 is where the two pathways, which encode information about the amplitude and the timing of the sensory stimuli, are recombined. Lamina 8 contains some neurons that respond when a fish's own signal frequency is higher than that from a neighbour and others that respond if the fish's signal frequency is lower.

This comparison is sharpened in the nucleus sensorius, which is part of the optic tectum and receives axons from all the laminae of the torus. The nucleus sensorius does not contain a somatotopic map of the body surface, but it does contain neurons that are extremely sensitive to the sign of the difference between two electric wave signals, able to respond to time differences as small as 1 μs, ten times less than any neurons in the torus. The response properties of these neurons are very similar to those of the jamming avoidance response, and their removal abolishes the response, which suggests they are involved in causing it.

This idea is borne out by experiments in which local regions of the nucleus sensorius were excited by using a small pipette to deliver the neurotransmitter glutamate, which normally has an excitatory action on neurons. One region of the nucleus sensorius causes an increase in the frequency of a fish's electric organ discharge, and another causes a decrease. The frequency of waves is set by a pacemaker nucleus in the brain. Two different kinds of neuron are found in the pacemaker nucleus: pacemaker neurons, which are primarily responsible for setting the rhythm; and relay neurons, which convey the rhythm to the electric organ motor neurons. Each area of the nucleus sensorius connects through a separate pathway to the pacemaker nucleus. Increases in frequency are caused by increasing excitation to the pacemaker neurons, whereas decreases in frequency are caused by decreases in excitation of the relay neurons, which has the effect of slowing down the rhythm generated in the pacemaker neurons because they are connected through electrical synapses with the relay neurons.

The design of neuronal pathways involved in jamming avoidance

The jamming avoidance response is, therefore, controlled by an extensive network of neurons that gathers information from different types of sensory neurons distributed over the body surface. Rather than a few principal neurons playing the most significant roles, the control process is democratically shared among many neurons at each stage. It may seem strange that this rather roundabout system is used to control the jamming avoidance behaviour, because it might be simpler for a fish's brain to use a corollary discharge of the commands to the electric organ to compare its own electric organ discharge with the sensory signal it receives on its skin. Electric fish do use corollary discharges; for example, they are important in sensory processing used by pulse-producing fish.

We need to bear in mind that the jamming avoidance reaction has not been engineered to achieve an ultimate aim, but is a behaviour that has an evolutionary history, shaped step by step through natural selection. The response has originated at least twice in evolution, in different orders of fish. The wave-producing African fish *Gymnarchus niloticus* achieves its jamming avoidance reaction in a similar way to *Eigenmannia*, by comparing amplitude and timing information from different skin surface regions, although details of the neuronal processing pathways differ. Study of relative species of *Eigenmannia* in South America indicates that the evolution of the jamming avoidance reaction proceeded with a number of progressive steps. *Sternopygus*, which discharges at 50–190 Hz, does not show a jamming avoidance reaction, but does have sensory pathways that separately process information about electrical signal amplitude and timing, before combining them in a similar manner to *Eigenmannia*. *Apteronotus*, in which the electric organ is made from modified axons rather than muscle cells and has a discharge rate that can exceed 1 kHz, shows a simpler jamming avoidance reaction than *Eigenmannia*. *Apteronotus* can only increase its electric organ discharge frequency in response to a neighbouring fish; it will not decrease it if the neighbour's frequency is a little higher than its own. So it is possible to envisage an evolutionary scenario in which fish first acquired the ability to process and distinguish different electric wave signals, then acquired the ability to increase their electric organ discharge rate to avoid jamming, and finally acquired the ability both to increase and to decrease discharge rate.

Bird song

Although the term 'bird brained' is commonly used as an insult, it is clear that birds' brains are at least as complex as those of many mammals. Many bird species display behaviours that need considerable cognitive ability, such as storing food in caches for later retrieval or learning to imitate sounds. A new interpretation of bird brain anatomy (Jarvis, 2005) pointed out that a classical view of bird brain anatomy hindered our understanding of the evolution, behaviour and neurobiology of birds. One behaviour that is characteristic of many species of birds is song. Almost all birds produce calls of various kinds but most passerines – 4000 species called song birds or oscines – produce extended and complex songs. A defining characteristic of these songs is that their full expression requires learning by a juvenile bird from mature, singing adults that act as tutors. The ability to learn extended vocalisations is rare in the animal kingdom – song birds share it with parrots and hummingbirds, and in mammals, only cetaceans and elephants in addition to humans are known to be able to learn to produce series of particular sounds. There is great interest in discovering whether birds and mammals that can learn vocalisations share particular brain characteristics that enable them to learn to sing and speak.

Song birds are believed to have evolved from one common ancestor, and include finches, warblers and thrushes. The most complex songs are usually produced by males during the breeding season, and the functions of song include establishing territories, attracting potential mates, and maintaining pair bonds (Catchpole and Slater, 2008; Marler and Slabberkoorn, 2004). To attract a female, a male needs to outcompete his neighbours, and this process of sexual selection must drive diversity in song between individuals of the same species. Usually, a bird does not make a perfect copy of tutor songs, so develops a unique repertoire with which to compete with others for mates and territory. Because birds incorporate parts of the songs of other birds into their own songs, local dialects develop in some species.

For a neuroethologist, bird song provides some general lessons about the way a complex type of behaviour is controlled by a central nervous system. Within the brain an interconnected network of discrete areas, or **nuclei**, have been shown to be involved in song. Only song birds plus others that learn vocalisations, such as parrots and hummingbirds, have these nuclei, and they are generally larger in males than in females. Each nucleus contains cell bodies and dendrites of a number of types of neuron. Some of the neurons

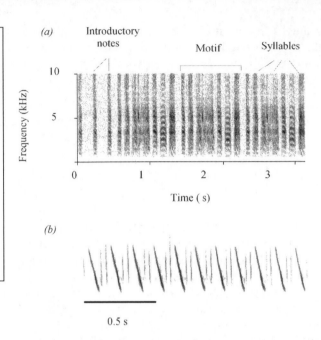

Fig. 9.13 Sonagraphs of excerpts from the songs of a zebra finch *Taeniopygia guttata* (*a*) and of a swamp sparrow, *Melospiza georgiana* (*b*), which show how the frequency content of the sound changes with time. The zebra finch song includes a few introductory notes, followed by three repetitions of the same motif. This motif contained six syllables, each one of which was a particular sequence of notes. The swamp sparrow song has one of the simplest structures of any bird. (Part (*a*) sonagraph kindly supplied by Dr D. Margoliash; (*b*) redrawn from Marler and Slabberkoorn, 2004.)

have axons that run in tracts to other nuclei, while others participate in processing information within their nucleus. One group of nuclei is responsible for generating the song in the adult, and another is involved in developing motor programmes for singing. A further group is involved in processing sounds.

A number of species are commonly studied. One is the zebra finch, *Taeniopygia guttata*, an Australian species that breeds readily in captivity and develops to maturity in only 100 days. An individual male produces song that is more easy to characterise than the songs of many other birds because his song is relatively stereotyped. In contrast, individual brown thrashers, from the eastern USA, use over a thousand different syllables in their songs. Another bird that has been important for research on neuronal mechanisms is the American swamp sparrow, *Melospiza georgiana*, in which an individual's repertoire is four or five songs composed of a series of trills, a relatively simple structure.

A lot of information about songs can be expressed in sonagraphs, which are particularly useful for comparing the songs of different individuals, or of one bird at different stages in its development. The sonagraphs in Fig. 9.13 illustrate different levels of organisation within the songs of a zebra finch and a swamp sparrow. In the zebra finch, series of notes are linked together into discrete syllables, and a series of syllables are linked together in a unit called a motif. When birds are interrupted during singing, they almost always finish a syllable, which means that a syllable is a basic unit in the organisation of song. The motif that a particular male zebra finch sings is fixed in form, although the number of motifs in a bout of singing varies from song to song, as do the brief introductory notes that precede the song and separate successive motifs. Each note of the

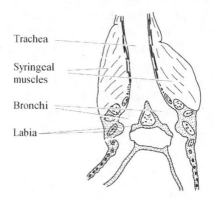

Trachea

Syringeal
muscles

Bronchi

Labia

Fig. 9.14 Main structures associated with the syrinx of a song bird. (Redrawn from Suthers, 1990.)

zebra finch song is composed of sounds of many different frequencies. Notes of most other species, including the swamp sparrow, contain more restricted tones and, as a result, have a more musical quality to a human listener.

The organ responsible for producing sounds during song is the **syrinx**, located where the trachea joins the bronchi of the two lungs (Fig. 9.14). Four to six muscles on either side are attached to the syrinx, and sound is produced when air is expelled through it. The flow of air induces parts of the syrinx wall to vibrate and generate sound, in a similar way to the operation of the human larynx. In song birds it is thought that fleshy labia at the mouth of each bronchus are primarily responsible. Some syrinx muscles determine the tone of sound which is produced and others control the timing of sounds by opening and closing the bronchi.

Respiratory muscles generate the force for expelling air through the syrinx, and so control the volume of sound. Each syllable of a song is produced by contraction of muscles that expel air from the interclavicular air sac. The lungs of birds are rigid and air is moved through them in one direction by the action of large air sacs that act as bellows. Electromyograms and pressure measurements have shown that, in canaries, each syllable is co-ordinated with a cycle of inspiration and expiration even at rates in excess of 20 per second. It is usual for the muscles on the left and right of the syrinx to act independently of each other, and zebra finches sing mostly by using muscles on the right. In some species, one side of the syrinx produces higher pitched notes than the other, and one thing a bird learns to do is to switch seamlessly between the two sides of its syrinx while singing. This asymmetry in the syrinx is matched by asymmetries between the left and right sides of the brain, a tantalising similarity to the asymmetrical arrangement of brain areas concerned with human speech.

Song development

The development of song in most species follows three distinct phases. In the first, the sensory phase, the bird hears songs produced by tutor birds around him. In temperate species, this usually occurs soon after hatching in the spring or summer, almost a year before the bird begins to sing himself. The bird is able to store memories of tutor songs at least until he starts to practise his own song, nearly a year later. The requirement for tutor songs was first shown experimentally by Thorpe (1958), who found that chaffinches (*Fringilla coelebs*) that had been reared in the laboratory without hearing the songs of adults produced very abnormal songs when they matured. The syllables of their songs were simple and disorganised. However, chaffinches that had heard tape recordings of adult chaffinch song as fledglings produced normal songs when they matured. Young chaffinches did not learn other sounds, including the songs of most other bird species, so a mechanism for recognising the song of members of its own species must be built into the brain of the young bird. The timing of exposure to tutor song is very important and a young chaffinch must hear adult songs before the middle of its first summer if it is to sing effectively the next year. This is one of the clearest examples of a critical period which, as described in Chapter 6 for the alignment of auditory with visual maps in a juvenile owl, is a restricted time window in an animal's life when it is able to learn something particular. A young bird can memorise songs he heard just after hatching for at least a year before using them as a template to compose his own song.

The second period of song development, the sensorimotor phase, begins early in the spring of a chaffinch's second year. He starts to sing spontaneously, at first producing quiet and variable songs. Gradually song becomes louder and it is called plastic because it is variable. Plastic song is often a mixture of different songs, and includes repeated syllables. Mark Konishi (1965a, b) showed, in different species of bird, that a bird must be able to hear its own plastic song for normal development. The timing of this is important, so it is a second critical period in song development. Finally, in late spring song crystallises to its mature form, probably triggered by a rise in the level of the steroid hormone testosterone in the blood.

This sequence of three phases is found in most species of song bird from temperate regions, but there are many variations in detail. Some species, including starlings and canaries, are called open-ended learners because they go through the processes of song practice and crystallisation every season. Zebra finches can breed throughout the year. Song development follows the same sequence as birds that breed seasonally except that young zebra finch males can still hear the songs of tutors while they start to produce plastic song. This is associated with their rapid development and social lifestyle. Their sensitive period for song learning starts at 20 days old, when they fledge, and lasts until they are 40 days old. Young zebra finches begin to sing 25 days after hatching and song crystallises 90–110 days after hatching, at sexual maturity. Although the sensory and sensorimotor phases of song development overlap in this species, it has been shown that a zebra finch is able to remember songs for several months without itself singing.

In most species of song birds, learning from tutors is essential for an individual to develop a song that fulfils its prime function of attracting mates. But a young bird will hear all kinds of songs, including the songs of other species, so is it predisposed to learn the most appropriate songs? The answer seems to be 'yes', which means that there must be a genetically encoded song recognition system in the bird's brain that allows it to tune in to males of its own species, and ignore songs of others. If a newly hatched white-crowned sparrow (*Zonotrichia leucophrys*) is kept in isolation and hears recordings of various species, including those of adult male white-crowned sparrows, he will later sing a typical white-crowned sparrow song. But this recognition is not a simple, hard-and-fast one that means a young bird will only pay attention to the correct species song. Other cues in addition to sounds are important, because if the young white-crowned sparrow has a live tutor of a song sparrow, he will learn the song sparrow song in preference to recordings of white-crowned sparrow songs. A feature of white-crowned sparrow song, as distinct from swamp sparrows *Melospiza georgiana* or song sparrows *Melospiza melodia*, is that the song starts with a whistle of constant tone, and this may be the most distinctive feature that young white-crowned sparrows usually use to recognise the song of their own species. But there is enormous variation in what different species of bird will learn and copy; zebra finches are relatively fussy and normally only copy songs of members of their own species, whereas lyre birds will copy into their songs almost any noise they hear including car engines and the sounds of cameras.

Neural centres for hearing and singing

Three different groups of nuclei, which are repeated on either side of the brain, are known to play roles in the production of song. One group of brain nuclei is involved in processing auditory information, a second in producing the motor programme for song (Fig. 9.15a) and a third in learning and developing the song pattern (Fig. 9.16). One nucleus, HVC, is a member of all three groups, and individual HVC neurons play key roles in generating adult song. Although some scientists use HVC as an abbreviation for the term 'higher vocal centre', we shall follow what others do and use HVC as a name because it can be misleading to attach functional names to brain regions (the anatomical term that HVC stands for is no longer used). Adult birds in which HVC has been destroyed cannot sing, although they still court females by adopting the same postures as during singing (Nottebohm *et al.*, 1976) and they produce alarm cries and some other calls. There is a good correlation between the size of HVC and the complexity of song in different species. In an adult male zebra finch, HVC contains about 35 000 neurons of three different

Fig. 9.15 Birdsong: pathways and neuronal activity in adult song birds. (*a*) A summary diagram of major brain nuclei and their connections. Nucleus HVC sends axons to RA, which in turn sends axons to motor centres that control muscles of the syrinx and air sacs. Field L is a brain area that contains neurons involved in recognising and processing sounds; it is similar to the auditory cortex in mammals. HVC receives auditory information through two small nuclei that are not drawn here. (*b*, *c*) Spikes recorded from a neuron in HVC (*b*) and from a neuron in RA (*c*) in singing zebra finches. The instantaneous spike rate was recorded from each neuron, and a sonogram for the song the bird sang is given below. (Redrawn from Yu and Margoliash, 1996.)

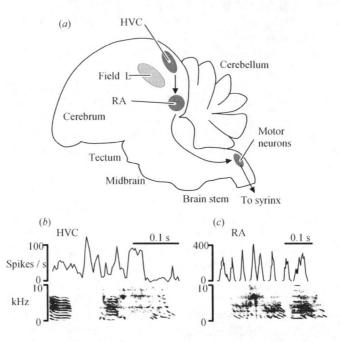

types: one type that is confined to HVC; one type that sends an axon to RA; and one type that sends an axon to Area X.

HVC exerts direct control over the production of song through connections it makes with nucleus RA (robust nucleus of the arcopallium). From RA, axons project to groups of motor neurons that innervate the syrinx and some respiratory muscles. Anthony Yu and Daniel Margoliash (1996) implanted fine wire electrodes into the brains of zebra finches and recorded spikes from single neurons while the birds were moving around their cages and singing normally. Many of the neurons in HVC started to spike before the bird began to sing, and remained excited until just before the end of the song (Fig. 9.15*b*). During a song, spike rate increased and decreased in a characteristic manner whenever the bird sang a particular syllable. An individual neuron was most excited when a particular syllable was sung, and each syllable would be associated with excitation of a unique population of HVC neurons. In contrast, individual neurons in RA produced discrete and intense bursts of spikes coinciding with a specific series of notes that occurred in a number of different syllables throughout the song (Fig. 9.15*c*).

In slightly earlier experiments, Eric Vu and colleagues (Vu *et al.*, 1994) used similar electrodes to stimulate neurons in singing birds with a short train of pulses that lasted less than the duration of a syllable. Stimuli to small regions of HVC disrupted the syllable the bird was singing and interfered with the order of subsequent syllables within a motif. For example, if a bird's motif consisted of syllables ABCD, a stimulus delivered just after syllable B might alter the motif's structure to ABG, combining syllables C and D to an unusual, abbreviated form, G. The electrical stimuli neither stopped the song, nor interfered with the pattern of syllables during

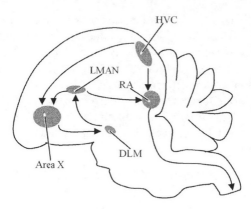

Fig. 9.16 A diagram to show the principal brain nuclei involved in the development of bird song. Notice that both HVC and LMAN send axons to RA and to Area X. Area X connects via DLM back onto LMAN.

following motifs. Stimuli delivered to RA had a more restricted effect, disrupting the current syllable without affecting later syllables within the motif. A stimulus delivered to RA just after B in the sequence ABCD would cause the bird to sing ABC'D, where C' is an altered form of syllable C.

These two kinds of experiment suggest that individual neurons in HVC control particular sequences of syllables within a motif by recruiting groups of neurons in RA in a specific order. The RA neurons each excite a pool of motor neurons that control muscles of the syrinx to produce a particular combination of notes. This correlates well with the anatomy of RA which is arranged topographically, so that a particular region in it corresponds with a particular group of motor neurons and muscles (Vicario, 1991). HVC is not organised topographically.

HVC has sensory as well as motor properties. In an anaesthetised zebra finch, many HVC neurons respond well to recordings of the bird's own song, but not to the songs of other birds. The neurons recognise both the order of notes within a syllable and the order of syllables within a motif. Specificity for the order of syllables in a motif is particularly interesting because it means that HVC neurons collect information over some time, so that, for example, they can distinguish ABCD from ACBD.

Auditory signals reach HVC though a series of processing areas, which include some that are part of the basic sound processing system that all birds have and others that are specialised features of song birds. As in other birds, the axons of auditory neurons travel from the ear on either side to its cochlear nucleus, which distributes information to a number of different processing stations. Much of this distributed information is brought together at the bird equivalent of the inferior colliculus – in an earlier chapter we described how this happens in the sound localisation system of owls. From here, information is relayed towards a number of distinct regions that together are called field L, which is just below HVC in song birds.

In early stages of the auditory pathway, individual neurons respond to simple stimuli such as tones or clicks, whereas in field L, many neurons respond to more complex features of natural sounds, including songs or cries of other birds. In song birds, a region called the auditory lobule, made up of two regions either side of field L, seems to be associated with recognising particular songs. In adult starlings, which have been trained to recognise and distinguish between different songs, individual neurons here respond to particular features of the learned songs (Gentner and Margoliash, 2003). These regions are likely to be very important in song recognition both by females and males, and in the development of song.

Individual HVC neurons: 'mirror neurons'

To understand the function of individual HVC neurons in singing behaviour, Jonathan Prather, Richard Mooney and colleagues (Prather *et al.*, 2008) managed to record responses from individual neurons of HVC in awake, moving swamp sparrows, a species in which individuals have a repertoire of a few different songs, all with a simple structure. In these remarkable experiments, they fixed a small and lightweight electrode manipulator onto the head of a bird. Gently advancing an electrode enabled them to record spikes from single neurons, and the electrode stayed in place for several days. They made recordings from neurons that carried spikes in their axons from HVC to Area X and found that about a third of the neurons they sampled responded to recordings of songs. Almost all of these HVC_X neurons were very selective in responding only to recordings of one of the songs from the bird's own repertoire, and this song is referred to as that neuron's primary song. Swamp sparrows often indulge in territorial singing matches with their neighbours, and playing recordings to the birds was a good way to induce them to sing during experiments. When a sparrow was played a recording of

A male swamp sparrow, *Melospiza georgiana*. These birds have been found to have mirror neurons in the system of brain nuclei that controls singing. A mirror neuron is one that responds when an animal performs a particular action, or when a neighbour performs the same action, so are important in tying together sensory and motor systems. (Photograph by Rob Lachlan, Duke University.)

its own song, an HVC_X neuron would respond both to the recording and when the bird itself sang. This differed from the responses of neurons that send axons to RA from HVC, which did not respond at all to songs played to the bird in these experiments. The particular song an HVC_X neuron responded to varied between neurons, so each song would be associated with its own group of HVC_X neurons. Responses were very brief, usually only one or two spikes, but consistently followed the start of each syllable in the preferred song. This kind of spike coding is called **sparse**; the neuron is usually silent, but each of the spikes it does produce says something significant.

HVC_X neurons clearly have auditory properties, and they could simply be responding to the sound the bird makes when it sings. The delay between a syllable and a spike in these neurons was the same whether the syllable was played from a recording or produced by the bird itself, which is consistent with this idea. But further investigation pointed to another explanation, which has interesting implications for the mechanisms of song construction in bird brains. This is that, when a bird sings, these neurons carry a copy of the commands that HVC issues to control the song.

There are a number of different lines of evidence that the neurons are not simply responding to sounds while the bird is singing. First, for half a second or so before and after a bout of singing, the neurons become deaf to sound recordings played to the bird. Second, when a recording of the same song type was played during a bout of singing, the responses to the ongoing song were not affected. In contrast, when a recording of the preferred song was played to a silent sparrow and a recording of another song by the bird was then superimposed, the sensory responses to the preferred song were suppressed. Third, it was quite common for the neurons to show some responses to a second song type when the bird sang, but not to recordings of the second song. These observations were not a peculiar feature of swamp sparrows because Prather and colleagues also found them in Bengalese finches, which have a more complex song structure.

In summary, when a bird is not singing, the HVC_X neurons are driven by auditory pathways and respond selectively to a particular song pattern; but when the bird is singing, the neurons are driven by a copy of the motor programme for singing. So, when a bird sings, HVC issues commands to neurons in RA that then organise activity in groups of motor neurons for the syringeal and respiratory muscles. The signals in HVC_X neurons may well be copies of the commands issued to RA, so using the terminology introduced earlier in the chapter, the signals in the HVC_X neurons are a corollary discharge of the motor commands. The corollary discharge is delayed relative to the commands sent to RA, probably by routing it through some of the local interneurons in HVC. The delay means that the corollary discharge is timed to coincide with responses of auditory neurons that report the results of the commands by responding to the sounds the bird is making. It is possible that HVC_X neurons are themselves a site of comparison between the motor commands and auditory feedback, or they may send their copies of the motor commands to another brain region where this comparison is made. In juvenile zebra finches during late stages of song development, there is good evidence that such a comparison is made in field L and another, neighbouring, auditory area, which contain neurons that seem to be able to report errors in the sound the bird makes while singing (Keller and Hahnloser, 2009). Although, in the experiments by Prather *et al.* (2008), the HVC_X neurons did not respond to external sounds while the bird was singing, it may have been that there was little difference between the song the bird intended to sing and what it actually did sing because the birds in the experiments were experienced singers.

A second function of the HVC_X neurons could be in enabling birds to interpret each other's songs. In a listening swamp sparrow, although HVC_X neurons responded best to recordings of one of the bird's own songs, they did also respond if a part of another bird's song closely resembled part of the neuron's primary song. Each HVC_X neuron is sensitive to the particular order of notes within a syllable. For example, if a neuron responded just after note C in a syllable with note order ABCD, it would not respond if the order was reversed to DCBA, or to C played alone. Thus it needs to be stimulated by at least a pair of syllables, and they must be in the correct order. Synthetic pairs of notes caused good responses if they corresponded with the appropriate pair of notes in a real song; and songs of other swamp sparrows that included the same sequential pair of notes in a syllable also caused good responses. So a particular HVC_X neuron will respond if a bird sings a song that contains a specific note sequence, and it will also respond if the bird hears one of its neighbours sing a song containing the same note sequence. When the neighbour sings a song, the HVC neurons could enable a swamp sparrow to match the song he hears to his own repertoire of songs, and then to select an appropriate response to the neighbour. So, in addition to a role in shaping a bird's song, HVC_X neurons might play a role in decoding the meaning of a neighbour's song.

Neurons whose excitation reflects the same action whether it is performed by an animal itself or while the animal watches a neighbour doing it were first found in monkeys. Gallese and colleagues in 1996 showed that a region of the brain's cortex contains neurons that respond both when a monkey extends its arm in a particular way and when it sees another monkey make the same movement. They have been named mirror neurons and they could have significant roles in perception, learning and perhaps empathy. The study on HVC_X neurons of swamp sparrows and Bengalese finches reported for the first time that there are mirror neurons with auditory properties.

Development of song and song nuclei

During development, where is memory of tutor song stored; and how is it used to instruct the bird as it develops its own song? It may not be possible to pinpoint one particular location in the brain that stores the memories of tutor birds; the memory could be distributed among different regions. But molecular biology techniques have provided evidence that one particular area of the auditory lobule, which is next to field L, is important in memorising and later recognising particular songs. The first evidence that part of the auditory lobule is involved in song memory came from experiments in which activity of different brain regions was detected by expression of a particular gene known as zenk (Mello et al., 1992). Zenk is known to be expressed in mammals and birds by a variety of different cell types, including neurons, when they are activated. So detecting regions that express zenk provides a way to suggest which particular brain areas are most active when an animal does a particular activity or has a particular kind of sensory experience, such as listening to song. The zenk protein is a transcription factor that regulates expression of other genes and, in neurons, it could lead to changes in the circuitry of a brain region by causing growth of new processes and synapses. In adult zebra finches or canaries, a part of the auditory lobule known as NCM, but not other parts of the auditory pathway or song nuclei, expressed zenk when the birds were played recordings of songs, particularly those of their own species. When birds sing songs, zenk and other genes are expressed in the forebrain nuclei that are involved in developing song (Jarvis and Nottebohm, 1997; Wada et al., 2006). Later experiments (Gentner et al., 2001) showed that in female birds NCM also expresses zenk when these females heard songs they found particularly attractive.

Although these observations label NCM as a possible site for song memory, they are far from conclusive, and do not show that NCM is involved in laying down the memory in a young bird. One way of investigating whether zenk production in NCM is important for a young bird to memorise songs is to block zenk production, and that is the approach used by Sarah London and David Clayton (2008). They perfused into the left and right NCMs of young zebra finches a drug that interferes with part of the molecular cascade regulating zenk expression. The birds were played recordings of tutor songs for an hour and a half every day between days 40 and 50 after hatching. When they matured, finches that received the drug during each tutoring session produced poor imitations of their tutor songs, whereas finches that received an inactive analogue of the drug copied the tutor songs very well.

For a bird to incorporate memories of tutor songs into its own adult song and so acquire a normal song pattern, nuclei LMAN, DLM and Area X are necessary (Bottjer et al., 1984). If LMAN is removed from a fledgling bird, the bird will produce an incomplete pattern of song, and if Area X is removed, song remains plastic. In the adult zebra finch, neurons in Area X and LMAN, like those in HVC, respond specifically to tape recordings of the bird's own song. This selectivity arises gradually during development. Neurons in LMAN and Area X of young birds are excited by many kinds of sounds, including songs played in reverse, and do not prefer particular songs (Doupe, 1997; Solis and Doupe, 1997). In 60-day-old zebra finches, some of the neurons prefer the bird's own song, while others respond best to tutor songs. When song crystallises, all the neurons that are recorded prefer the bird's own song.

Memories of alternative songs can displace each other during song learning. This was shown in zebra finches by using two different tutors: one early on during development, and the second 60 days after hatching, at the time when the finches were starting to sing (Yazaki-Sugiyama and Mooney, 2004). When the bird started to sing, it included elements that resembled the song of the first tutor, but when it matured, its own song was clearly modelled on the song of the second tutor. In some experiments, the first tutor was a Bengalese finch rather than a zebra finch to enhance the difference between songs. In 60-day-old finches, neurons in LMAN responded very selectively to recordings of tutor 1 or of the bird's own juvenile song. However, in adult finches, no LMAN neurons were found that responded to either of those songs, but many responded to recordings of songs made by either the second tutor or by the bird itself. So LMAN does not retain memory of all songs experienced, but does keep a record of the most relevant songs, including the current bird's own song.

Changing from babbling to effective song

When a young bird starts to sing, its song is a kind of babbling that includes poorly defined syllables and, in zebra finches, variable motifs unlike the fixed motif structure that is characteristic of adult song. It is quite usual for complex motor programmes, such as song, to be relatively poorly organised when an animal first performs them, but the animal learns by a process of exploring alternative movements to find out gradually which ones work best. It is generally assumed that learning a motor programme involves improving the pathways and networks that will be responsible for controlling the behaviour when it has been perfected. So one possibility is that LMAN, Area X and DLM are responsible for shaping the motor commands that HVC issues as a bird develops its song.

It was surprising, therefore, when Dmitriy Aronov, Aaron Andalman and Michale Fee reported (Aronov et al., 2008) that LMAN plays a direct role in controlling babbling by young zebra finches that is similar to that played by HVC in song by adults. HVC is not needed for a young finch to babble. When a juvenile finch babbles, particular neurons in LMAN spike just before defined features of the song, such as a particular syllable type, suggesting that these neurons have a role in driving aspects of the babbling behaviour. Like HVC, LMAN sends axons both to RA and to Area X (Fig. 9.16). In a young zebra finch in the sensorimotor phase of song development, if LMAN is inactivated on both sides, the finch no longer babbles. However, if its left and right HVC nuclei are removed, it will still babble. Furthermore, in an adult zebra finch removing HVC does not completely block singing, but instead of singing its adult song the finch sings a poorly defined and variable song. Analysing the acoustic features of this song showed it has all the characteristics of babbling that juveniles produce, and is not just a poor version of the adult song.

In one type of experiment, Aronov and colleagues temporarily inactivated HVC in adult finches by dialysing a solution of the drug tetrodotoxin, which blocks spikes, into it. The songs of these birds lost their adult organisation so the birds babbled rather than sang, but the properly organised adult song returned within 20 minutes as the drug wore off. These results suggest that LMAN is capable of generating experimental song programmes, so the juvenile finch tries out a number of different ways of singing. HVC stores the programme for the final, well-crafted adult song.

Significant structural changes occur in HVC, RA, LMAN, DLM and Area X during the development of a zebra finch, and some of these changes are probably associated with the switch in the dominant nucleus controlling song from LMAN to HVC. HVC is first recognisable as a distinct nucleus 10–15 days after hatching and new neurons are added to it throughout sensorimotor learning. These new neurons originate from a proliferative layer in the ventricle above HVC; in canaries, which learn new songs annually, new neurons are added to HVC every year, and this was the first example in a vertebrate where new neurons were shown to be born during adult life (Nottebohm, 1989). RA also increases in size during sensorimotor learning, but LMAN loses about half its neurons. To begin with, 37% of the synapses in RA are from LMAN neurons, but this drops to 4% by the time song crystallises, and an increasing proportion of synapses come from HVC (Herrmann and Arnold, 1991). In finches that are raised without exposure to adult songs the extent of these changes is less than normal (Wallhausser-Franke *et al.*, 1995).

Conclusions

The examples of signalling behaviour in this chapter illustrate how it is possible to understand the roles of individual nerve cells in controlling relatively complex behaviours. All three of the behaviours are clearly defined activities in which specialised morphological structures produce and receive signals; and because the areas of the nervous system associated with the behaviours are themselves specialised, unambiguous associations can be made between activity in their neurons and particular behaviours. In each of the behaviours, there are examples of studies where the roles of individual neurons have been defined by correlating neuronal signals with behaviour that the animal is performing, by stimulating the neuron to observe effects on behaviour, and by removing neurons (or, at least, brain nuclei). One of our main aims in the book is to show how the control of behaviour can be understood by following the flow of information through networks of neurons that interact with each other in various ways. There are several examples in this chapter of basic operations performed by nervous systems, such as sensory filtering and motor pattern generation. A new concept introduced here is corollary discharge. In the cricket, where a corollary discharge interneuron has been identified, sending motor commands to the auditory system enables the cricket to hear sounds around him as soon as he has stopped making a sound himself. In birds, both when an adult sings and during the development of song, it is quite clear that comparisons are made between templates for song stored in the brain and the sound the bird actually makes while singing. The HVC_X neurons are probably part of a corollary discharge system, or something very similar, that enables these comparisons to be made. *Eigenmannia* does not use corollary discharge to control its jamming avoidance response, although an engineer given the job of designing a control system for a jamming avoidance response would probably do so. However, *Eigenmannia*'s strategy, of using the electrical stimuli that impinge on its skin, has developed in a number of steps that can be traced in a possible evolutionary series of electric fish. As in other areas of biology, the function of nervous systems can only be fully appreciated in an evolutionary context.

Questions

What commands the command neuron for cricket singing?

What are the relative advantages to electric fish in using waves or pulses as signals?

Why is vocal learning restricted to only a few groups of birds and mammals?

Summary

- Signals have clearly defined functions in behaviour, and both their generation and reception involve clear specialisations in nervous systems.

Cricket song

- A male cricket sings by stridulation: rubbing a toothed file on one forewing against a scraper on the other.
- The tone of song is characteristic of species, and resonance amplifies the sound.
- Sound is produced in syllables that can be organised into chirps or trills.
- The calling song attracts receptive females; courtship is quieter and sung when a female is nearby; competing males sing rivalry songs.

- Flight muscles move the wings during singing, with elevators closing the wings and causing the loudest sounds.
- A command neuron in the brain switches on calling song, and also plays a role in setting chirp repetition rate.
- The basic song pattern is generated by a central pattern generator, but functional song needs proprioceptive feedback.
- Different interneurons are involved in generating the patterns for singing and flying.
- Receptive females run or fly towards the sound of a male's calling song. To release this phonotaxis, the sound must have the correct tone and be pulsed into syllables. Syllable rate is a significant parameter.
- Each syllable at normal song rate evokes a turn by a running female, controlled by a neuronal network in the thorax. A likely function for brain neurons is to recognise songs with appropriate syllable rates and enhance the thoracic networks that control phonotaxis.
- Auditory receptor neurons transduce vibrations of the tympanum just below the front knees and send axons to the thoracic ganglion. They contact local auditory interneurons as well as those that project to the brain.
- Most receptors are tuned to sound frequencies close to the tone of the song of its species.
- Directional sensitivity is enhanced by the physical design of the ears and tracheae, and by interactions between neurons such as the omega neurons.
- When a cricket sings, a corollary discharge interneuron reduces the responsiveness of many auditory neurons so that they are not deafened by the cricket's own sounds.
- Some of the same genes influence both the pattern generators for singing in males and the pattern recognition mechanisms in females.

Electric fish jamming avoidance

- Several aquatic vertebrates are sensitive to electric fields around their bodies, and some freshwater fish generate their own weak electric fields for communication and electrolocation.
- Electric fish have ampullary and tuberous electroreceptors that monitor electric fields across the skin.
- *Eigenmannia* produces waves at 240–630 Hz, with frequency tightly controlled in an individual.
- The jamming avoidance reaction enables a fish to maintain its own private frequency. If it senses a fish with frequency similar to its own, it will raise or lower its own frequency.
- A fish's own electric organ discharge creates electric fields that cross its skin at right angles all around its body. The discharge of a neighbour creates an electric field that crosses the skin at right angles nearest to the neighbour, but glances off skin in other regions.

- Where the two fish's fields sum, beats are generated with a frequency equal to the difference in the frequencies of the two fish.
- At a patch of skin, electric fields are encoded in two separate pathways in different electroreceptors: P-units code strength and T-units code timing.
- In the brain, electroreceptors project to the electrosensory lobe, which projects to the torus semicircularis.
- Information about electric signal amplitude and timing is recombined in one lamina of the torus semicircularis. Subsequently, in an area of the tectum, some neurons respond to extreme time differences in electric wave signals. Their removal abolishes a fish's jamming avoidance reaction, and stimulating or inhibiting them alters the fish's electric organ frequency.
- Electric organ discharge is controlled by a pacemaker nucleus in the hindbrain that connects via relay neurons to motor neurons.
- The way that the jamming avoidance reaction is controlled by comparing amplitude and timing of signals across different skin areas can be understood from tracing an evolutionary series through living species of fish.

Bird song

- Most passerines produce extended and complex songs that, to be fully functional, involve a young bird learning sounds in the songs of mature, tutor birds.
- Song is produced by the syrinx, which contains several muscles and fleshy lobes that vibrate as air passes over them.
- In temperate birds, song development has three phases:
  - sensory, in which a newly hatched bird hears songs produced by adults nearby
  - sensorimotor in the spring of year 2, when the bird makes broken variable or plastic songs
  - crystallisation to the adult song, triggered by a rise in testosterone.
- In producing and developing bird song, one group of brain nuclei are involved in producing song in the adult, another group in developing song, and a third group in processing auditory information. The nucleus called HVC is a member of all three groups.
- In adult song, experiments using electrodes for stimulation and for recording indicate that there is a hierarchical arrangement in which HVC neurons organise activity in groups of neurons in RA, which then drive sets of motor neurons.
- In anaesthetised or sleeping birds, individual HVC neurons respond to particular features of recordings of a bird's own song.
- In the auditory pathway, neurons in field L and in the auditory lobule respond selectively to particular songs, and may be a site where a juvenile compares remembered songs of tutors with his own plastic songs.
- In adult swamp sparrows and Bengalese finches, some neurons that connect HVC with Area X each respond to recordings of one

particular song from the bird's repertoire – the neuron's primary song. They also spike when the bird itself sings that song, in this case carrying as a corollary discharge a copy of part of the commands within the brain that control that song. They could be part of a system ensuring a match between the song the bird intends to sing and the song he actually produces.

- These HVC_X neurons also spike if another bird includes in its song notes that resemble notes in the neuron's primary song. They are similar to mirror neurons in monkeys: neurons that are excited when either an animal itself or else a neighbour performs a particular act. They may play a role in birds' interpreting each others' songs.
- Activity-related genes are particularly expressed in field L and neighbouring regions when a young bird is learning tutors' songs, indicating that these regions are important in recognising and remembering tutor songs.
- Nuclei LMAN and DLM as well as Area X are important for a juvenile bird to develop song. These regions contain neurons that become increasingly selective for recordings of a bird's own song rather than other songs during development, and the regions undergo considerable anatomical reorganisation.
- In a juvenile zebra finch, LMAN plays a similar role to HVC in the adult by controlling immature, babbling song. LMAN may be the site where the final, adult song pattern is assembled.

Further reading

Brenowitz, E. A. and Beecher, M. D. (2005). Song learning in birds: diversity and plasticity, opportunities and challenges. *Trends Neurosci.* **28**, 127–135. A discussion of how the approach of comparing different species with the more commonly studied ones, especially the zebra finch, would be fruitful in understanding how birds learn their songs.

Hedwig, B. (2005). Pulses, patterns and paths: neurobiology of acoustic behaviour in crickets. *J. Comp. Physiol. A*, **192**, 677–689. A review that covers significant discoveries in the neurobiology of cricket song after 2000.

Rose, G. J. (2004). Insights into neural mechanisms and evolution of behaviour from electric fish. *Nat. Rev. Neurosci.* **5**, 943–951. A good review of the jamming avoidance reaction and other aspects of the neurobiology of electric fish.

References

Amaral, D. G. and Witter, M. P. (1995). Hippocampal formation. In *The Rat Nervous System* (ed. G. Paxinos), pp. 443–493. London: Academic Press.

Anstey, M. L., Rogers, S. M., Ott, S. R., Burrows, M. and Simpson, S. J. (2008). Serotonin mediates behavioral gregarization underlying swarm formation in desert locusts. *Science* **323**, 627–630.

Antonsen, B. and Edwards, D. (2007). Mechanisms of serotonergic facilitation of a command neuron. *J. Neurophysiol.* **98**, 3494–3504.

Aronov, D., Andalman, A. S. and Fee, M. S. (2008). A specialized forebrain circuit for vocal babbling in the juvenile song bird. *Science* **320**, 630–634.

Bacon, J. and Möhl, B. (1983a). The tritocerebral commissure giant (TCG) wind-sensitive interneurone in the locust. 1. Its activity in straight flight. *J. Comp. Physiol.* **150**, 439–452.

Bacon, J. and Möhl, B. (1983b). The tritocerebral commissure giant (TCG) wind-sensitive interneurone in the locust. 2. Directional sensitivity and role in flight stabilisation. *J. Comp. Physiol.* **150**, 453–465.

Bacon, J. and Tyrer, N. M. (1978). The tritocerebral commissure giant (TCG): a bimodal interneurone in the locust, *Schistocerca gregaria. J. Comp. Physiol.* **126**, 317–325.

Barlow, R. B., Hitt, J. M. and Dodge, J. A. (2001). *Limulus* vision in the marine environment. *Biol. Bull.* **200**, 169–176.

Barnett, P. D., Nordström, K. and O'Carroll, D. C. (2007). Retinotopic organization of small-field-target-detecting neurons in the insect visual system. *Curr. Biol.* **17**, 569–578.

Bartelmez, G. W. (1915). Mauthner's cell and the nucleus motorius tegmenti. *J. Comp. Neurol.* **25**, 87–128.

Bentley, D. R. and Hoy, R. R. (1972). Genetic control of the neuronal network generating cricket (*Teleogryllus*) song patterns. *Anim. Behav.* **20**, 478–492.

Bicker, G. and Pearson, K. G. (1983). Initiation of flight by stimulation of a single identified wind sensitive neurone (TCG) in the locust. *J. Exp. Biol.* **104**, 289–294.

Bitterman, M. E., Menzel, R., Fietz, A. and Schäfer, S. (1983). Classical conditioning of proboscis extension in honeybees (*Apis mellifera*). *J. Comp. Psychol.* **97**, 107–119.

Bliss, T. V. P. and Lømo, T. (1973). Long-lasting potentiation of synaptic transmission in the dentate area of the anaesthetized rabbit following stimulation of the perforant path. *J. Physiol.* **232**, 331–356.

Bodenhamer, R., Pollak, G. D. and Marsh, D. S. (1979). Coding of fine frequency information by echoranging neurons in the inferior colliculus of the Mexican free-tailed bat. *Brain Res.* **171**, 530–535.

Boice, R. (1977). Burrows of wild and albino rats: effects of domestication, outdoor raising, age, experience, and maternal state. *J. Comp. Physiol. Psychol.* **91**, 649–661.

Bolhuis, J. and Verhulst, S. (2008). *Tinbergen's Legacy: Function and Mechanism in Behavioural Biology*. Cambridge: Cambridge University Press.

Borst, A. (2007). Correlation versus gradient type motion detectors: the pros and cons. *Phil. Trans. Royal Soc. B* **362**, 369–374.

Borst, A. and Haag, J. (2002). Neural networks in the cockpit of the fly. *J. Comp. Physiol.* **188**, 419–437.

Borst, A. and Haag, J. (2007). Optic flow processing in the cockpit of the fly. In *Invertebrate Neurobiology* (ed. G. North and R. J. Greenspan), pp. 101–122. Cold Spring Harbor, NY: Cold Spring Harbor Laboratory Press.

Bottjer, S. W., Miesner, E. A. and Arnold, A. P. (1984). Forebrain lesions disrupt development but not maintenance of song in passerine birds. *Science* **224**, 901–903.

Brainard, M. S. and Knudsen, E. I. (1993). Experience-dependent plasticity in the inferior colliculus: a site for visual calibration of the neural representation of auditory space in the barn owl. *J. Neurosci.* **13**, 4589–4608.

Brun, V. H., Otnæss, M. K., Molden, S., *et al.* (2002). Place cells and place recognition maintained by direct entorhinal-hippocampal circuitry. *Science* **296**, 2243–2246.

Bruns, V. and Schmieszek, E. (1980). Cochlear innervation in the greater horseshoe bat: demonstration of an acoustic fovea. *Hearing Res.* **3**, 27–43.

Bullock, T. H. and Horridge, G. A. (1965). *Structure and Function in the Nervous Systems of Invertebrates.* San Francisco, CA: W.H. Freeman.

Burrows, M. (1975). Monosynaptic connexions between wing stretch receptors and flight motoneurones of the locust. *J. Exp. Biol.* **62**, 189–219.

Burrows, M. (1989). Processing of mechanosensory signals in local reflex pathways of the locust. *J. Exp. Biol.* **146**, 209–227.

Burrows, M. (1992). Reliability and effectiveness of transmission from exteroceptive sensory neurons to spiking local interneurons in the locust. *J. Neurosci.* **12**, 1477–1499.

Burrows, M. (1996). *The Neurobiology of an Insect Brain.* Oxford: Oxford University Press.

Burrows, M. and Newland, P. L. (1993). Correlation between the receptive fields of locust interneurons, their dendritic morphology, and the central projections of mechanosensory neurons. *J. Comp. Neurol.* **329**, 412–426.

Burrows, M. and Siegler, M. V. S. (1978). Graded synaptic transmission between local interneurones and motor neurones in the metathoracic ganglion of the locust. *J. Physiol. (London)* **285**, 231–255.

Burrows, M. and Siegler, M. V. S. (1985). The organization of receptive fields of spiking local interneurons in the locust with inputs from hair afferents. *J. Neurophysiol.* **53**, 1147–1157.

Burton, B. G., Tatler, B. W. and Laughlin, S. B. (2001). Variations in photoreceptor response dynamics across the fly retina. *J. Neurophysiol.* **86**, 950–960.

Camhi, J. M. and Tom, W. (1978). The escape system of the cockroach *Periplaneta americana*. I. The turning response to wind puffs. *J. Comp. Physiol.* **128**, 193–201.

Canfield, J. G. and Rose, G. J. (1993). Activation of Mauthner neurons during prey capture. *J. Comp. Physiol. A* **172**, 611–618.

Carew, T. J. (2001). *Behavioral Neurobiology: The Cellular Organization of Natural Behavior.* Sunderland, MA: Sinauer Associates.

Carr, C. E. and Konishi, M. (1990). A circuit for detection of interaural time differences in the brain stem of the barn owl. *J. Neurosci.* **10**, 3227–3246.

Catania, K. C. (1999). A nose that looks like a hand and acts like an eye: the unusual mechanosensory system of the star-nosed mole. *J. Comp. Physiol. A* **185**, 367–372.

Catania, K. C. and Kaas, J. H. (1996). The unusual nose and brain of the star-nosed mole. *BioScience* **46**, 578–586.

Catania, K. C. and Kaas, J. H. (1997). Somatosensory fovea in the star-nosed mole: behavioral use of the star in relation to innervation patterns and cortical representation. *J. Comp. Neurol.* **387**, 215–233.

Catania, K. C. and Remple, F. E. (2005). Asymptotic prey profitability drives star-nosed moles to the foraging speed limit. *Nature* **442**, 519–522.

Catchpole, C. K. and Slater, P. J. B. (2008). *Bird Song: Biological Themes and Variations.* 2nd edition. Cambridge: Cambridge University Press.

Clemens, S. and Katz, P. S. (2001). Identified serotonergic neurons in the *Tritonia* swim CPG activate both ionotropic and metabotropic receptors. *J. Neurophysiol.* **85**, 476–479.

Clyne, J. D. and Miesenböck, G. (2008). Sex-specific control and tuning of the pattern generator for courtship song in *Drosophila. Cell* **133**, 354–363.

Collett, T. S. (2007). Insect navigation: visual panoramas and the sky compass. *Curr. Biol.* **18**, R1058–R1060.

Collett, T. S. and Land, M. F. (1975). Visual control of flight behaviour in the hoverfly *Syritta pipiens*, L. *J. Comp. Physiol.* **99**, 1–66.

Comer, C. (1985). Analyzing cockroach escape behavior with lesions of individual giant interneurons. *Brain Res.* **335**, 342–346.

Comer, C. M. and Dowd, J. P. (1993). Multisensory processing for movement: antennal and cercal mediation of escape turning in the cockroach. In *Biological Neural Networks in Invertebrate Neuroethology and Robotics* (ed. R. D. Beer, R. E. Ritzmann and T. McKenna), pp. 89–112. Boston, MA: Academic Press.

Dagan, D. and Camhi, J. M. (1979). Responses to wind recorded from the cercal nerve of the cockroach *Periplaneta americana*. II. Directional selectivity of the sensory nerves innervating single columns of filiform hairs. *J. Comp. Physiol. A* **133**, 103–110.

Dambach, M. and Rausche, G. (1985). Low frequency airborne vibrations in crickets and feedback control of calling song. In *Acoustic Vibrational Communication in Insects* (ed. K. Kalmring and N. Elsner), pp. 177–182. Berlin: Paul Parey.

Davis, R. L. (2005). Olfactory memory formation in *Drosophila*: from molecular to systems neuroscience. *Annu. Rev. Neurosci.* **28**, 275–302.

Dawson, J. W., Kutsch, W. and Robertson, R. M. (2004). Auditory-evoked evasive manoeuvres in free-flying locusts and moths. *J. Comp. Physiol. A* **190**, 69–84.

de Ruyter van Steveninck, R. and Laughlin, S. B. (1996). The rate of information transfer at graded-potential synapses. *Nature* **379**, 642–645.

Demir, E. and Dickson, B. J. (2005). *fruitless* splicing specifies male courtship behavior in *Drosophila. Cell* **121**, 785–794.

Diamond, J. (1968). The activation and distribution of gaba and L-glutamate receptors on goldfish Mauthner neurons: an analysis of dendritic remote inhibition. *J. Physiol.* **194**, 669–723.

Douglass, J. K. and Strausfeld, N. J. (2003). Anatomical organization of retino-topic motion-sensitive pathways in the optic lobes of flies. *Microsc. Res. Techniq.* **62**, 132–150.

Doupe, A. J. (1997). Song- and order-selective neurons in the songbird anterior forebrain and their emergence during vocal development. *J. Neurosci.* **17**, 1147–1167.

Dowling, J. and Boycott, B. (1966). Organization of the primate retina: electron microscopy. *Proc. Roy. Soc. Lond. B* **166**, 80–111.

Dowling, J. E. and Werblin, F. S. (1969). Organization of retina of the mud-puppy, *Necturus maculosus*. I. Synaptic structure. *J. Neurophysiol.* **32**, 315–338.

Dvorak, D. R., Bishop, L. G. and Eckert, H. E. (1975). On the identification of movement detectors in the fly optic lobe. *J. Comp. Physiol.* **100**, 5–23.

Eaton, R. C. and Emberley, D. S. (1991). How stimulus direction determines the angle of the Mauthner initiated response in teleost fish. *J. Exp. Biol.* **161**, 469–487.

Eaton, R. C., Lavender, W. A. and Wieland, C. M. (1981). Identification of Mauthner-initiated response patterns in goldfish: evidence from simultaneous cinematography and electrophysiology. *J. Comp. Physiol.* **144**, 521–531.

Eaton, R., Lavender, W. and Wieland, C. (1982). Alternative neural pathways initiate fast-start responses following lesions of the Mauthner neuron in goldfish. *J. Comp. Physiol. A* **145**, 485–496.

Eccles, J. C. (1957). *The Physiology of Nerve Cells*. Baltimore, MD: Johns Hopkins Press.

Eccles, J. C. (1977). *The Understanding of the Brain*. New York: McGraw Hill.

Edwards, D., Yeh, S.-R. and Krasne, F. (1998). Neuronal coincidence detection by voltage-sensitive electrical synapses. *Proc. Natl. Acad. Sci. USA* **95**, 1745–1750.

Egelhaaf, M. (1985). On the neuronal basis of figure-ground discrimination by relative motion in the visual system of the fly. I. Behavioural constraints imposed by the neuronal network and the role of the optomotor system. *Biol. Cybern.* **52**, 123–140.

Egelhaaf, M. and Borst, A. (1993). Motion computation and visual orientation in flies. *Comp. Biochem. Physiol.* **104A**, 659–673.

Elyada, Y. M., Haag, J. and Borst, A. (2009). Different receptive fields in axons and dendrites underlie robust coding in motion-sensitive neurons. *Nat. Neurosci.* **12**, 327–332.

Erber, J., Masuhr, T. and Menzel, R. (1980). Localization of short-term memory in the brain of the bee, *Apis mellifera*. *Physiol. Entomol.* **5**, 343–358.

Espinoza, S., Breen, L., Varghese, N. and Faulkes, Z. (2006). Loss of escape-related giant neurons in a spiny lobster, *Panulirus argus*. *Biol. Bull.* **211**, 223–231.

Ewert, J.-P. (1980). *Neuroethology*. Berlin: Springer-Verlag.

Ewert, J.-P. (1985). Concepts in vertebrate neuroethology. *Anim. Behav.* **33**, 1–29.

Ewert, J.-P. (1987). Neuroethology of releasing mechanisms: prey-catching in toads. *Behav. Brain Sci.* **10**, 337–403.

Farooqui, T., Robinson, K., Vaessin, H. and Smith, B. H. (2003). Modulation of early olfactory processing by an octopaminergic reinforcement pathway in the honeybee. *J. Neurosci.* **23**, 5370–5380.

Finkenstädt, T. and Ewert, J. (1988). Stimulus-specific long-term habituation of visually guided orienting behavior toward prey in toads: a ^{14}C-2DG study. *J. Comp. Physiol. A* **163**, 1–11.

Fischer, H. and Kutsch, W. (2000). Relationships between body mass, motor output and flight variables during free flight of juvenile and mature adult locusts, *Schistocerca gregaria*. *J. Exp. Biol.* **203**, 2723–2735.

Fortune, E. S. (2006). The decoding of electrosensory systems. *Curr. Opin. Neurobiol.* **16**, 474–480.

Franceschini, N., Hardie, R. C., Ribi, W. and Kirschfeld, K. (1981). Sexual dimorphism in a photoreceptor. *Nature* **291**, 241–244.

Franceschini, N., Riehle, A. and Le Nestour, A. (1989). Directionally selective motion detection by insect neurons. In *Facets of Vision* (ed. D. G. Stavenga and R. C. Hardie), pp. 360–390. Berlin: Springer.

Frost, W. N. and Katz, P. S. (1996). Single neuron control over a complex motor pattern. *Proc. Natl. Acad. Sci. USA* **93**, 422–426.

Fry, S. N., Sayaman, R. and Dickinson, M. H. (2003). The aerodynamics of free-flight maneuvers in *Drosophila*. *Science* **300**, 495–498.

Furshpan, E. J. and Potter, D. D. (1959). Transmission at the giant motor synapses of the crayfish. *J. Physiol.* **145**, 289–325.

Gahtan, E., Sankrithi, N., Campos, J. and O'Malley, D. (2002). Evidence for a widespread brain stem escape network in larval zebrafish. *J. Neurophysiol.* **87**, 608–614.

Gallese, V., Fadiga, L., Fogassi, L. and Rizzolatti, G. (1996). Action recognition in the premotor cortex. *Brain* **119**, 593–609.

Gentner, T. Q. and Margoliash, D. (2003). Neuronal populations and single cells representing learned auditory objects. *Nature* **424**, 669–674.

Gentner, T. Q., Hulse, S. H., Duffy, D. and Ball, G. F. (2001). Response biases in auditory forebrain regions of female song birds following exposure to sexually relevant variation in male song. *J. Neurobiol.* **46**, 48–58.

Grinnell, A. D. and Hagiwara, S. (1972). Studies of auditory neurophysiology in non-echolocating bats, and adaptations for echolocation in one genus, *Rousettus*. *Z. vergl. Physiol.* **76**, 82–96.

Grinnell, A. D. and Schnitzler, H.-U. (1977). Directional sensitivity of echolocation in the horseshoe bat, *Rhinolophus ferrumequinum*. II. Behavioural directionality of hearing. *J. Comp. Physiol.* **116**, 63–76.

Grothe, B. (2003). New roles for synaptic inhibition in sound localization. *Nat. Rev. Neurosci.* **4**, 540–550.

Haag, J. and Borst, A. (2004). Neural mechanism underlying complex receptive field properties of motion-sensitive interneurons. *Nature Neurosci.* **7**, 628–634.

Habersetzer, J. and Vogler, B. (1983). Discrimination of surface-structured targets by the echolocating bat *Myotis myotis* during flight. *J. Comp. Physiol.* **152**, 275–282.

Hafting, T., Fyhn, M., Molden, S., Moser, M. B. and Moser, E. (2005). Microstructure of a spatial map in the entorhinal cortex. *Nature* **436**, 801–806.

Hagedorn, M. and Heiligenberg, W. (1985). Court and spark: electric signals in the courtship and mating of gymnotoid fish. *Anim. Behav.* **33**, 254–265.

Hammer, M. (1993). An identified neuron mediates the unconditioned stimulus in associative olfactory learning in honeybees. *Nature* **366**, 59–63.

Hammer, M. and Menzel, R. (1995). Learning and memory in the honeybee. *J. Neurosci.* **15**, 1617–1630.

Hanson, A. (2004) History of the Norway rat (*Rattus norvegicus*). 'Rat behavior' and 'Rat biology'. Online www.ratbehavior.org/history.htm. Accessed 25 August 2009.

Hardie, R. C. (1986). The photoreceptor array of the dipteran retina. *Trends Neurosci.* **9**, 419–23.

Harrow, I. D., Hue, B., Pelhate, M. and Sattelle, D. B. (1980). Cockroach giant interneurones stained by cobalt-backfilling of dissected axons. *J. Exp. Biol.* **84**, 341–343.

Hartline, H. K., Wagner, H. G. and Ratliff, F. (1956). Inhibition in the eye of *Limulus*. *J. Gen. Physiol.* **39**, 651–673.

Hausen, K. and Egelhaaf, M. (1989). Neural mechanisms of visual course control in insects. In *Facets of Vision* (ed. D. G. Stavenga and R. C. Hardie), pp. 391–424. Berlin: Springer.

Hedwig, B. (2000). Control of cricket stridulation by a command neuron: efficacy depends on the behavioral state. *J. Neurophysiol.* **83**, 712–722.

Hedwig, B. (2006). Pulses, patterns and paths: neurobiology of acoustic behaviour in crickets. *J. Comp. Physiol.* A **192**, 677–689.

Hedwig, B. and Heinrich, R. (1997). Identified descending brain neurons control different stridulatory motor patterns in an acridid grasshopper. *J. Comp. Physiol.* A **180**, 285–294.

Hedwig, B. and Pearson, K. G. (1984). Patterns of synaptic input to identified flight motoneurons in the locust. *J. Comp. Physiol.* A **154**, 745–760.

Hedwig, B. and Poulet, J. (2004). Complex auditory behaviour emerges from simple reactive steering. *Nature* **430**, 781–785.

Heiligenberg, W. (1991). *Neural Nets in Electric Fish*. Boston, MA: MIT Press.

Heiligenberg, W. and Partridge, B. L. (1981). How electroreceptors encode JAR-eliciting stimulus regimes: reading trajectories in a phase-amplitude plane. *J. Comp. Physiol.* A **142**, 295–308.

Heiligenberg, W. and Rose, G. (1985). Phase and amplitude computations in the midbrain of an electric fish: intracellular studies of neurons participating in the jamming avoidance response of *Eigenmannia*. *J. Neurosci.* **5**, 515–531.

Heiligenberg, W., Baker, C. and Matsubara, J. (1978). The jamming avoidance response in *Eigenmannia* revisited: the structure of a neuronal democracy. *J. Comp. Physiol.* A **127**, 267–286.

Heitler, W. J. and Fraser, K. (1993). Thoracic connections between crayfish giant fibres and motor giant neurones reverse abdominal patterns. *J. Exp. Biol.* **181**, 329–333.

Heitler, W., Fraser, K. and Ferrero, E. (2000). Escape behaviour in the stomatopod crustacean *Squilla mantis*, and the evolution of the caridoid escape reaction. *J. Exp. Biol.* **203**, 183–192.

Henneman, E., Somjen, G. and Carpenter, D. O. (1965). Functional significance of cell size in spinal motoneurons. *J. Neurophysiol.* **28**, 560–580.

Hennig, R. M. (1990). Neuronal control of the forewings in two different behaviours: stridulation and flight in the cricket, *Teleogryllus commodus*. *J. Comp. Physiol.* A **167**, 617–627.

Hensler, K. (1992). Neuronal co-processing of course deviation and head movements in locusts. I. Descending deviation detectors. *J. Comp. Physiol.* A **171**, 257–271.

Herberholz, J., Issa, F. and Edwards, D. (2001). Patterns of neural circuit activation and behavior during dominance hierarchy formation in freely behaving crayfish. *J. Neurosci.* **21**, 2759–2767.

Herberholz, J., Sen, M. and Edwards, D. (2004). Escape behavior and escape circuit activation in juvenile crayfish during prey–predator interactions. *J. Exp. Biol.* **207**, 1855–1863.

Herrmann, K. and Arnold, A. P. (1991). The development of afferent projections to the robust archistriatal nucleus in male zebra finches: a quantitative electron microscopic study. *J. Neurosci.* **11**, 2063–2074.

Higgins, C. M., Douglass, J. K. and Strausfeld, N. J. (2004). The computational basis of an identified neuronal circuit for elementary motion detection in dipterous insects. *Visual Neurosci.* **21**, 567–586.

Hill, K. G. and Boyan, G. S. (1977). Sensitivity to frequency and direction of sound in the auditory system of crickets (Gryllidae). *J. Comp. Physiol.* A **121**, 79–97.

Hopkins, C. D. (1999). Design features for electric communication. *J. Exp. Biol.* **202**, 1217–1228.

Horsman, U., Heinzel, H.-G. and Wendler, G. (1983). The phasic influence of self-generated air current modulations on the locust flight motor. *J. Comp. Physiol.* **150**, 427–438.

Huston, S. J. and Krapp, H. G. (2008). Visuomotor transformation in the fly gaze stabilization system. *PLoS Biol.* **6**, 1468–1478.

Hyde, P. S. and Knudsen, E. I. (2002). The optic tectum controls visually guided adaptive plasticity in the owl's auditory space map. *Nature* **415**, 73–76.

Issa, F., Adamson, D. and Edwards, D. (1999). Dominance hierarchy formation in juvenile crayfish *Procambarus clarkii*. *J. Exp. Biol.* **202**, 3497–3506.

Jacobs, L. F. (2003). The evolution of the cognitive map. *Brain Behav. Evol.* **62**, 128–139.

Jarvis, E. (2005). Avian brains and a new understanding of vertebrate brain evolution. *Nat. Rev. Neurosci.* **6**, 151–159.

Jarvis, E. D. and Nottebohm, F. (1997). Motor-driven gene expression. *Proc. Natl. Acad. Sci. USA* **94**, 4097–4102.

Jeffery, K. J. and Burgess, N. (2006). A metric for the cognitive map: found at last? *Trends Cogn. Sci.* **10**, 1–3.

Jones, G. and Holderied, M. W. (2007). Bat echolocation calls: adaptation and convergent evolution. *Proc. Roy. Soc. Lond. B* **274**, 905–912.

Judge, S. J. and Rind, F. C. (1997). The locust DCMD, a movement-detecting neurone tightly tuned to collision trajectories. *J. Exp. Biol.* **200**, 2209–2216.

Kalko, E. K. V. and Schnitzler, H.-U. (1998). How echolocating bats approach and acquire food. In *Bat Biology and Conservation* (ed. T. H. Kunz and P. A. Racey), pp. 197–204. Washington, DC: Smithsonian Institution Press.

Kandel, E. R. (1979). *The Behavioral Biology of Aplysia*. San Francisco, CA: Freeman.

Katz, P. S., Getting, P. A. and Frost, W. N. (1994). Dynamic neuromodulation of synaptic strength intrinsic to a central pattern generator circuit. *Nature* **367**, 729–731.

Keil, T. (1997). Functional morphology of insect mechanoreceptors. *Microsc. Res. Techniq.* **39**, 506–531.

Keller, G. B. and Hahnloser, H. R. (2009). Neural processing of auditory feedback during vocal practice in a song bird. *Nature* **457**, 187–190.

Kern, R., van Hateren, J. H., Michaelis, C., Lindemann, J. P. and Egelhaaf, M. (2005). Eye movements during natural flight shape the function of a blowfly motion sensitive neuron. *PLoS Biol.* **6**, 1131–1138.

Kern, R., van Hateren, J. H. and Egelhaaf, M. (2006). Representation of behaviourally relevant information by blowfly motion-sensitive visual interneurons requires precise compensatory head movements. *J. Exp. Biol.* **209**, 1251–1260.

Kimchi, T., Xu, J. and Dulac, C. (2007). A functional circuit underlying male sexual behaviour in the female mouse brain. *Nature* **448**, 1009–1014.

Kimmel, C. B. and Eaton, R. C. (1976). Development of the Mauthner cell. In *Simpler Networks and behavior* (ed. J. C. Fentress), pp. 186–202. Sunderland, MA: Sinauer Associates.

Kimmerle, B., Warzecha, A.-K. and Egelhaaf, M. (1997). Object detection in the fly during simulated translatory flight. *J. Comp. Physiol. A* **181**, 247–255.

Kirchner, W. H. and Srinivasan, M. V. (1989). Freely flying honeybees use image motion to estimate object distance. *Naturwissenschaften* **76**, 281–282.

Knudsen, E. I. (1981). The hearing of the barn owl. *Sci. Am.* **245**, 83–91.

Knudsen, E. I. (2002). Instructed learning in the auditory localization pathway of the barn owl. *Nature* **417**, 322–328.

Knudsen, E. I. and Knudsen, P. F. (1990). Sensitive and critical periods for visual calibration of sound localization by barn owls. *J. Neurosci.* **10**, 222–232.

Knudsen, E. I. and Konishi, M. (1979). Mechanisms of sound localisation in the barn owl (*Tyto alba*). *J. Comp. Physiol* **133**, 13–21.

Kolton, L. and Camhi, J. M. (1995). Cartesian representation of stimulus direction: parallel processing by two sets of giant interneurons in the cockroach. *J. Comp. Physiol. A* **176**, 691–702.

Konishi, M. (1965a). Effects of deafening on song development in American robins and black-headed grosbeaks. *Z. Tierpsychol.* **22**, 584–599.

Konishi, M. (1965b). The role of auditory feedback in the control of vocalization in the white-crowned sparrow. *Z. Tierpsychol.* **22**, 770–783.

Konishi, M. (1992). The neural algorithm for sound localisation in the owl. *Harvey Lect.* **86**, 47–64.

Konishi, M. (1993). Listening with two ears. *Sci. Am.* **268**, 34–41.

Konishi, M. (2006). Behavioral guides for sensory neurophysiology. *J. Comp. Physiol.* **192**, 671–676.

Koppl, C., Gleich, O. and Manley, G. A. (1993). An auditory fovea in the barn owl cochlea. *J. Comp. Physiol. A.* **171**, 695–704.

Krapp, H. G., Hengstenberg, B. and Hengstenberg, R. (1998). Dendritic structure and receptive-field organization of optic flow processing interneurons in the fly. *J. Neurophysiol.* **79**, 1902–1917.

Krasne, F. B. (1969). Excitation and habituation of the crayfish escape reflex: the depolarising response in lateral giant fibres of the isolated abdomen. *J. Exp. Biol.* **50**, 29–46.

Krasne, F. B. and Lee, S. (1988). Response-dedicated trigger neurons as control points for behavioral actions: selective inhibition of lateral giant command neurons during feeding in crayfish. *J. Neurosci.* **8**, 3703–3712.

Krasne, F. B. and Wine, J. J. (1975). Extrinsic modulation of crayfish escape and behaviour. *J. Exp. Biol.* **63**, 433–450.

Krasne, F. B. and Wine, J. J. (1977). Control of crayfish escape behavior. In *Identified Neurons and Behavior of Arthropods* (ed. G. Hoyle), pp. 275–292. New York: Plenum.

Kupferman, I. and Weiss, K. R. (1978). The command neuron concept. *Brain Behav. Sci.* **1**, 3–39.

Kutsch, W. (1969). Neuromuskuläre Aktivität bei verschiedenen Verhaltensweisen von drei Grillenarten. *Z. vergl. Physiol.* **63**, 335–378.

Kutsch, W., Schwarz, G., Fischer, H. and Kautz, H. (1993). Wireless transmission of muscle potentials during free flight of a locust. *J. Exp. Biol.* **185**, 367–373.

Lambert, T. D., Howard, J., Plant, A., Soffe, S. R. and Roberts, A. (2004). Mechanisms and significance of reduced activity and responsiveness in resting frog tadpoles. *J. Exp. Biol.* **207**, 1113–1125.

Laughlin, S. B. (1981). Neural principles in the peripheral visual systems of invertebrates. In *Handbook of Sensory Physiology*. Vol.VII/6B. *Comparative Physiology and Evolution of Vision in Invertebrates: Invertebrate Visual Centers and Behavior* (ed. H. Autrum), pp. 133–280. Berlin: Springer Verlag.

Laughlin, S. B. and Hardie, R. C. (1978). Common strategies for light adaptation in the peripheral visual system of fly and dragonfly. *J. Comp. Physiol.* **128**, 319–340.

Laughlin, S. B. and Weckström, M. (1993). Fast and slow photoreceptors: a comparative study of the functional diversity of coding and conductances in the Diptera. *J. Comp. Physiol. A* **172**, 593–609.

Laughlin, S. B., Howard, J. and Blakeslee, B. (1987). Synaptic limitation to contrast coding in the retina of the blowfly *Calliphora*. *Proc. Roy. Soc. Lond. B* **231**, 437–467.

Lee, R. and Eaton, R. C. (1991). Identifiable reticulospinal neurons of the adult zebrafish, *Brachydanio rerio. J. Comp. Neurol.* **304**, 34–52.

Lee, R., Eaton, R. and Zottoli, S. (1993). Segmental arrangement of reticulospinal neurons in the goldfish hindbrain. *J. Comp. Neurol.* **329**, 539–556.

Letzkus, P., Ribi, W. A., Wood, J. T., Zhu, H., Zhang, S.-W. and Srinivasan, M. V. (2006). Lateralization of olfaction in the honeybee *Apis mellifera. Curr. Biol.* **16**, 1471–1476.

Levi, R. and Camhi, J. M. (2000a). Population vector coding by the giant interneurons of the cockroach. *J. Neurosci.* **20**, 3822–3829.

Levi, R. and Camhi, J. M. (2000b). Wind direction coding in the cockroach escape response: winner does not take all. *J. Neurosci* **20**, 3814–3821.

Lewen, G. D., de Ruyter van Steveninck, R. and Bialek, W. (2001). Neural coding of naturalistic motion stimuli. *Network* **12**, 317–329.

Li, W.-C., Soffe, S. R., Wolf, E. and Roberts, A. (2006). Persistent responses to brief stimuli: feedback excitation among brainstem neurons. *J. Neurosci.* **26**, 4026–4035.

Liebenthal, E., Uhlman, O. and Camhi, J. M. (1994). Critical parameters of the spike trains in a cell assembly: coding of turn direction by the giant interneurons of the cockroach. *J. Comp. Physiol. A* **174**, 281–296.

Lillywhite, P. G. (1977). Single photon signals and transduction in an insect eye. *J. Comp. Physiol.* **122**, 189–200.

Link, A., Marimuthu, G. and Neuweiler, G. (1986). Movement as a specific stimulus for prey catching behaviour in rhinolophid and hipposiderid bats. *J. Comp. Physiol. A* **159**, 403–413.

Liu, K. S. and Fetcho, J. R. (1999). Laser ablations reveal functional relationships of segmental hindbrain neurons in zebrafish. *Neuron* **23**, 325–335.

Livingstone, M., Harris-Warrick, R. and Kravitz, E. (1980). Serotonin and octopamine produce opposite postures in lobsters. *Science* **208**, 76–79.

London, S. E. and Clayton, D. F. (2008). Functional identification of sensory mechanisms required for developmental song learning. *Nat. Neurosci.* **11**, 579–586.

Lorenz, K. and Tinbergen, N. (1938). Taxis und Instinkthandlung in der Eirollbewegung der Graugans. *Z. Tierpysychol.* **2**, 1–29.

Manley, G. A., Koppl, C. and Konishi, M. (1988). A neural map of interaural intensity differences in the brain stem of the barn owl. *J. Neurosci.* **8**, 2665–2676.

Manoli, D. S. and Baker, B. S. (2004). Median bundle neurons coordinate behaviours during *Drosophila* male courtship. *Nature* **430**, 564–569.

Manoli, D., Foss, M., Villella, M., *et al.* (2005). Male-specific *fruitless* specifies the neural substrates of *Drosophila* courtship behaviour. *Nature* **436**, 395–400.

Manoli, D., Meissner, G. and Baker, B. (2006). Blueprints for behavior: genetic specification of neural circuitry for innate behaviors. *Trends Neurosci.* **29**, 444–451.

Marler, P. and Slabberkoorn, H. (2004). *Nature's Music: The Science of Birdsong.* San Diego, CA: Elsevier.

Masino, T. and Knudsen, E. I. (1990). Horizontal and vertical components of head movement are controlled by distinct neural circuits in the barn owl. *Nature* **345**, 434–437.

Matheson, T., Rogers, S. M. and Krapp, H. G. (2004). Plasticity in the visual system is correlated with a change in lifestyle of solitarious and gregarious locusts. *J. Neurophysiol.* **91**, 1–12.

Mauelshagen, J. (1993). Neural correlates of olfactory learning paradigms in an identified neuron in the honeybee brain. *J. Neurophysiol.* **69**, 609–625.

McLean, D. L., Fan, J., Higashijima, S., Hale, M. E. and Fetcho, J. R. (2007). A topographic map of recruitment in spinal cord. *Nature* **446**, 71–75.

McLean, D. L., Masino, M. A., Koh, I. Y. Y., Lindquist, W. B. and Fetcho, J. R. (2008). Continuous shifts in the active set of spinal interneurons during changes in locomotor speed. *Nat. Neurosci.* **11**, 1419–1429.

Mello, C. V., Vicario, D. S. and Clayton, D. F. (1992). Song presentation induces gene expression in the song bird forebrain. *Proc. Natl. Acad. Sci. USA* **89**, 6818–6822.

Menzel, R. (1999). Memory dynamics in the honeybee. *J. Comp. Physiol. A* **185**, 323–340.

Menzel, R. and Erber, J. (1978). Learning and memory in bees. *Sci. Am.* **239**, 80–87.

Menzel, R. and Giurfa, M. (2001). Cognitive architecture of a mini-brain: the honeybee. *Trends Cogn. Sci.* **5**, 62–71.

Menzel, R., De Marco, R. J. and Greggers, U. (2006). Spatial memory, navigation and dance behaviour in *Apis mellifera. J. Comp. Physiol. A.* **192**, 889–903.

Metzner, W. (1993). The jamming avoidance response in *Eigemannia* is controlled by two separate motor pathways. *J. Neurosci.* **13**, 1862–1878.

Metzner, W. (1999). Neural circuitry for communication and jamming avoidance in gymnotiform electric fish. *J. Exp. Biol.* **202**, 1365–1375.

Meyrand, P., Simmers, A. J. and Moulins, M. (1991). Construction of a pattern generating circuit with neurons of different networks. *Nature* **351**, 60–63.

Meyrand, P., Simmers, A. J. and Moulins, M. (1994). Dynamic construction of a neural network from multiple pattern generators in the lobster stomatogastric nervous system. *J. Neurosci.* **14**, 630–644.

Michelsen, A. and Nocke, H. (1974). Biophysical aspects of sound communication in insects. *Adv. Insect Physiol.* **10**, 247–296.

Michelsen, A., Anderson, B. B., Kirchner, W. H. and Lindauer, M. (1989). Honeybees can be recruited by a mechanical model of a dancing bee. *Naturwissenschaften* **76**, 277–280.

Miller, J. P. and Selverston, A. I. (1982) Mechanisms underlying pattern generation in lobster stomatogastric ganglion as determined by selective inactivation of identified neurons. IV. Network properties of pyloric system. *J. Neurophysiol.* **48**, 1416–1432.

Mizunami, M., Weibrecht, J. M. and Strausfeld, N. J. (1998). Mushroom bodies of the cockroach: their participation in place memory. *J. Comp. Neurol.* **402**, 520–527.

Möhl, B. (1985). The role of proprioception in locust flight control. II. Information relayed by forewing stretch receptors during flight. *J. Comp. Physiol. A* **156**, 103–116.

Möhl, B. (1988). Short-term learning during flight control in *Locusta migratoria. J. Comp. Physiol. A* **163**, 803–812.

Möhl, B. (1993). The role of proprioception for motor learning in locust flight. *J. Comp. Physiol. A* **172**, 325–332.

Moiseff, A. and Konishi, M. (1981). Neuronal and behavioural sensitivity to binaural time differences in the owl. *J. Neurosci.* **1**, 40–48.

Moller, P. (1995). *Electric Fishes: History and Behavior*. London: Chapman and Hall.

Morris, R. M. (1981). Developments of a water-maze procedure for studying spatial learning in the rat. *J. Neurosci. Methods* **11**, 47–60.

Morris, R., Anderson, E., Lynch, G. S. and Baudry, M. (1986). Selective impairment of learning and blockade of longterm potentiation by an N-methyl-d-aspartate receptor antagonist, ap5. *Nature* **319**, 774–776.

Moser, E. I. and Moser, M. B. (2008). A metric for space. *Hippocampus* **18**, 1142–1156.

Moser, E. I., Kropff, E. and Moser, M.-B. (2008). Place cells, grid cells, and the brain's spatial representation system. *Ann. Rev. Neurosci.* **31**, 69–89.

Muller, R. U., Kubie, J. L. and Ranck, J. B. J. (1987). Spatial firing patterns of hippocampal complex-spike cells in a fixed environment. *J. Neurosci.* **7**, 1935–1950.

Nabatiyan, A., Poulet, J. F. A., De Polavieja, G. G. and Hedwig, B. (2003). Temporal pattern recognition based on instantaneous spike rate coding in a simple auditory system. *J. Neurophysiol.* **90**, 2484–2493.

Nakayama, H. and Oda, Y. (2004). Common sensory inputs and differential excitability of segmentally homologous reticulospinal neurons in the hindbrain. *J. Neurosci.* **24**, 3199–3209.

Neuweiler, G. (1983). Echolocation and adaptivity to ecological constraints. In *Neuroethology and Behavioural Physiology* (ed. F. Huber and H. Markl), pp. 280–302. Berlin: Springer Verlag.

Neuweiler, G. (2000). *The Biology of Bats.* New York: Oxford University Press.

Neuweiler, G., Bruns, V. and Schuller, G. (1980). Ears adapted for the detection of motion, or how echolocating bats have exploited the capacities of the mammalian auditory system. *J. Acoust. Soc. Am.* **68**, 741–753.

Neuweiler, G., Singh, S. and Sripathi, K. (1984). Audiograms of a South Indian bat community. *J. Comp. Physiol.* **154**, 133–142.

Neuweiler, G., Metzner, W., Heilmann, U., *et al.* (1987). Foraging behaviour and echolocation in the rufous horseshoe bat (*Rhinolophus rouxi*) of Sri Lanka. *Behav. Ecol. Sociobiol.* **20**, 53–67.

Neves, G., Cooke, S. F. and Bliss, T. V. P. (2008). Synaptic plasticity, memory and the hippocampus: a neural network approach to causality. *Nat. Rev. Neurosci.* **9**, 65–75.

Newland, P. L. (1991). Morphology and somatotopic organisation of the central projections of afferents from tactile hairs on the hind leg of the locust. *J. Comp. Neurol.* **311**, 1–16.

Nicholl, R. A., Kauer, J. A. and Malenka, R. C. (1988). The current excitement in long term potentiation. *Neuron* **1**, 97–103.

Nicol, D. and Meinertzhagen, I. A. (1982). An analysis of the number and composition of synaptic populations formed by photoreceptors of the fly. *J. Comp. Neurol.* **207**, 29–44.

Nissanov, J., Eaton, R. C. and DiDomenico, R. (1990). The motor output of the Mauthner cell, a reticulospinal command neuron. *Brain Res.* **517**, 88–98.

Nolen, T. G. and Hoy, R. R. (1984). Initiation of behavior by single neurons: the role of behavioral context. *Science* **226**, 992–994.

Norberg, R. A. (1970). Hunting technique of Tengmalm's owl, *Aegolius funereus* (L.). *Ornis Scand.* **1**, 51–64.

Norberg, R. A. (1977). Occurrence and independent evolution of bilateral ear asymmetry in owls and implications on owl taxonomy. *Phil. Trans. R. Soc. Lond. B* **280**, 375–408.

Nordström, K., Barnett, P. D. and O'Carroll, D. C. (2006). Insect detection of small targets moving in visual clutter. *PLoS Biol.* **4**, 378–386.

Nottebohm, F. (1989). From bird song to neurogenesis. *Sci. Am.* **260**, 74–79.

Nottebohm, F., Stokes, T. and Leonard, C. (1976). Central control of song in the canary. *J. Comp. Neurol.* **165**, 457–486.

Okada, R., Rybak, J., Manz, G. and Menzel, R. (2007). Learning-related plasticity in PE1 and other mushroom body-extrinsic neurons in the honeybee brain. *J. Neurosci.* **27**, 11 736–11 747.

O'Keefe, J. and Dostrovsky, J. (1971). The hippocampus as a spatial map: preliminary evidence from unit activity in the freely-moving rat. *Brain Res.* **34**, 171–175.

Olson, G. C. and Krasne, F. B. (1981). The crayfish lateral giants are command neurons for escape behavior. *Brain Res.* **214**, 89–100.

O'Malley, D., Kao, Y.-H. and Fetcho, J. R. (1996). Imaging the functional organization of zebrafish hindbrain segments during escape behaviours. *Neuron* **17**, 1145–1155.

O'Neill, W. E., and Suga, N. (1982). Encoding of target range and its representation in the auditory cortex of the moustached bat. *J. Neurosci.* **2**, 17–31.

O'Shea, M. and Rowell, C. H. F. (1976). The neuronal basis of a sensory analyser, the acridid movement detector system. II. Response decrement, convergence and the nature of the excitatory afferents to the fan-like dendrites of the LGMD. *J. Exp. Biol.* **65**, 289–308.

Payne, R. S. (1971). Acoustic location of prey by barn owls (*Tyto alba*). *J. Exp. Biol.* **54**, 535–573.

Pearson, K. G. and Ramirez, J.-M. (1990). Influence of input from the forewing stretch receptors on motoneurones in flying locusts. *J. Exp. Biol.* **151**, 317–340.

Pearson, K. G. and Wolf, H. (1987). Comparison of motor patterns in the intact and deafferented flight motor system of the locust. *J. Comp. Physiol. A* **160**, 259–268.

Pearson, K. G. and Wolf, H. (1988). Connections of hindwing tegulae with flight neurones in the locust, *Locusta migratoria. J. Exp. Biol.* **135**, 381–409.

Pearson, K. G., Reye, D. N., Parsons, D. W. and Bicker, G. (1985). Flight-initiating interneurons in the locust. *J. Neurophysiol.* **53**, 910–923.

Pereda, A., Bell, T. and Faber, D. (1995). Retrograde synaptic communication via gap junctions coupling auditory afferents to the Mauthner cell. *J. Neurosci.* **15**, 5943–5955.

Peron, S. and Gabbiani, F. (2009). Spike frequency adaptation mediates looming stimulus selectivity in a collision-detecting neuron. *Nat. Neurosci.* **12**, 318–326.

Pires, A. and Hoy, R. R. (1992). Temperature coupling in cricket acoustic communication. 1. Field and laboratory studies of temperature effects on calling song production and recognition in *Gryllus firmus. J. Comp. Physiol. A* **171**, 68–79.

Pittenger, C. and Kandel, E. R. (2003). In search of general mechanisms for long-lasting plasticity: *Aplysia* and the hippocampus. *Phil. Trans. Roy. Soc. Lond. B* **358**, 757–763.

Plummer, M. and Camhi, J. M. (1981). Discrimination of sensory signals from noise in the escape system of the cockroach: the role of wind acceleration. *J. Comp. Physiol.* **142**, 347–357.

Pollack, G. S. (1988). Selective attention in an insect auditory neuron. *J. Neurosci.* **8**, 2635–2639.

Pollak, G. D. (1980). Organizational and encoding features of single neurons in the inferior colliculus of bats. In *Animal Sonar Systems* (ed. R. G. Busnel, and J. F. Fish), pp. 549–587. New York: Plenum Press.

Pollak, G. D. and Schuller, G. (1981). Tonotopic organization and encoding features of single units in inferior colliculus of horseshoe bats: functional implications for prey identification. *J. Neurophysiol.* **45**, 208–226.

Pollak, G. D., Marsh, D., Bodenhamer, R. and Souther, A. (1977). Characteristics of phasic-on neurons in the inferior colliculus of unanaesthetised bats with observations relating to mechanisms of echo ranging. *J. Neurophysiol.* **40**, 926–942.

Poulet, J. F. A. and Hedwig, B. (2003). Corollary discharge inhibition of ascending auditory neurons in the stridulating cricket. *J. Neurosci.* **23**, 4717–4725.

Poulet, J. F. A. and Hedwig, B. (2005). Auditory orientation in crickets: pattern recognition controls reactive steering. *Proc. Natl. Acad. Sci. USA* **102**, 15 665–15 669.

Poulet, J. F. A. and Hedwig, B. (2006). The cellular basis of a corollary discharge. *Science* **311**, 518–522.

Poulet, J. F. A. and Hedwig, B. (2007). New insights into corollary discharges mediated by identified neural pathways. *Trends Neurosci.* **30**, 14–21.

Prather, J. F., Peters, S., Nowicki, S. and Mooney, R. (2008). Precise auditory-vocal mirroring in neurons for learned vocal communication. *Nature* **451**, 305–310.

Pringle, J. W. S. (1975). *Insect Flight*. Oxford Biology Readers 52. Glasgow: Oxford University Press.

Prugh, J. I., Kimmel, C. B. and Metcalfe, W. K. (1982). Noninvasive recording of the Mauthner neurone action potential in larval zebrafish. *J. Exp. Biol.* **101**, 83–92.

Ramirez, J.-M. and Pearson, K. G. (1991). Octopaminergic modulation of interneurons in the flight system of the locust. *J. Neurophysiol.* **66**, 1522–1537.

Ramon y Cajal, S. (1909). Histologie du système nerveux de l'homme et des vertébrés. Paris: Maloine.

Reichert, H. and Wine, J. J. (1983). Coordination of lateral giant and non-giant escape systems in crayfish escape behaviour. *J. Comp. Physiol.* **153**, 3–15.

Reichert, H., Wine, J. J. and Hagiwara, G. (1981). Crayfish escape behaviour: behavioural analysis of phasic extension reveals dual systems for motor control. *J. Comp. Physiol.* **142**, 281–294.

Rind, F. C. (1984). A chemical synapse between two motion detecting neurones in the locust brain. *J. Exp. Biol.* **110**, 143–167.

Rind, F. C. (1996). Intracellular characterization of neurons in the locust brain signalling impending collision. *J. Neurophysiol.* **75**, 986–995.

Rind, F. C. and Bramwell, D. I. (1996). A neural network based on the input organisation of an identified neuron signalling impending collision. *J. Neurophysiol.* **75**, 967–985.

Rind, F. C. and Simmons, P. J. (1992). Orthopteran DCMD neuron: a reevaluation of responses to moving objects. I. Selective responses to approaching objects. *J. Neurophysiol.* **68**, 1654–1666.

Rind, F. C. and Simmons, P. J. (1998). A local circuit for the computation of object approach by an identified visual neuron in the locust. *J. Comp. Neurol.* **395**, 405–415.

Rind, F. C., Santer, R. D. and Wright, G. A. (2008). Arousal facilitates collision avoidance mediated by a looming sensitive visual neuron in a flying locust. *J. Neurophysiol.* **100**, 670–680.

Riquimaroux, H., Gaioni, S. J. and Suga, N. (1991). Cortical computational maps control auditory perception. *Science* **251**, 565–568.

Rister, J., Pauls, D., Schnell, B., *et al.* (2007). Dissection of the peripheral motion channel in the visual system of *Drosophila melanogaster*. *Neuron* **56**, 155–170.

Ritzmann, R. E. (1993). The neural organization of cockroach escape and its role in context-dependent orientation. In *Biological Neuronal Networks in Invertebrate Neuroethology and Robotics* (ed. R. D. Beer, R. E. Ritzmann and T. McKenna), pp. 113–137. New York: Academic Press.

Robert, D. (1989). The auditory behaviour of flying locusts. *J. Exp. Biol.* **147**, 279–301.

Roberts, A. (1990). How does a nervous system produce behaviour? A case study in neurobiology. *Sci. Progr.* **74**, 31–51.

Roberts, A. (2000). Early functional organization of spinal neurons in developing lower vertebrates. *Brain Res. Bull.* **53**, 585–593.

Roberts, A. and Tunstall, M. J. (1990). Mutual re-excitation with post-inhibitory rebound: a simulation study of the mechanisms or locomotor rhythm generation in the spinal cord of *Xenopus* embryos. *Eur. J. Neurosci.* **2**, 11–23.

Roberts, A., Li, W.-C. and Soffe, S. R. (2008a). Roles for inhibition: studies on networks controlling swimming. *J. Comp. Physiol. A* **194**, 185–193.

Roberts, A., Li, W.-C., Soffe, S. R. and Wolf, E. (2008b). Origin of excitatory drive to a spinal locomotor network. *Brain Res. Rev.* **57**, 22–28.

Roberts, B. L. (1969). Spontaneous rhythms in the motoneurones of spinal dogfish (*Scyliorhinus canicule*). *J. Mar. Biol. Assoc. UK.* **49**, 3349.

Robertson, R. M. and Pearson, K. G. (1982). A preparation for the intracellular analysis of neuronal activity during flight in the locust. *J. Comp. Physiol.* **146**, 311–320.

Robertson, R. M. and Pearson, K. G. (1983). Interneurons in the flight system of the locust: distribution, properties and resetting properties. *J. Comp. Neurol.* **215**, 33–50.

Robertson, R. M. and Pearson, K. G. (1985). Neural circuits in the flight system of the locust. *J. Neurophysiol.* **53**, 110–128.

Rodríguez, F., Lópeza, J. C., Vargasa, J. P., *et al.* (2002). Spatial memory and hippocampal pallium through vertebrate evolution: insights from reptiles and teleost fish. *Brain Res. Bull.* **57**, 499–503.

Roeder, K. D. (1962). The behaviour of free flying moths in the presence of artificial ultrasonic pulses. *Anim. Behav.* **10**, 300–304.

Roessingh, P., Simpson, S. J. and James, S. (1993). Analysis of phase-related changes in behaviour of desert locust nymphs. *Proc. Roy. Soc. Lond. B* **252**, 43–49.

Roessingh, P., Bouaïchi, A. and Simpson, S. J. (1998). Effects of sensory stimuli on the behavioural phase state of the desert locust, *Schistocerca gregaria*. *J. Insect Physiol.* **44**, 883–893.

Rogers, S. M., Matheson, T., Despland, E., *et al.* (2003). Mechanosensory-induced behavioural gregarization in the desert locust *Schistocerca gregaria*. *J. Exp. Biol.* **206**, 3991–4002.

Rogers, S. M., Krapp, H. G., Burrows, M. and Matheson, T. (2007). Compensatory plasticity at an identified synapse tunes a visuomotor pathway. *J. Neurosci.* **27**, 4621–4633.

Rose, G. J. (2004). Insights into neural mechanisms and evolution of behaviour from electric fish. *Nat. Rev. Neurosci.* **5**, 943–951.

Rose, G. and Heiligenberg, W. (1985). Structure and function of electrosensory neurons in the torus semicircularis of *Eigenmannia*: morphological correlates of phase and amplitude sensitivity. *J. Neurosci.* **5**, 2269–2280.

Rose, G. J., Kawasaki, M. and Heiligenberg, W. (1988). 'Recognition units' at the top of a neuronal hierarchy? – prepacemaker neurons in *Eigenmannia* code the sign of frequency differences unambiguously. *J. Comp. Physiol. A* **162**, 759–772.

Rossell, S. (1979). Regional differences in photoreceptor performance in the eye of the praying mantis. *J. Comp. Physiol.* **131**, 95–112.

Rowell, C. H. F. (1971). The orthopteran descending movement detector (DMD) neurones: a characterisation and review. *J. Comp. Physiol.* **73**, 167–194.

Russell, J. C., Towns, D. R., Anderson, S. H. and Clout, M. N. (2005). Intercepting the first rat ashore. *Nature* **437**, 1107.

Rydqvist, B., Lin, J.-H. and Swerup, C. (2007). Mechanotransduction and the crayfish stretch receptor. *Physiol. Behav.* **92**, 21–28.

Sales, G. and Pye, D. (1974). *Ultrasonic Communication by Animals.* London: Chapman and Hall.

Santer, R. D., Simmons, P. J. and Rind, F. C. (2005) Gliding behaviour elicited by lateral looming stimuli in flying locusts. *J. Comp. Physiol. A* **191**, 61–73.

Santer, R. D., Rind, F. C., Stafford, R. and Simmons, P. J. (2006). Role of an identified looming-sensitive neuron in triggering a flying locust's escape. *J. Neurophysiol.* **95**, 3391–3400.

Sautois, B., Soffe, S. R., Li, W.-C. and Roberts, A. (2007). Role of type-specific neuron properties in a spinal cord motor network *J. Comput. Neurosci.* **23**, 59–77.

Schilstra, C. and van Hateren, J. H. (1998). Stabilizing gaze in flying blowflies. *Nature* **395**, 654.

Schilstra, C. and van Hateren, J. H. (1999). Blowfly flight and optic flow. I. Thorax kinematics and flight dynamics. *J. Exp. Biol.* **202**, 1481–1490.

Schmitz, B., Scharstein, H. and Wendler, G. (1983). Phonotaxis in *Gryllus campestris* L. (Orthoptera, Gryllidae): II. Acoustic orientation of female crickets after occlusion of single sound entrances. *J. Comp. Physiol. A* **152**, 257–264.

Schnitzler, H.-U. and Kalko, E. K. V. (2001). Echolocation by insect eating bats. *BioScience* **51**, 557–569.

Schramek, J. E. (1970). Crayfish swimming: alternating motor output and giant fiber activity. *Science* **169**, 698–700.

Schuller, G. (1984). Natural ultrasonic echoes from wing beating insects are encoded by collicular neurons in the CF-FM bat, *Rhinolophus ferrumequinum*. *J. Comp. Physiol.* **154**, 121–128.

Selverston, A. I. and Miller, J. P. (1980). Mechanisms underlying pattern generation in lobster stomatogastric ganglion as determined by selective inactivation of identified neurons. I. Pyloric system. *J. Neurophysiol.* **44**, 1102–1121.

Sherman, A. and Dickinson, M. H. (2003). A comparison of visual and haltere-mediated equilibrium reflexes in the fruit fly *Drosophila melanogaster*. *J. Exp. Biol.* **206**, 295–302.

Shettleworth, S. J. (2003). Memory and hippocampal specialization in food-storing birds: challenges for research on comparative cognition *Brain Behav. Evol.* **62**, 108–116.

Sillar, K., Wedderburn, J. F. S. and Simmers, A. J. (1991). The postembryonic development of locomotor rhythmicity in *Xenopus laevis* tadpoles. *Proc. Roy. Soc. Lond. B.* **246**, 147–153.

Silvey, G. and Wilson, I. (1979). Structure and function of the lateral giant neurone of the primitive crustacean *Anaspides tasmaniae*. *J. Exp. Biol.* **78**, 121–136.

Simmons, P. J. (1980). A locust wind and ocellar brain neurone. *J. Exp. Biol.* **85**, 281–294.

Simmons, P. J. (2002). Signal processing in a simple visual system: the locust ocellar system and its synapses. *Microsc. Res. Techniq.* **56**, 270–280.

Simmons, P. J. and Rind, F. C. (1992). Orthopteran DCMD neuron: a reevaluation of responses to moving objects. II. Critical cues for detecting approaching objects. *J. Neurophysiol.* **68**, 1667–1682.

Simmons, P. J. and Young, D. (1978). The tymbal mechanism and song patterns of the bladder cicada, *Cystosoma saundersii. J. Exp. Biol.* **76**, 27–45.

Simpson, S. J., Despland, E., Hägele, B. F. and Dodgson, T. (2001). Gregarious behavior in desert locusts is evoked by touching their back legs. *Proc. Natl. Acad. Sci. USA* **98**, 3895–3897.

Snodgrass, R. E. (1935). *Principles of Insect Morphology.* New York: McGraw-Hill.

Sokolowski, M. L. (2001). *Drosophila*: genetics meets behaviour. *Nat. Rev. Genet.* **2**, 879–892.

Solis, M. M. and Doupe, A. J. (1997). Anterior forebrain neurons develop selectivity by an intermediate stage of birdsong learning. *J. Neurosci.* **17**, 6447–6462.

Srinivasan, M. V. (1992). How bees exploit optic flow: behavioural experiments and neural models. *Phil. Trans. Roy. Soc. B* **337**, 253–259.

Srinivasan, M. V., Laughlin, S. B. and Dubs, A. (1982). Predictive coding: a fresh view of inhibition in the retina. *Proc. Roy. Soc. Lond. B.* **216**, 427–459.

Stockinger, P., Kvitsiani, D., Rotkopf, S., Tirian, L. and Dickson, B. J. (2005). Neural circuitry that governs *Drosophila* male courtship behavior. *Cell* **121**, 795–807.

Suga, N., Neuweiler, G. and Moller, J. (1976). Peripheral auditory tuning for fine frequency analysis by the CF-FM bat, *Rhinolophus ferrumequinum*. IV. Properties of peripheral auditory neurons. *J. Comp. Physiol.* **106**, 111–125.

Sullivan, W. E. (1982). Neural representation of target distance in auditory cortex of the echolocating bat, *Myotis lucifugus. J. Neurophysiol.* **48**, 1011–1032.

Suthers, R. A. (1990). Contributions to birdsong from the left and right sides of the intact syrinx. *Nature* **347**, 473–477.

Takahashi, T., Moiseff, A. and Konishi, M. (1984). Time and intensity cues are processed independently in the auditory system of the owl. *J. Neurosci.* **4**, 1781–1786.

Taube, J. S. (2007). The head direction signal: origins and sensory-motor integration. *Ann. Rev. Neurosci.* **30**, 181–207.

Taube, J. S., Muller, R. U. and Ranck, J. B. J. (1990a). Head-direction cells recorded from the postsubiculum in freely moving rats. I. Description and quantitative analysis. *J. Neurosci.* **10**, 420–435.

Taube, J. S., Muller, R. U. and Ranck, J. B. J. (1990b). Head-direction cells recorded from the postsubiculum in freely moving rats. II. Effects of environmental manipulations. *J. Neurosci.* **10**, 436–447.

Thompson, L. T. and Best, P. J. (1990). Long-term stability of the place-field activity of single units recorded from the dorsal hippocampus of freely behaving rats. *Brain Res.* **509**, 299–308.

Thorpe, W. H. (1958). The learning of song patterns by birds, with especial references to the song of the chaffinch, *Fringilla coelebs. Ibis* **100**, 535–570.

Tinbergen, N. (1951). *The Study of Instinct.* Oxford: Clarendon Press.

Tinbergen, N. (1963). On aims and methods in Ethology. *Z. für Tierpsychol.* **20**, 410–433.

Tolman, E. C. (1948). Cognitive maps in rats and men. *Psychol. Rev.* **55**, 189–208.

Tyrer, N. M., Turner, J. D. and Altman, J. (1984). Identifiable neurons in the locust central nervous system that react with antibodies to serotonin. *J. Comp. Neurol.* **227**, 313–333.

Vater, M., Feng, A. S. and Betz, M. (1985). An HRP-study of the frequency-place map of the horseshoe bat cochlea: morphological correlates of the sharp tuning to a narrow frequency band. *J. Comp. Physiol. A* **157**, 671–686.

Vicario, D. S. (1991). Organization of the zebrafinch song control system. II. Functional organization of the output from the nucleus robustus archistri-atalis. *J. Comp. Neurol.* **309**, 486–494.

von Frisch, K. (1967). *The Dance Language and Orientation of Bees.* Cambridge, MA: Harvard University Press.

Vrontou, E., Nilsen, S. P., Demir, E., Kravitz, E. A. and Dickson, B. J. (2006). *fruitless* regulates aggression and dominance in *Drosophila*. *Nat. Neurosci.* **9**, 1469–1471.

Vu, E. and Krasne, F. (1993). The mechanism of tonic inhibition of crayfish escape behavior: distal inhibition and its functional significance. *J. Neurosci.* **13**, 4394–4402.

Vu, E. T., Mazurek, M. E. and Kuo, Y.-C. (1994). Identification of a forebrain motor programming network for the learned song of zebra finches. *J. Neurosci.* **14**, 6924–6934.

Wada, K. *et al.* (2006). A molecular neuroethological approach for identifying and characterizing a cascade of behaviorally regulated genes. *Proc. Natl. Acad. Sci. USA* **103**, 15 212–15 217.

Wagner, H. (1986). Flight performance and visual control of flight of the free-flying housefly (*Musca domestica* L.). I. Organization of the flight motor. *Phil. Trans. Roy. Soc. B* **312**, 527–551.

Wallhausser-Franke, E., Nixdorf-Bergweiler, B. E. and DeVoogd, T. J. (1995). Song isolation is associated with maintaining high spine frequencies on zebrafinch lMAN neurons. *Neurobiol. Learn. Mem.* **64**, 25–35.

Warzecha, A.-K., Egelhaaf, M. and Borst, A. (1993). Neural circuit tuning fly visual neurons to motion of small objects. I. Dissection of the circuit by pharmacological and photoinactivation techniques. *J. Neurophysiol.* **69**, 329–339.

Werblin, F. S. and Dowling, J. E. (1969). Organization of the retina of the mudpuppy, *Necturus maculosus*. II. Intracellular recording. *J. Neurophysiol.* **32**, 339–355.

Wiersma, C. A. G. (1947). Giant nerve fiber system of the crayfish: a contribu-tion to comparative physiology of the synapse. *J. Neurophysiol.* **10**, 23–38.

Wiersma, C. A. G. and Ikeda, K. (1964). Interneurons commanding swimmeret movements in the crayfish, *Procambarus clarkii* (Girard). *Comp. Biochem. Physiol.* **12**, 509–525.

Willows, A. O. D., Dorsett, D. A. and Hoyle, G. (1973). The neuronal basis of behavior in *Tritonia*. III. Neuronal mechanism of a fixed action pattern. *J. Neurobiol.* **4**, 255–285.

Wilson, D. M. (1960). The central nervous control of flight in a locust. *J. Exp. Biol.* **38**, 471–490.

Wilson, D. M. (1968). The flight control system of the locust. *Sci. Am.* **218**, 83–90.

Wilson, M. (1978). The functional organization of locust ocelli. *J. Comp. Physiol.* **124**, 297–316.

Wilson, M., Garrard, P. and McGiness, S. (1978). The unit structure of the locust compound eye. *Cell Tiss. Res.* **195**, 205–226.

Wine, J. (1984). The structural basis of an innate behavioural pattern. *J. Exp. Biol.* **112**, 283–319.

Wine, J. J. and Krasne, J. B. (1982). The cellular organization of crayfish escape behavior. In *The Biology of Crustacea*, vol. 4 (ed. E. D. Bliss), pp. 241–292. New York: Academic Press.

Wine, J. J. and Mistick, D. C. (1977). Temporal organization of crayfish escape behavior: delayed recruitment of peripheral inhibition. *J. Neurophysiol.* **40**, 904–925.

Witten, I. B., Bergan, J. F. and Knudsen, E. I. (2006). Dynamic shifts in the owl's auditory space map predict moving sound location. *Nat. Neurosci.* **11**, 1439–1445.

Wohlers, D. W. and Huber, F. (1982). Processing of sound signals by six types of neurons in the prothoracic ganglion of the cricket, *Gryllus campestris* L. *J. Comp. Physiol. A* **146**, 161–173.

Wong, D. (2004). The auditory cortex of the little brown bat, *Myotis lucifugus*. In *Echolocation in Bats and Dolphins* (ed. J. A. Thomas, C. F. Moss and M. Vater), pp. 185–189. Chicago, IL: Chicago University Press.

Wong, D., Maekawa, M. and Tanaka, H. (1992). The effect of pulse repetition rate on the delay sensitivity of neurons in the auditory cortex of the FM bat, *Myotis lucifugus*. *J. Comp. Physiol.* **170**, 393–402.

Yazaki-Sugiyama, Y. and Mooney, R. (2004). Sequential learning from multiple tutors and serial retuning of auditory neurons in a brain area important to birdsong learning. *J. Neurophysiol.* **92**, 2771–2788.

Yeh, S., Musolf, B. and Edwards, D. (1997). Neuronal adaptations to changes in the social dominance status of crayfish. *J. Neurosci.* **17**, 697–708.

Young, D. and Ball, E. (1974). Structure and development of the tracheal organ in the mesothoracic leg of the cricket *Teleogryllus commodus* (Walker). *Z. Zellforsch.* **147**, 325–334.

Young, D. and Bennett–Clark, H. C. (1995). The role of the tymbal in cicada song production. *J. Exp. Biol.* **198**, 1001–1019.

Yu, A. C. and Margoliash, D. (1996). Temporal hierarchical control of singing in birds. *Science* **273**, 1871–1875.

Zottoli, S. J. (1977). Correlation of the startle reflex and Mauthner cell auditory responses in unrestrained goldfish. *J. Exp. Biol.* **66**, 243–254.

Zottoli, S. J. (1978). Comparative morphology of the Mauthner cell in fish and amphibians. In *Neurobiology of the Mauthner Cell* (ed. D. S. Faber, and H. Korn), pp. 13–45. New York: Raven Press.

Index

Page numbers in **bold** indicate places where the meaning of a word or phrase set in **bold** in the text is explained.